Operational Amplifier Characteristics and Applications

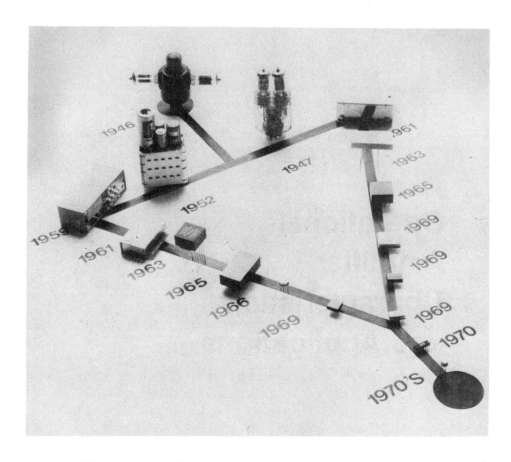

This photograph illustrates the progression of development that has taken place in the operational amplifier from the twin-triode version to the monolithic integrated circuit version. (Photo courtesy of Teledyne Philbrick Corporation.)

Operational Amplifier Characteristics and Applications

Second Edition

ROBERT G. IRVINE

California State Polytechnic University, Pomona

Prentice-Hall, Inc., Englewood Cliffs, New Jersey 07632

Library of Congress Cataloging-in-Publication Data

IRVINE, ROBERT G., (date)
  Operational amplifier characteristics and applications.

  Bibliography: p.
  Includes index.
  1. Operational amplifiers.   I. Title.
TK7871.58.06I78 1987   621.3815'35   86-25483
ISBN 0-13-637661-4

Editorial/production supervision
  and interior design: PATRICK WALSH
Cover design: EDSAL ENTERPRISES
Manufacturing buyer: RHETT CONKLIN

Printed in the United States of America

10   9   8   7   6   5   4   3   2

ISBN 0-13-637661-4   025

Prentice-Hall International (UK) Limited, *London*
Prentice-Hall of Australia Pty. Limited, *Sydney*
Prentice-Hall Canada Inc., *Toronto*
Prentice-Hall Hispanoamericana, S.A., *Mexico*
Prentice-Hall of India Private Limited, *New Delhi*
Prentice-Hall of Japan, Inc., *Tokyo*
Prentice-Hall of Southeast Asia Pte. Ltd., *Singapore*
Editora Prentice-Hall do Brasil, Ltda., *Rio de Janeiro*

To my wife
Joan

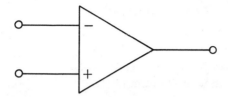

Contents

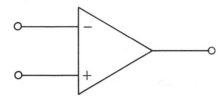

Preface

The operational amplifier was developed during the late 1930's in response to a government research grant to perform integrations and differential equations electronically. The first devices were crude vacuum tube models, but worked. Later, devices were developed that were self-contained and plugged into octal sockets. The op amp was a strange device, not understood by many engineers and certainly not taught in the universities.

In 1962 Robert J. Widlar, working for Fairchild Semiconductor, took some digital circuit monolithic ideas and applied them to a linear op amp. The first device, a μA702, was not very successful. But his second device, the μA709, became an instant success and became integrated into almost every new design. The 709 and Mr. Widlar became famous. Mr. Widlar left Fairchild and helped form National Semiconductor. The story from there is history.

The op amp now enjoys a place in our technology as ubiquitous as the transistor. The inner workings of the op amp remain misunderstood today even though many texts and references are available describing its features and functions. This prompted the writing of the first edition of this book, which appeared in 1981. The author felt that many things had not been said about the characteristics of the op amp and set forth to correct that. This second edition refines and updates many of the ideas presented in the first edition. A completely new problem set has been written for the book; the reader is provided with answers to selected problems. The book is again written in such a manner that self-study is possible even though it is a teaching text; the book is also useful as a reference.

The first edition is being taught in both Engineering Technology and Electrical Engineering programs. The math level is directed toward the four-year Engineering

Technologists, and yet, some calculus is used so as not to be demeaning to the Electrical Engineering student.

The organization of this edition is the same as in the first edition and is taught by the author in two quarters; Chapters 2 through 6 are taught in the first quarter and Chapters 7 through 11, excluding 10, are taught in the second quarter, both at junior level. The text could be used for a semester course by either including Chapter 7 for the more comprehensive course or by beginning at Chapter 2 and proceeding through Chapter 11, excluding Chapter 9 (active filters). Chapter 9 should be used only in the more comprehensive course. It takes almost three weeks to complete Chapter 9. Section 2.6 presents the four types of feedback from an op amp point of view; the author has found this to be a very valuable addition to the text, particularly for Electrical Engineering students.

Chapter 9 is presented in such a manner as to enhance the student's understanding without diminishing the quality of presentation. The six major filter types are all presented, and examples and problems using these ensure a feeling of ease by the student after finishing the chapter. This chapter and presentation have proved to be very valuable to the reader. A special method of performing the required mathematics has been developed to prevent the student from becoming bogged down in math and not seeing the function being analyzed. In every case, the pole-zero diagram is shown opposite the filter's frequency response to give the student a "feeling" for the numbers in the equations and pole locations for a particular filter type.

Chapter 11 has some new material in it, primarily in the area of high frequency op amps. The high frequency op amp is emerging as the primary emphasis area for the late 80's and early 90's. Many companies have broken the 1 MHz barrier for high frequency performance and are now producing devices into the hundreds of MHz and some devices now break the 1 GHz barrier. This will surely enhance the analog area and keep it a viable option to the ever advancing digital market and techniques. In one area of high frequency response, current feedback is used to eliminate the constant -20 dB/decade gain-bandwidth frequency barrier. An engineer, David Nelson, has invented and patented an idea for breaking the gain-bandwidth barrier by using current feedback. He has formed a company called Comlinear in Ft. Collins, Colorado that produces a product line using these techniques. This is a process and idea that should be followed closely by the reader as it could emerge as the new wave of the future. The author has a patent application pending on an idea similar to that of David Nelson's. Both ideas are presented in Chapter 11 as the author feels that this process will become a viable alternative to the other methods of high frequency op amp construction.

The book has been written to give students and practicing engineers the tools with which to analyze and understand unfamiliar or special op amp circuits that will be encountered during the course of a career. Almost all circuits encountered by the author in industry were combinations or variations of the classical circuits presented in this book. A solutions manual, with all problems worked in detail, is available to instructors and those practicing engineering.

Acknowledgments

I wish to acknowledge the efforts of Kathleen I. Zeller who has rewritten the problem set and corresponding answers to the second edition. Her efforts in organizing the material for submission are also appreciated. I am grateful as well to Adolfo Jimenez for his help with the preparation of the Index and Answers to Selected Problems and to John Witte for pointing out problem areas.

My thanks to Richard Cockrum, my mentor and critic, who has always helped with the book; to William Donovan, who critiqued the first six chapters' problem sets, and is also a mentor and critic; and to Dr. Lyle McCurdy, for providing inspiration to upgrade the level of the book.

Dr. Jea K. Park helped with derivations and Owen K. Skousen provided the lead to Comlinear. Richard Cockrum and Earl Schoenwetter provided helpful critiques and suggestions for the second edition. My wife has provided encouragement and patience throughout the effort.

Robert G. Irvine
Claremont, California

As is the policy begun in the first edition, I am recognizing an individual who is an innovator in the area of analog engineering. This by no means suggests that this person stands above others who are also innovators in this field. I believe that students and young engineers should be inspired to achieve to their maximum ability. This motivates me to provide an image for them. David Nelson is an example of an innovator who saw a need, found a solution and made it available to the world.

David A. Nelson was born in 1952, raised and educated in Utah, and graduated from Brigham Young University with a BS in Physics in 1975. He earned his MSEE at UCLA in 1978 and was the recipient of the Hughes Masters Fellow award while working for Hughes Aircraft Company. At Hughes, and later at Hewlett-Packard, he found that the lack of fast settling, wideband op amps often meant that many design problems could not always be solved simply and inexpensively.

Designing a better op amp became his hobby. Op amps hadn't changed much in 20 years since the first monolithic devices appeared. He found that performance improvements due to faster transistors and exotic processes had obscured the high speed limitations inherent in op amp architectures. In 1982, after patenting a new circuit architecture and forming Comlinear using $50,000 invested by relatives, he succeeded in introducing the first op amp with truly high speed performance. It was easy to use, in contrast to earlier high speed op amps, because bandwidth and signal fidelity were independent of gain, and no compensation was required for optimum performance. Today Comlinear continues to manufacture innovative products for a widening variety of analog problems.

David wishes readers to find motivation in his experience, and to encourage them to leave the relative security of a large company and take a small risk on their own early in their careers.

"I have found that nothing teaches success like failure, and that the only real failure is not having tried."

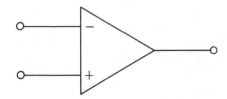

1 Construction of the Operational Amplifier

1.1 Introduction

The operational amplifier is a special-purpose amplifier intended to be used with external feedback components for proper operation. Only the operational amplifier itself will be discussed in this chapter. Its use with feedback elements will be reserved for later chapters, where an in-depth analysis will be presented.

1.2 Notation

The notation used in this text will be as follows:

1. Average (dc), maximum, and effective (root-mean-square or rms) values are represented using uppercase letters (V for voltage, I for current, and P for power).

2. Instantaneous values of quantities that vary with time are represented using lowercase letters (v, i, or p).

3. Instantaneous total values and average (dc) values are represented using an uppercase subscript letter for the appropriate terminal of the device being described (v_E).

4. Varying component values are indicated using a lowercase subscript letter for the appropriate device terminal (i_b).

5. A double subscript is used in two cases:

   a. Where the power-supply voltage (V_{CC} or V_{SS}) is referenced.

   b. Where the voltage being represented is not ground-referenced; that is, it is the potential difference *between* two points in the circuit neither of which is at ground potential (zero volts). An example is V_{BE} for a transistor circuit with an emitter resistor.

6. *E* or *e* will be used in place of *V* or *v* for voltage sources. Power-supply sources are excluded from this notation.

7. The zero-volts reference plane for circuits will be represented using the terms "ground" and "common" interchangeably. These only differ in systems employing a chassis. There, the chassis is *ground* and the circuit *common* is a bus wire that is attached to the chassis at one point.

1.3   SI Metric Units

The use of SI (International System of Units) metric units will be followed. As a result, the reader will encounter a few units that differ from the previous MKS or CGS systems. The symbols, together with the SI metric units, are listed in Table 1-1.

1.4   Differential Amplifier Operation

Much emphasis and time will be spent on the differential amplifier. Many of the good and most of the bad characteristics of the operational amplifier originate in the differential amplifier.

A basic differential amplifier is shown in Fig. 1-1. It functions in such a manner that voltages common to both inputs (v_{I_1} and v_{I_2}) cause both collector voltages to move in unison, whereas voltages differing on the two inputs cause the two collector voltages to move in opposition. The basic rules of transistor action still apply; the collector voltage moves in opposition to the base voltage direction, regardless of whether the base change was caused by common or differing voltages.

Table 1-1   SI Metric Units

Quantity	Symbol	Unit Name	SI Metric Units
Voltage	*VE*	(V) Volt	MV, kV, V, mV, μV
Current	*I*	(A) Ampere	kA, A, mA, μA, nA, pA
Power	*P*	(W) Watt	MW, kW, W, mW, μW
Energy (work)	*W*	(J) Joule	MJ, kJ, J, mJ, μJ
Resistance	*R*	(Ω) Ohm	MΩ, kΩ, Ω, mΩ
Conductance	*G*	(S) Siemen[b]	kS, S, mS, μS
Capacitance	*C*	(F) Farad	F, mF, μF, nF, pF
Inductance	*L*	(H) Henry	H, mH, μH
Electric charge	*Q*	(C) Coulomb	kC, C, mC, μC, nC, pC
Frequency	*f*	(Hz) Hertz	THz, GHz, MHz, kHz, Hz
Force	*F*	(N) Newton	MN, kN, N, mN, μN
Length	*l*	(m) Meter	km, m, mm, μm
Mass	*M*	(kg) Kilogram	Mg, kg, g, mg, μg
Time	*t*	(s) Second	s, ms, μs, ns, ps
Angular velocity	ω	(ω) rad/s	ω
Damping force[a]	*d*	(N) Newton	N
Spring constant[a]	*K*	(N/m) $\dfrac{\text{Newtons}}{\text{Meter}}$	N/m

[a] Not actually SI units but represented as their SI equivalents.
[b] Mho (v) has been replaced by Siemen (S).

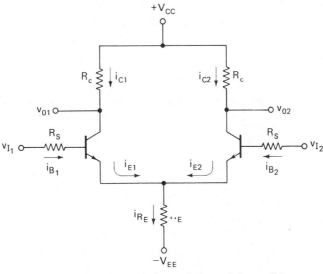

Fig. 1-1 Bipolar transistor differential amplifier.

1.4.1 Common-Mode Voltages

The voltage, common to both inputs, is labeled *common-mode input voltage* ($v_{I_{CM}}$). If v_{I_1} is the voltage on base 1 and v_{I_2} is the voltage on base 2, the common-mode input voltage is the average of these two voltages, as expressed by

$$v_{I_{CM}} = \frac{v_{I_1} + v_{I_2}}{2} \qquad (1\text{-}1)$$

If $v_{I_1} = v_{I_2}$, then $v_{I_{CM}} = v_{I_1} = v_{I_2}$.

The voltage, common to both outputs, is labeled *common-mode output voltage* ($v_{O_{CM}}$). If v_{O_1} is the voltage on collector 1 and v_{O_2} is the voltage on collector 2, the common-mode output voltage is the average of these two voltages at any instant in time, as expressed by

$$v_{O_{CM}} = \frac{v_{O_1} + v_{O_2}}{2} \qquad (1\text{-}2)$$

With the two inputs connected together, v_{O_1} should be equal to v_{O_2}, presuming matched components, but that is seldom the case for actual circuits. Example 1-1 will help to clarify the use of Eqs. (1-1) and (1-2).

EXAMPLE 1-1

A differential amplifier has the characteristics that cause the common-mode output voltage to be $+5.5$ V when the input common-mode voltage is zero and to have an output common-mode voltage of $+5.0$ V when the input common-mode voltage is $+1$ V. Determine the input and output voltages for the following set of voltages:

Case	V_{I_1}	V_{I_2}	V_{O_1}	V_{O_2}	Find
1	0	0	+5.5		v_{O_2}
2	+1.1	+0.9	+2.0	+8.0	$v_{I_{CM}}, v_{O_{CM}}$
3	+0.01	−0.01		+6.1	$v_{I_{CM}}, v_{O_1}$

Solution

Case 1

$$v_{I_{CM}} = \frac{0 + 0}{2} = 0 \quad \text{thus} \quad v_{O_{CM}} = +5.5 \text{ V}$$

$$+5.5 = \frac{+5.5 + v_{O_2}}{2} \qquad v_{O_2} = +5.5 \text{ V}$$

Case 2

$$v_{I_{CM}} = \frac{+1.1 + 0.9}{2} = +1.0$$

$$v_{O_{CM}} = \frac{2.0 + 8.0}{2} = +5.0$$

Case 3

$$v_{I_{CM}} = \frac{+0.01 + (-0.01)}{2} = 0 \text{ V}$$

$$v_{O_{CM}} = +5.5 = \frac{v_{O_1} + 6.1}{2} \qquad v_{O_1} = +4.9 \text{ V}$$

In the most frequently encountered case for common-mode voltages, the two input voltages are varying at different rates, thus producing a changing dc or an ac common-mode voltage. In this situation Eq. (1-1), between two instants in time, becomes

$$\Delta v_{I_{CM}} = \frac{\Delta v_{I_1} + \Delta v_{I_2}}{2} \tag{1-3}$$

and Eq. (1-2), between two instants in time becomes

$$\Delta v_{O_{CM}} = \frac{\Delta v_{O_1} + \Delta v_{O_2}}{2} \tag{1-4}$$

where the individual v_I's or v_O's must be established at two separate instants of time to determine the variation over the time interval.

EXAMPLE 1-2

A differential amplifier has v_{I_1} grounded while v_{I_2} is connected to an ac voltage of $v_{I_2} = 6$ mV $\sin(2\pi \cdot 0.1t)$. The output common-mode voltage is +5.0 V when $v_{I_{CM}} = 0$, +5.001 V when $v_{I_{CM}} = -3$ mV, and +4.999 V when $v_{I_{CM}} = +3$ mV. Determine the input and output common-mode voltages.

Solution

$$\Delta v_{I_{CM}} = \frac{0 + 6 \text{ mV} \sin(2\pi \cdot 0.1t)}{2} = 3 \text{ mV} \sin(2\pi \cdot 0.1t)$$

which is a sine wave with peak instantaneous voltages of ± 3 mV. The collector voltage decreases as the base voltage increases, which results in a phase reversal on the output. Thus $\Delta v_{O_{CM}} = +5 \text{ V} - 1 \text{ mV} \sin(2\pi \cdot 0.1t)$. [*Note:* The $(-)$ sign indicates the phase reversal.]

EXAMPLE 1-3

A differential amplifier has the following input and output voltages at different instants of time:

Instant	v_{I_1}	v_{I_2}	v_{O_1}	v_{O_2}
t_1	-3.4	-3.6	$+4.2$	$+4.4$
t_2	$+1.8$	$+1.2$	$+2.6$	$+3.2$

Determine the change in the input and output common-mode voltage.

Solution

Δv_{I_1} is v_{I_1} at t_2 minus v_{I_1} at t_1; Δv_{I_2} is v_{I_2} at t_2 minus v_{I_2} at t_1. The same is true for the output voltages.

$$\Delta v_{I_{CM}} = \frac{[1.8 - (-3.4)] + [1.2 - (-3.6)]}{2} = \frac{5.2 + 4.8}{2}$$

$$= +5.0 \text{ V}$$

$$\Delta v_{O_{CM}} = \frac{(2.6 - 4.2) + (3.2 - 4.4)}{2} = \frac{-1.6 + (-1.2)}{2}$$

$$= -1.4 \text{ V}$$

which illustrates that, as the input common-mode voltages increases by 5 V, the output common-mode voltage decreases by 1.4 V.

The *common-mode gain* is defined as the change in common-mode output voltage divided by the change of input common-mode voltage, or

$$A_{CM} = \frac{\Delta v_{O_{CM}}}{\Delta v_{I_{CM}}} \tag{1-5}$$

where A_{CM} becomes negative when numerical values are applied.

EXAMPLE 1-4

Using the values determined in Example 1-3, find A_{CM}.

Solution

$$A_{CM} = \frac{-1.4 \text{ V}}{+5.0 \text{ V}} = -0.28$$

Common-mode gains are typically less than unity (1). The most desirable case is for the common-mode gain to be zero, and, as will be seen in subsequent sections, this is the motivation for particular circuit selections.

1.4.2 Differential Voltages

The differential voltages must be considered for two cases: (1) where the deviation voltage is swinging around a zero common-mode voltage, and (2) where the deviation voltage is swinging around a nonzero common-mode voltage. The first case should only be considered for the inputs, as the collector voltages of a differential amplifier generally swing around a nonzero common-mode voltage.

The differential voltage swing about any common-mode voltage, including zero volts, is called the *deviation voltage*. The input deviation voltage,* at any instant in time, is given by one-half the difference voltage at the input or

$$v_{I_{DEV}} = \pm \frac{v_{I_1} - v_{I_2}}{2} \tag{1-6}$$

where $(v_{I_1} - v_{I_2})$ is called the *differential input voltage* and is given an individual label (v_{Id}); thus

$$v_{Id} = v_{I_1} - v_{I_2} \tag{1-7}$$

and

$$v_{Id} = \pm 2 v_{I_{DEV}} \tag{1-8}$$

The output deviation voltage, at any instant in time, is given by

$$v_{O_{DEV}} = \pm \frac{v_{O_1} - v_{O_2}}{2} \tag{1-9}$$

which is the voltage deviation around the common-mode output voltage at any instant in time on each collector.

EXAMPLE 1-5

A differential amplifier has zero input and output common-mode voltages (not physically possible). The input and output voltages are:

v_{I_1}	v_{I_2}	v_{O_1}	v_{O_2}
-0.01	$+0.01$	$+1.4$	-1.4

Determine the input and output deviation voltages and the differential input voltage.

Solution

$$v_{I_{DEV}} = \pm \frac{-0.01 - 0.01}{2} = \mp 0.01 \text{ V}$$

$$v_{O_{DEV}} = \pm \frac{1.4 - (-1.4)}{2} = \pm 1.4 \text{ V}$$

$$v_{Id} = \pm 2(\mp 0.01 \text{ V}) = -0.02 \text{ V}^\dagger$$

The two input voltages are usually varying at different rates, thus producing a changing dc or ac deviation voltage on the input and output of the differential

---

*The ($\pm$) sign indicates that the deviation voltage swings to each side of the common-mode voltage. This ($\pm$) indicator is dropped when discussing v_{Id}, and this seems to be a mathematical error, but the differential voltage has only one algebraic sign, because it is a single voltage value.

†Note: $(\pm)(\pm) = +$, $(\pm)(\mp) = -$, $(\pm)/(\pm) = +$, $(\pm)/(\mp) = -$, $(\mp)/(\pm) = -$.

amplifier. In this situation, Eq. (1-6) becomes

$$\Delta v_{I_{DEV}} = \pm \frac{\Delta v_{I_1} - \Delta v_{I_2}}{2} \qquad (1\text{-}10)$$

Eq. (1-8) becomes*

$$\Delta v_{Id} = \pm 2\Delta v_{I_{DEV}} \qquad (1\text{-}11)$$

and Eq. (1-9) becomes

$$\Delta v_{O_{DEV}} = \pm \frac{\Delta v_{O_1} - \Delta v_{O_2}}{2} \qquad (1\text{-}12a)$$

Then the deviation gain is*

$$A_{DEV} = + \frac{\Delta v_{O_{DEV}}}{\Delta v_{I_{DEV}}} \qquad (1\text{-}12b)$$

where the individual v_I's or v_O's must be established at two separate instants of time to determine the variation over the time interval $(t_2 - t_1)$. The deviation gain in Example 1-5 is $\pm 1.4 \text{ V} / \mp 0.01 \text{ V} = -140$.

EXAMPLE 1-6

A differential amplifier has a varying dc voltage on the input, which has the values at two instants in time given below ($v_{I_{CM}}$ and $v_{O_{CM}}$ are both zero):

Instant	v_{I_1}	v_{I_2}	v_{O_1}	v_{O_2}
t_1	+0.04	−0.04	−8	+8
t_2	−0.02	+0.02	+4	−4

Determine the change in input deviation voltage, differential input voltage, and output deviation voltage.

Solution

$$
\begin{array}{cccc}
 & v_{i_1} & & v_{i_2} \\
 & t_2 \quad t_1 & & t_2 \quad t_1
\end{array}
$$

$$\Delta v_{I_{DEV}} = \pm \frac{(-0.02 - 0.04) - [0.02 - (-0.04)]}{2}$$

$$= \pm \frac{-0.06 - 0.06}{2} = \mp 0.06 \text{ V}$$

$$\Delta v_{Id} = \pm 2(\mp 0.06 \text{ V}) = -0.12 \text{ V}$$

$$\Delta v_{O_{DEV}} = \pm \frac{[4 - (-8)] - (-4 - 8)}{2} = \pm \frac{12 + 12}{2}$$

$$= \pm 12 \text{ V}$$

*Note: $(\pm)(\pm) = +, (\pm)(\mp) = -, (\pm)/(\pm) = +, (\pm)/(\mp) = -, (\mp)/(\pm) = -$.

Differential Amplifier Operation

7

1.4.3 Differential and Common-Mode Voltages

The differential gain is defined as the change in output deviation voltage divided by the change in differential input voltage given by

$$A_D = \pm \frac{\Delta v_{O_{DEV}}}{\Delta v_{Id}} \qquad (1\text{-}13)$$

Because the input *differential* and not the input deviation voltage is used in the denominator, a factor of 2 is introduced which is not, at first, apparent. This causes A_D to be half the expected value. The differential gain in Example 1-6 is $\pm[(\pm 12 \text{ V})/(-0.12 \text{ V})] = -100$.

EXAMPLE 1-7

A differential amplifier has a sine-wave voltage applied to the two inputs. The dc average of the sine wave is zero.

$$\Delta v_{Id} = 0.04 \sin(2\pi \cdot 0.1t)$$

if the differential gain is -150, determine the output deviation voltage on each collector.

Solution

$$\Delta v_{O_{DEV}} = \pm A_d (\Delta v_{Id})$$

$$\Delta v_{O_{DEV}} = \pm(-150)[0.04 \sin(2\pi \cdot 0.1t)] = -6 \sin(2\pi \cdot 0.1t)$$

Thus each collector voltage is changing ± 6 V around the common-mode collector voltage. Collector 2 is swinging in an equal but opposite direction to collector 1.

The most general case of input and output differential voltages occurs when the deviation voltage swings around a common-mode voltage. The combination of these two is the voltage measured by a voltmeter attached to a base or collector lead. It is not apparent when making these measurements that two separate voltages actually exist. The actual voltages on the differential amplifier inputs are given by*

$$v_{I_1} = v_{I_{CM}} \pm v_{I_{DEV}} \qquad (1\text{-}14)$$

$$v_{I_2} = v_{I_{CM}} \mp v_{I_{DEV}} \qquad (1\text{-}15)$$

The voltages on the output are given by

$$v_{O_1} = v_{O_{CM}} \pm v_{O_{DEV}} \qquad (1\text{-}16)$$

$$v_{O_2} = v_{O_{CM}} \mp v_{O_{DEV}} \qquad (1\text{-}17)$$

The combination of common-mode and deviation voltages is shown as input and output voltages to a differential amplifier in Fig. 1-2. The common-mode voltage is the average of the two voltages in the bases (inputs) and collectors (outputs). The collector common-mode voltage is out of phase with that in the base and the amplitude is less, indicating a common-mode gain of less than unity. The deviation voltage is a peak excursion, around the common-mode voltage, where the deviation

---

*A $\mp$ voltage is 180° out of phase with a $\pm$ voltage.

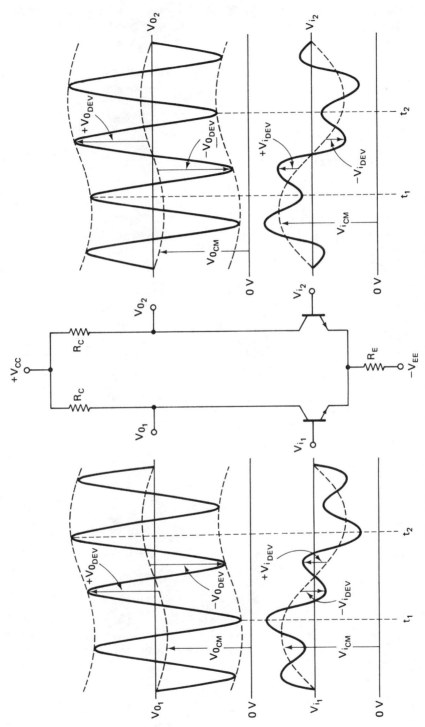

Fig. 1-2 Representation of common-mode and deviation voltages.

9

voltages on the two bases or two collectors are out of phase with each other. The deviation voltage on a particular transistor's collector is out of phase with the deviation voltage on its base and the amplitude is significantly greater, indicating a large deviation gain.

When both the common-mode and deviation voltages are changing independently, the equations for the input voltages become

$$\Delta v_{I_1} = \Delta v_{I_{CM}} + \Delta v_{I_{DEV}} \tag{1-18}$$

$$\Delta v_{I_2} = \Delta v_{I_{CM}} - \Delta v_{I_{DEV}} \tag{1-19}$$

and those for the output voltages become

$$\Delta v_{O_1} = \Delta v_{O_{CM}} + \Delta v_{O_{DEV}} \tag{1-20}$$

$$\Delta v_{O_2} = \Delta v_{O_{CM}} - \Delta v_{O_{DEV}} \tag{1-21}$$

An example will help clarify the use of these equations.

EXAMPLE 1-8

A differential amplifier has a common-mode gain of -0.8 and a differential gain of -10. Base 2 is grounded and base 1 has the following voltage applied:

$$\Delta v_{I_1} = 1.0 \sin(2\pi \cdot 10t)$$

Determine the maximum and minimum voltages that appear on collectors 1 and 2.

Solution

$$\Delta v_{I_{CM}} = \frac{\Delta v_{I_1} + \Delta v_{I_2}}{2} = \frac{1.0 \sin(2\pi \cdot 10t) + 0}{2}$$

$$= 0.5 \sin(2\pi \cdot 10t)$$

$$\Delta v_{I_{DEV}} = \pm \frac{\Delta v_{I_1} - \Delta v_{I_2}}{2} = \frac{1.0 \sin(2\pi \cdot 10t) - 0}{2}$$

$$= 0.5 \sin(2\pi \cdot 10t)$$

$$\Delta v_{Id} = 1.0 \sin(2\pi \cdot 10t)$$

$$\Delta v_{O_{CM}} = (A_{CM})(\Delta v_{I_{CM}}) = (-0.8)[0.5 \sin(2\pi \cdot 10t)]$$

$$= -0.4 \sin(2\pi \cdot 10t) = +0.4 \sin(2\pi \cdot 10t + 180°)$$

$$\Delta v_{O_{DEV}} = (A_D)(\Delta v_{Id}) = (-10)[1.0 \sin(2\pi \cdot 10t)]$$

$$= -10 \sin(2\pi \cdot 10t) = +10 \sin(2\pi \cdot 10t + 180°)$$

The common-mode voltage on both collectors is out of phase with respect to that on base 1. The deviation voltage on collector 1 is out of phase with respect to the voltage on base 1 and the voltage on collector 2 is in phase with respect to that on base 1. Thus the voltage on collector 1 is

$$\Delta v_{O_1} = \Delta v_{O_{CM}} + \Delta v_{O_{DEV}}$$

$$= -0.4 \sin(2\pi \cdot 10t) - 10 \sin(2\pi \cdot 10t)$$

$$= -10.4 \sin(2\pi \cdot 10t)$$

and the voltage on collector 2 is

$$\Delta v_{O_2} = \Delta v_{O_{CM}} - \Delta v_{O_{DEV}}$$

$$= -0.4\sin(2\pi \cdot 10t) + 10\sin(2\pi \cdot 10t)$$

$$= +9.6\sin(2\pi \cdot 10t)$$

Thus the ac voltage on collector 1 has an amplitude of 10.4 V peak while that on collector 2 has an amplitude of 9.6 V peak.

Example 1-8 illustrates a phenomenon that can be used as a measurement tool on differential amplifiers; when the two collector voltages do not have the same ac amplitude, a common-mode voltage shift is occurring. The amount of common-mode shift can be determined using Eq. (1-4) and the amount of deviation swing can be determined using Eq. (1-12) as illustrated in Example 1-9.

EXAMPLE 1-9

Determine the maximum common-mode shift and the deviation swing for the collector voltages in Example 1-8.

Instant	v_{O_1}	v_{O_2}
t_1	-10.4	$+9.6$
t_2	$+10.4$	-9.6

Solution

$$\Delta v_{O_{CM}} = \frac{[10.4 - (-10.4)] + (-9.6) - 9.6}{2}$$

$$= \frac{20.8 - 19.2}{2} = \frac{1.6}{2} = 0.8 \text{ V p-p (or 0.4 V peak)}$$

which is the voltage excursion of the common-mode voltage from t_2 to t_1.

$$\Delta v_{O_{DEV}} = \frac{[10.4 - (-10.4)] - [(-9.6) - 9.6]}{2}$$

$$= \frac{20.8 + 19.2}{2} = 20 \text{ V p-p (or 10 V peak)}$$

which is the voltage deviation on the two collectors over the time interval between t_1 and t_2.

A more realistic situation occurs with differential amplifiers when the two collectors are slightly unbalanced and are at a positive dc voltage level. The equations for common-mode and deviation gain are still valid, as illustrated in Example 1-10 and Fig. 1-3.

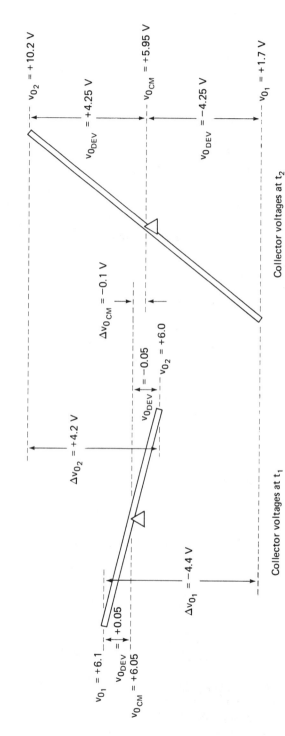

Fig. 1-3 Teeter-totter representation of differential amplifier operation on output (Example 1-10).

EXAMPLE 1-10

A differential amplifier has the following voltages on the collectors:

Instant	v_{O_1}	v_{O_2}
t_1	+6.1	+6.0
t_2	+1.7	+10.2

Find the instantaneous common-mode and deviation voltages at t_1 and t_2. Also find the change in common-mode and deviation voltages over the interval $(t_2 - t_1)$. Check your answers (see Fig. 1-3).

Solution

$$v_{O_{CM}}(\text{at } t_1) = \frac{6.1 + 6.0}{2} = +6.05 \text{ V}$$

$$v_{O_{CM}}(\text{at } t_2) = \frac{1.7 + 10.2}{2} = +5.95 \text{ V}$$

A downward shift in common-mode voltage occurred between t_1 and t_2. This shift is determined by

$$\Delta v_{O_{CM}}(\text{between } t_2 \text{ and } t_1) = \frac{(1.7 - 6.1) + (10.2 - 6.0)}{2}$$

$$= \frac{-4.4 + 4.2}{2} = -0.1 \text{ V}$$

$$v_{O_{DEV}}(\text{at } t_1) = \pm\frac{6.1 - 6.0}{2} = \pm 0.05 \text{ V}$$

$$v_{O_{DEV}}(\text{at } t_2) = \pm\frac{1.7 - 10.2}{2} = \mp 4.25 \text{ V}$$

The deviation has increased in absolute value and has changed phase. The amount of change is given by

$$\Delta v_{O_{DEV}}(\text{between } t_2 \text{ and } t_1) = \frac{(10.2 - 6.0) - (1.7 - 6.1)}{2}$$

$$= \frac{+4.2 + 4.4}{2} = +4.3 \text{ V}$$

Answer check: The voltage on each collector can be found at t_1 from

$$v_{O_1}(\text{at } t_1) = v_{O_{CM}} + v_{O_{DEV}} = 6.05 + 0.05 = 6.1 \text{ V}$$

$$v_{O_2}(\text{at } t_1) = v_{O_{CM}} - v_{O_{DEV}} = 6.05 - 0.05 = 6.0 \text{ V}$$

Voltages at t_2 are

$$v_{O_1}(\text{at } t_2) = 5.95 + (-4.25) = +1.7 \text{ V}$$

$$v_{O_2}(\text{at } t_2) = 5.95 - (-4.25) = +10.2 \text{ V}$$

The change in v_{O_1} between t_1 and t_2 is found from

$$\Delta v_{O_1} = \Delta v_{O_{CM}} + \Delta v_{O_{DEV}} = -0.1 + (-4.3) = -4.4 \text{ V}$$

$$\Delta v_{O_2} = -0.1 - (-4.3) = +4.2 \text{ V}$$

These voltages can be seen to be correct when checked against the original collector voltages.

1.4.4 Common-Mode Rejection Quotient

The common-mode rejection quotient (CMRQ) [1] is a measure of the amount of differential gain present with respect to the amount of common-mode gain in a differential amplifier. Thus the CMRQ is

$$\text{CMRQ} = \frac{A_D}{A_{\text{CM}}} \qquad (1\text{-}22)$$

which in the case of Example 1-8 is $-10/-0.8 = 12.5$. This means that the differential gain is 25 times as large as the common-mode gain. Since, in the ideal case, the differential amplifier has only differential gain, the CMRQ is ideally infinite and A_{CM} is ideally zero. Typical operational amplifiers have differential gains in the first differential amplifier stage ranging between 100 and 10^6. They have common-mode gains ranging between 10^{-4} and 10^{-5}. Thus typical CMRQ's range between $\times 10^6$ and $\times 10^{11}$.

Operational amplifiers use multiple cascaded differential amplifiers for large voltage gains (to be discussed later). One would normally expect this to increase the CMRQ as stages are added. According to Burr-Brown [2], an established operational amplifier manufacturer, the law of diminishing returns prevents this from occurring. Thus the CMRQ obtained in the first differential amplifier stage is usually all that can be expected. The CMRQ is one of the good features of the operational amplifier which occurs in the differential amplifier.

1.5 Bipolar Transistor Differential Amplifier

The equations developed in Section 1.5 can be developed using bipolar transistor parameters and circuit values.

1.5.1 Common-Mode Gain Equation

In Fig. 1-1 it can be seen that if the two bases are connected together (to assure common-mode operation), the two transistors act together as though they were one (presuming matched transistors and resistors). Both collector voltages rise and fall in unison as the base voltage is varied. Since both emitter currents pass through a common-emitter resistor and the voltage drop across that resistor is twice the value as if only one transistor existed, Fig. 1-1 can be redrawn as shown in Fig. 1-4a.

The effect of twice the voltage being dropped across the emitter resistor is simulated by separating the two emitter resistors and making each one twice the resistance value of the original emitter resistor. Since the two transistors act in unison, the effects of only one need be analyzed as shown in Fig. 1-4b, to determine the common-mode gain equation.

The input impedance of each transistor of the differential amplifier is

$$R_{\text{in}} = h_{ie} + (h_{fe} + 1)(2R_E) \simeq 2h_{fe}R_E \qquad (1\text{-}23)$$

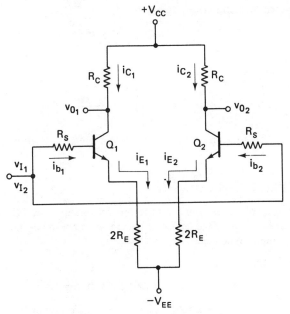

(a) Circuit with split emitter resistor

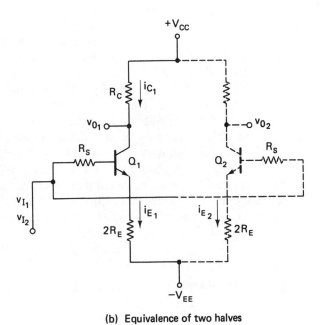

(b) Equivalence of two halves

Fig. 1-4 Circuit for common-mode rejection.

and it follows that the voltage gain is

$$A_v = \frac{-h_{fe}R_C}{R_{in}} \simeq \frac{-h_{fe}R_C}{2h_{fe}R_E} \simeq \frac{-R_C}{2R_E}$$

Thus

$$A_{CM} = \frac{-R_C}{2R_E} \tag{1-24}$$

This is the voltage gain from the two common inputs to *either* transistor collector output.

EXAMPLE 1-11

A differential amplifier uses 50 kΩ collector resistors and a 5 kΩ emitter resistor. $h_{fe} = 99$ and $h_{ie} = 25$ kΩ ($I_C = 100$ μA). Determine the common-mode gain using both the exact and approximate input impedance.

Solution

$$R_{in} = 25\ \text{k}\Omega + (100)(2)(5\ \text{k}\Omega) = 1025\ \text{k}\Omega\ (\text{exact})$$

$$= 2(99)(5\ \text{k}\Omega) = 990\ \text{k}\Omega\ (\text{approximate})$$

$$A_{CM} = -\frac{(99)(50\ \text{k}\Omega)}{1025\ \text{k}\Omega} = -4.8\ (\text{exact})$$

$$\simeq -\frac{50\ \text{k}\Omega}{2(5\ \text{k}\Omega)} \simeq -5\ (\text{approximate})$$

1.5.2   Differential Gain Equation

In calculating the deviation gain, it must be presumed that the common-mode gain is zero, so that it does not enter into the expression.

Figure 1-5 is similar to Fig. 1-1 except that it depicts the change in collector and emitter currents as signal varies. From Fig. 1-5a it can be seen that as one collector current rises, the other falls in a like amount. Since most of the emitter current is composed of collector current, the respective emitter currents also rise and fall in a like manner. Both emitter currents pass through the R_E resistor and as one current rises and the other falls, the nominal current passing through R_E exhibits no change. It thus appears, as in Fig. 1-5b, to the two transistors that the emitter is bypassed with a capacitor (C_E), since no signal current change exists across R_E.

The gain from either input to a collector output can now be determined. For the bypassed (grounded) emitter case, the deviation voltage gain is

$$\Delta v_{O_{DEV_1}} = -\frac{h_{fe}R_C}{R_s + h_{ie}}\Delta v_{I_{DEV}} \tag{1-25}$$

Substituting the differential input voltage, Eq. (1-11), for $\Delta v_{i_{DEV}}$, we get

$$\Delta v_{O_{DEV_1}} = -\frac{h_{fe}R_C}{R_s + h_{ie}}\left(\frac{\Delta v_{Id}}{2}\right) = -\frac{h_{fe}R_C}{2(R_s + h_{ie})}\Delta v_{Id}$$

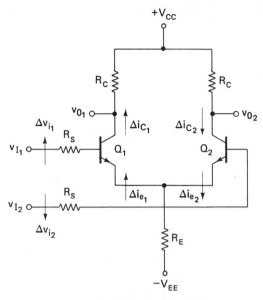

(a) Circuit showing signal phases

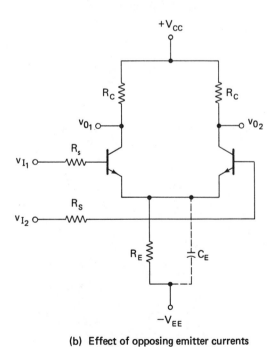

(b) Effect of opposing emitter currents

Fig. 1-5 Circuit for deviation-gain equation.

Then the *differential* voltage gain is

$$A_D = \frac{\Delta v_{O_{DEV_1}}}{\Delta v_{Id}} = -\frac{h_{fe}R_C}{2(R_s + h_{ie})} \tag{1-26}$$

Equation (1-26) also applies for the differential gain to collector 2. The only difference between the two equations (A_D to collectors 1 and 2) is the sign; the equation for A_D to collector 2 has a positive sign.

EXAMPLE 1-12

A differential amplifier has $R_S = 10$ kΩ, $R_C = 50$ kΩ, $R_E = 75$ kΩ, $h_{fe} = 80$, and draws 100 μA of current in each collector. Determine the differential gain.

Solution

$h_{ie} = 80(26 \text{ mV}/100 \ \mu\text{A}) = 20.8$ kΩ for each transistor of the differential amplifier.

$$A_D = -\frac{(80)(50 \text{ k}\Omega)}{2(10 \text{ k}\Omega + 20 \text{ k}\Omega)} = -67$$

Differential amplifiers used in operational amplifiers typically draw small collector currents. This causes the h_{ie} to be large, which in turn causes the differential gain to be relatively small (< 100).

1.5.3 Common-Mode Rejection Quotient

The common-mode rejection quotient [1] is found using Eqs. (1-24) and (1-26).

$$\text{CMRQ} = \frac{A_D}{A_{CM}} = \frac{h_{fe}}{R_S + h_{ie}}(R_E) \tag{1-27}$$

Since R_S is of the same order of magnitude as h_{ie}, it can be seen that only one controllable function exists in the equation; that function is R_E. It is thus R_E which sets the size of CMRQ. If we wish a large CMRQ, we must have a large R_E; but how large?

The following example will help in deciding the minimum value for R_E.

EXAMPLE 1-13

Let $R_S = 1$ kΩ, $h_{ie} = 1200$ Ω, $h_{fe} = 80$, and CMRQ = 31,600. What value must be selected for R_E to satisfy these criteria?

Solution

Rearranging Eq. (1-27), we have

$$R_E = \frac{(CMRQ)(R_S + h_{ie})}{h_{fe}}$$

$$= \frac{(31,600)(2200)}{80} = 870 \text{ k}\Omega$$

A physical resistor is out of the question because of the large negative voltage required to maintain a reasonable tail current.

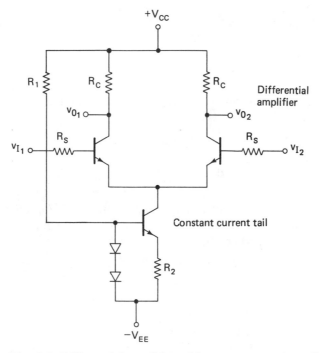

Fig. 1-6 Differential amplifier with constant current tail.

Thus the need for a constant current source exhibiting a large output impedance becomes apparent. One circuit that will yield the desired impedance while maintaining a reasonable current is the common-base circuit. This circuit has all the desirable characteristics for providing a large CMRQ while maintaining a reasonable tail current. The differential amplifier with the common-base constant current tail is shown in Fig. 1-6.

Since the output impedance of the common-base tail circuit is about 1 MΩ, it satisfies the criteria set forth in Example 1-13 of being in excess of 870 kΩ. If the current tail output impedance is 1 MΩ, the CMRQ under the previous conditions will be 36,364.

One added benefit gained using the constant current tail is that any nonlinear effects in the transistors as well as the common-mode gain are essentially eliminated. Only the differential gain exists with a constant current tail on the differential amplifier. This occurs because as one collector current rises, the other collector current must drop by exactly the same amount to maintain the tail current constant.

1.5.4 Constant Current Tail

A constant current tail is a common-base amplifier stage without a signal input lead. The whole purpose of this stage is to exhibit a high output impedance at a fixed collector current. The collector current is a function of the emitter current if β is high, which is usually the case.

To design the current tail, one first determines the desired collector current and the design then establishes parameters that fulfill this requirement. The circuit of a current tail is shown in Fig. 1-6. The two diodes between base and $-V_{EE}$ hold the base voltage constant at $-V_{EE} + 1.4$ V and this *ac grounds* the base to a specific dc voltage. The value of R_2 then sets the emitter current, which, in turn, sets the collector current. Writing the loop equation around the base, emitter, R_E, diode loop, reveals that

$$I_E = \frac{1.4 \text{ V} - V_{BE}}{R_2} \approx I_C \qquad (1\text{-}28)$$

The output impedance of the current tail is found using the equation for the output impedance of a common-emitter amplifier [3, p. 344] with an emitter resistor. The exact equation and the two succeedingly more simple (and less accurate) equations are given below. (*Note:* The output impedance is dependent on the source impedance on the base, which, because of the two diodes, is zero for our case.)

$$R_O = \left(\frac{1}{h_{oe}}\right) \frac{(1 + h_{fe})R_E + (R_S + h_{ie})(1 + h_{oe}R_e)}{R_E + R_S + h_{ie} - (h_{re}h_{fe}/h_{oe})} \quad \begin{matrix} \text{exact} \\ \text{equation} \end{matrix} \quad (1\text{-}29a)$$

$$R_O = \left(\frac{h_{fe}}{h_{oe}}\right) \frac{R_E + r_e}{(1 + h_{fe})r_e + R_E} \quad \begin{matrix} \text{first} \\ \text{approximation} \end{matrix} \quad (1\text{-}29b)$$

$$\left[\text{see Section 2.8.1, where } r_e \cong h_{ie}/(1 + h_{fe})\right]$$

$$R_O = \frac{h_{fe}}{h_{oe}} \quad \begin{matrix} \text{second} \\ \text{approximation} \end{matrix} \quad \left(\begin{matrix} \text{least} \\ \text{accurate} \end{matrix}\right) \quad (1\text{-}29c)$$

Equation (1-29b) is sufficiently accurate for any textbook problems; besides, h_{re} is rarely given in transistor specifications and it is not found from simple transistor measurements. An example will illustrate the use of Eq. (1-29b).

EXAMPLE 1-14

Find the value of R_2 and the output impedance of a current tail having a collector current of 200 μA, $h_{oe} = 16$ μS, $\beta = 80$.

Solution

$$R_2 = \frac{1.4 \text{ V} - 0.7 \text{ V}}{0.2 \text{ mA}} = 3.5 \text{ k}\Omega$$

$$r_e = \frac{26 \text{ mV}}{200 \text{ μ}A} = 130 \text{ }\Omega$$

$$R_O = \left(\frac{80}{16 \text{ μS}}\right) \frac{130 + 3500}{(81)130 + 3500} = 1.29 \text{ M}\Omega$$

Now CMRQ will have the needed high value.

Besides the common-base transistor circuit just described, three other circuits are used as current tails in monolithic operational amplifiers. They are the two- and three-transistor current mirrors and the Widlar current mirror. The two-transistor

     Construction of the Operational Amplifier

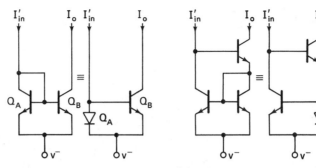

(a) Two transistor current mirror   (b) Three transistor current mirror

Fig. 1-7 Monolithic current mirrors.

current mirror illustrated in Fig. 1-7a has an output current that is related to the input current by the relationship

$$I_o = I_{in}\left(\frac{h_{fe}}{h_{fe} + 2}\right) \tag{1-30}$$

The three-transistor current mirror (Fig. 1-7b) has the relationship

$$I_o = I_{in}\left(\frac{h_{fe}^2 + 2h_{fe}}{h_{fe}^2 + 2h_{fe} + 2}\right) \tag{1-31}$$

In Eq. (1-30), I_o is within 2% of I_{in} when h_{fe} is 100; for Eq. (1-31), I_o is within 0.02% of I_{in} when h_{fe} is 100. Thus both current mirrors are quite accurate with reasonable h_{fe}'s. The two-transistor current mirror has an output impedance greater than 100 MΩ. The output impedance of a three-transistor current mirror is greater than 1000 MΩ.

EXAMPLE 1-15

The transistors in a linear monolithic circuit using current mirrors all have $h_{fe} = 80$.

   a. Find the output current for the two-transistor current mirror, if $I_{in} = 100\ \mu A$.
   b. Find the output current for the three-transistor current mirror if the input current is 1 mA.

Solution

   a. $I_o = 100\ \mu A\left(\dfrac{80}{80 + 2}\right) = 97.56\ \mu A$

   b. $I_o = 1000\ \mu A\left[\dfrac{80^2 + 2(80)}{80^2 + 2(80) + 2}\right] = 1000\ \mu A\ (0.9997) = 999.7\ \mu A$

The Widlar current mirror [4, p. 592] uses a resistor in the emitter of the *controlled* transistor. This permits the controlled transistor to have a smaller current than the controlling transistor. The method of finding R_2 is as follows (see Fig. 1-7):

$$V_{BE1} = V_T \ln\left(\frac{I_{ref}}{I_{S1}}\right)$$

$$V_{BE2} = V_T \ln\left(\frac{I_O}{I_{S2}}\right)$$

$$I_{S1} = I_{S2} \quad \text{then } V_{BE1} - V_{BE2} = \ln\left(\frac{I_{ref}}{I_O}\right)$$

and from Fig. 1-7 we see that the loop equation becomes

$$V_{BE1} - V_{BE2} = I_{E2}R_2 \quad \text{and} \quad I_{E2} = I_O \quad \text{then} \quad R_2 = \frac{V_T}{I_O} \ln\left(\frac{I_{ref}}{I_O}\right) \quad (1\text{-}32)$$

where V_T is as given in Sec. 2-81. An example will illustrate the use of Eq. (1-32).

EXAMPLE 1-16

A Widlar current tail draws 200 μA when the reference current is 1 mA. Find the value of R_2.

Solution

$$R_2 = \left(\frac{26\ \text{mV}}{0.2\ \text{mA}}\right) \ln\left(\frac{1\ \text{mA}}{0.2\ \text{mA}}\right) = 209\ \Omega$$

1.6 Common-Mode Input Voltage Range

The use of a constant current tail on the differential amplifier produces a circuit with remarkable characteristics. The two input leads can have the same voltage difference but vary over a wide common-mode voltage range without causing *any* voltage change to occur on the collectors. But over what range can the base common-mode voltage vary? The differential gain of the differential amplifier is usually quite large; this means that saturation on the collectors occurs for a small difference in the two base voltages. Thus the differential input voltage is usually quite small and the two bases are nearly at the same potential.

The emitter voltage of the differential amplifier follows the common-mode base voltages, just V_{BE} volts lower, and the emitter voltage has two hard limits. The first hard limit takes place when the current tail transistor saturates. This occurs at $-V_{EE} + V_{CE_{Sat}} + I_E R_2$ for the current tail, which is usually about 1 V above the $-V_{EE}$ voltage. The second hard limit occurs when the differential amplifier transistors saturate. This occurs at $+V_{CC} - I_C R_C + 0.2$ V, which depends on the collector current and the R_C value.

It is within this voltage range that the collectors will not vary when the input common-mode voltage changes. This means that the input common-mode voltage can vary anywhere inside these two hard limits and *no* appreciable change will occur on the collectors. This is, of course, dependent on a large CMRQ in the differential amplifier, which implies an insignificant common-mode gain. The differential amplifier for an operational amplifier is designed to have the dc collector voltage be as close to $+V_{CC}$ as possible, to increase the input common-mode voltage range.

EXAMPLE 1-17

The differential amplifier illustrated in Fig. 1-6 has the following components: $R_C = 50\text{ k}\Omega$, $R_2 = 3.5\text{ k}\Omega$, $R_1 = 10\text{ k}\Omega$, and $R_S = 1\text{ k}\Omega$. $+V_{CC} = +10$ V and $-V_{EE} = -10$ V. Determine the input common-mode voltage range for this differential amplifier if $h_{fe} = 80$.

Solution

The voltage across R_2 is 700 mV (V_{BE}). The tail current is found to be $I_2 = 700\text{ mV}/3.5$ $\text{k}\Omega = 200\ \mu$A. Each collector current in the differential amplifier is 100 μA and the dc voltage drop across the R_C's is (100 μA)(50 kΩ) = 5 V. The lower limit of the differential amplifier emitters is -10 V $+ 0.7$ V $+ 0.2$ V $= -9.1$ V. The upper limit of the differential amplifier emitters is $+10$ V $- 5$ V $- 0.2$ V $= +4.8$ V. The bases of the differential amplifier are 0.7 V above the emitter voltages, so the base voltage range is $+5.5$ to -8.4 V.

1.7 Operational Amplifier Realization

The operational amplifier has three basic circuit stages formed from monolithic circuit elements:

1. An input differential amplifier.
2. A voltage-level shifter with voltage gain.
3. An output stage with current gain.

There are almost infinite variations of these basic stages.

1.7.1 Monolithic Circuit Elements

Most operational amplifiers that the user will encounter today are monolithic; that is, they are formed on one integrated substrate through masking and deposition methods.

Circuit elements, unavailable to the discrete element circuit designer, can be realized on a monolithic substrate. These elements make the schematic of the IC operational amplifier appear to be drastically different from the classical circuits just presented. A brief discussion of some of these methods will be presented here to permit a better understanding of IC circuits (see Fig. 1-8).

A diode may be formed by using a transistor with the collector tied to the base. The base–emitter junction then forms a forward-biased diode when current is passed in the forward direction (Fig. 1-8a).

The same circuit forms a zener diode of quite high figure of merit [5] when the current is passed in the reverse direction. The avalanche voltage ranges from 6 to 8 V (Fig. 1-8b). A zener diode in series with a forward-biased diode forms a temperature-compensated zener diode (Fig. 1-8c).

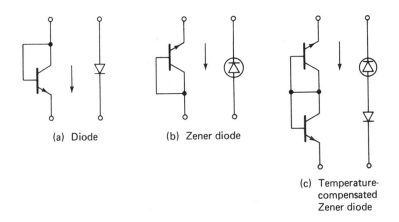

(a) Diode   (b) Zener diode

(c) Temperature-
compensated
Zener diode

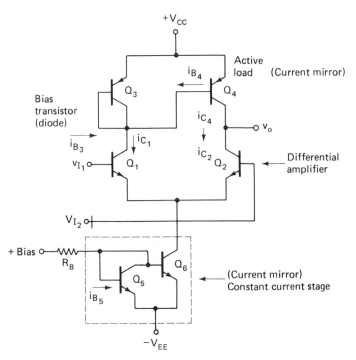

(d) Differential amplifier with current tail

Fig. 1-8 Monolithic realization of circuit components. (Reprinted with permission from *Analog Integrated Circuit Design*, Alan B. Grebene; Litton Publishing Company.

Transistors are less difficult to realize than resistors on a monolithic substrate; thus resistors are avoided at every opportunity. One of the most important circuits in the design of analog integrated circuits is the current mirror. In this circuit, the collector current of a transistor can be controlled precisely by the base and collector currents of another transistor. Thus to cause a specific collector current to flow, one need only hold the input current constant. The output impedence of the two-transistor current mirror is several hundred megohms; for the three-transistor mirror it is several thousand megohms. This circuit is used as both the constant current tail in a differential amplifier and as an "active load" in the collector of the differential amplifier. Through its use, both large differential gains and insignificant common-mode gains can be obtained. Thus a monolithic differential amplifier with a current mirror as the active collector load and current tail can have a differential gain of several thousand and a common-mode gain of 10^{-5}; this circuit is illustrated in Fig. 1-8d. It is through this means that designers of a monolithic op amps are able to secure large differential gains with few stages.

1.7.2 Differential Amplifier

The differential amplifier was discussed extensively in the preceding sections. Multiple differential amplifier stages are employed where the collectors of the first are used as sources for the bases of the second. This method usually produces a large-deviation voltage gain but no additional rejection of common-mode signals [2].

1.7.3 Level Shifter

As was illustrated in the examples of preceding sections, the nominal collector voltage on a differential amplifier output is offset from zero volts by a constant dc level. The operational amplifier requires a nominal zero dc output voltage for proper operation. Thus a voltage level shifter is employed between the differential amplifier and the output stage. Figure 1-9 illustrates several manifestations of a level shifter [5, sec. 4.4].

Circuits b, c, and d are of the emitter-follower type and thus have a voltage gain of nearly unity. Circuit a is a voltage divider, and thus a loss in gain is suffered while the dc voltage level is shifted.

1.7.4 Output Stage

The requirements of the output stage include a current gain, a low output impedance, and the ability to pass ac signals with large amplitudes and little distortion. Circuits that have all these qualities are basically forms of emitter followers. They can be either push-pull complementary-symmetry emitter followers or emitter followers with a current tail.

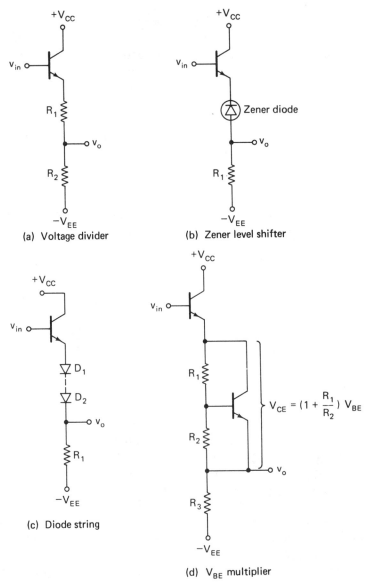

Fig. 1-9 Voltage-level shifters. (Reprinted with permission from *Analog Integrated Circuit Design*, Alan B. Grebene; Litton Publishing Company.

1.7.5 The Operational Amplifier

The schematic of a complete operational amplifier is illustrated in Fig. 1-10. An operational amplifier with a simple straightforward schematic was specifically chosen for this figure. One with multiple collectors and many additional circuits would have tended to mask the stages in the op amp. In Fig. 1-10 the input differential amplifier with its constant current tail is clearly visible. The second

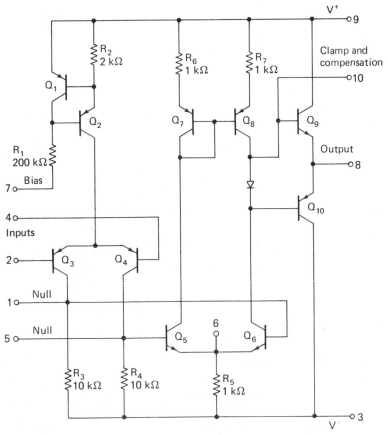

Fig. 1-10 LH0003/LH0003C wide-bandwidth operational amplifier. (Courtesy of National Semiconductor Corp., Santa Clara, Calif.)

differential amplifier, inverted from the first, serves a dual purpose; it is both the level shifter and an additional stage of voltage gain. The output stage with the two complementary-symmetry emitter followers has one diode between the bases to reduce crossover distortion. The amount of idling current drawn by the output stage is still low, as the op amp supply current is listed as being 3 mA.

Pin 7, bias input, is used to set the current through the first differential amplifier via its constant current tail; this permits adjustment of gain.

1.7.6 Operational Amplifier Symbol

The symbol for an operational amplifier is a triangle with two inputs and one output, as shown in Fig. 1-12a. The polarity signs of the inputs are placed inside the triangle adjacent to the appropriate leads.

Variations of the basic symbol also exist. These show the power-supply terminals, the offset "null" terminals, the frequency rolloff "compensation" terminals, and other miscellaneous terminals, such as the bias (7) and clamp (10) terminals in Fig.

Operational Amplifier Realization **27**

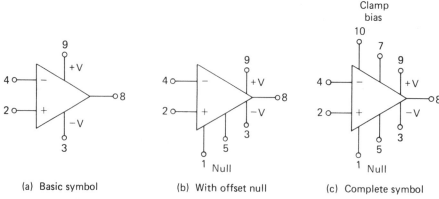

(a) Basic symbol   (b) With offset null   (c) Complete symbol

Fig. 1-11 Variations of the operational amplifier symbol.

1-10. Figure 1-12 illustrates the various configurations for the op amp symbol. The variations of the basic symbol are seldom used unless they add clarity to the understanding of the circuit operation.

1.7.7 Complementary MOSFET Operational Amplifier

An operational amplifier has been developed using a technology that is a fallout from a digital application. The *CO*mplementary—*S*ymmetry/*M*etal-*O*xide Semiconductor (COS/MOS) integrated-circuit principle is one where an *N*-channel and a *P*-channel MOS-FET are used together. An inverter is formed by connecting the gates and sources, respectively, of the two types of MOS-FETs. This circuit has remarkable characteristics: for use as the input stage of an operational amplifier, the input impedance is 10^{14} to 10^{16} Ω; for use as the output stage, each stage, when in saturation, will attain virtually the power-supply "rail" voltages. These two characteristics alone make the COS/MOS op amp have nearly ideal input and output characteristics. Several terms are used to replace the basic acronym by various manufacturers; some of these are CMOS, C-MOS, and McMOS.

Figure 1-11 illustrates a COS/MOS operational amplifier. The MOS-FETs on the input can clearly be seen, as can the COS/MOS inverter on the output. A combination of junction transistor and MOS-FET technology is used in constructing this operational amplifier.

1.8 Common-Mode Rejection Ratio

Up to this point, only the common-mode rejection quoteitn (CMRQ) has been discussed. We, as op amp users, never hear the term CMRQ used. Instead, we read in the manufacturer's specification sheet about the common-mode rejection ratio (CMRR). What, then, is the difference between the two? CMRQ has already been well defined in preceding sections as the ratio between differential and common-mode gains.

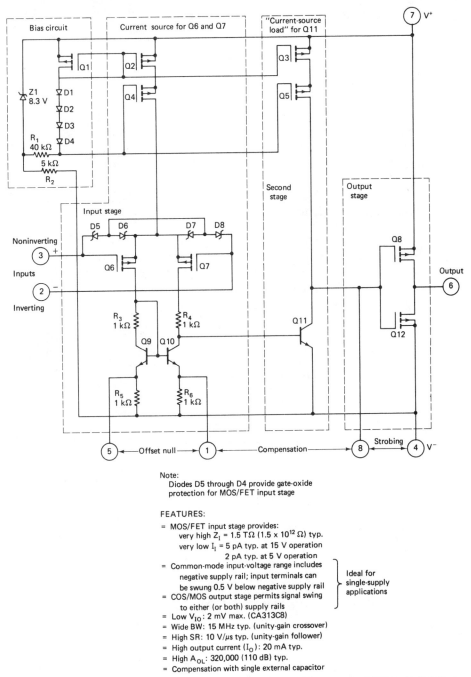

Note:
Diodes D5 through D4 provide gate-oxide
protection for MOS/FET input stage

FEATURES:

= MOS/FET input stage provides:
  very high Z_I = 1.5 TΩ (1.5 x 10^{12} Ω) typ.
  very low I_I = 5 pA typ. at 15 V operation
            2 pA typ. at 5 V operation
= Common-mode input-voltage range includes
  negative supply rail; input terminals can   Ideal for
  be swung 0.5 V below negative supply rail   > single-supply
= COS/MOS output stage permits signal swing   applications
  to either (or both) supply rails
= Low V_{IO}: 2 mV max. (CA313C8)
= Wide BW: 15 MHz typ. (unity-gain crossover)
= High SR: 10 V/μs typ. (unity-gain follower)
= High output current (I_O): 20 mA typ.
= High A_{OL}: 320,000 (110 dB) typ.
= Compensation with single external capacitor

Fig. 1-12 COS/MOS CA3130 operational amplifier. (Courtesy of RCA Solid State, Somerville, N.J.)

The IEEE definition [1, p. 98] for CMRR is

$$\text{CMRR} = \frac{\Delta V_{I_{CM}}}{\Delta V_{O_{CM}}}(A_{\text{VCL}}) = \frac{A_{\text{VCL}}}{A_{\text{CM}}} \tag{1-33}$$

where A_{VCL} is the "closed loop gain." Close examination of this equation reveals that it is the reciprocal of Eq. (1-5) for common-mode gain with the exception of the (A_{VCL}) term. Presuming that the (A_{VCL}) is unity, Eqs. (1-5) and (1-33) are reciprocals. Thus the common-mode rejection ratio (CMRR) for an operational amplifier is the reciprocal of the common-mode gain between the input leads and the output lead, presuming that the closed loop gain is unity. The meaning of CMRR for gains other than unity will be discussed in Chapter 4.

The common-mode rejection ratio is usually converted to dB and labeled common-mode rejection (dB) or CMR (dB), which is calculated using

$$\text{CMR(dB)} = 20\log_{10}(\text{CMRR}) \tag{1-34}$$

Typical operational amplifiers have CMR values ranging between 90 and 100 dB; this represents CMRR values ranging between 31,623 and 100,000.

The change in output voltage for a given change in common-mode input voltage is given by [1]

$$\Delta V_{O_{CM}} = \frac{-\Delta V_{I_{CM}}}{\text{CMRR}}(A_{\text{VCL}}) \tag{1-35}$$

Remember, common-mode input signals occur when the voltages on the two input leads move together in unison.

EXAMPLE 1-18

An operational amplifier circuit with $A_{\text{VCL}} = 1$ has 20 μV rms on the output when a common-mode voltage of 1 V rms is present on the input. Find the CMRR and the CMR(dB).

Solution

$$\text{CMRR} = \frac{1 \text{ V rms}}{20 \text{ } \mu\text{V rms}}(1) = 50,000$$

$$\text{CMR(dB)} = 20\log_{10}(50,000)$$
$$= 20(4.7) = 94 \text{ dB}$$

EXAMPLE 1-19

A specification sheet for an operational amplifier gives CMR(dB) as 91 dB. Determine the CMRR and the expected output voltage (rms) for a 5.6 V rms common-mode input signal.

Solution

$$91 \text{ dB} = 20\log_{10}(\text{CMRR})$$

$$\frac{91}{20} = 4.55 = \log_{10}(\text{CMRR})$$

$$\text{CMRR} = 10^{4.55} = 35,481$$

$$\Delta V_{O_{CM}} = \frac{-5.6}{35,481}(1) = -158 \text{ } \mu\text{V rms}$$

REFERENCES

1. *IEEE Standard Dictionary of Electrical and Electronics Terms* (New York: Wiley-Interscience, 1972), p. 98 (0-18E1).
2. **Jerald G. Graeme, Gene E. Tobey, and Lawrence P. Huelsman eds.,** *Operational Amplifiers Design and Applications* (New York: McGraw-Hill Book Company, 1971), p. 96.
3. **M. J. Hellstrom and W. R. Harden,** *An Integrated Low Voltage Operational Amplifier* (Int. Solid State Ckts. Conf., Digest Tech. Papers, 12, 1969), pp. 16–17.
4. **Jacob Millman and Christos C. Halkias,** *Electronic Devices and Circuits* (New York: McGraw-Hill Book Company, 1967).
5. **Alan B. Grebene,** *Analog Integrated Circuit Design* (New York: Van Nostrand Reinhold Co., 1972), pp. 157–158.

PROBLEMS

1. A bipolar transistor differential amplifier pair has $+1$ V on base 1 and $+2$ V on base 2. Determine the value of the common-mode input voltage.

2. A differential amplifier has -1.5 V on base 1 and -3 V on base 2. Determine the common-mode input voltage.

3. A bipolar transistor differential amplifier has $+10$ V on collector 1 and -5 V on collector 2. Determine the common-mode output voltage.

4. A differential amplifier has -5 V on collector 1 and $+2$ V on collector 2. Determine the common-mode output voltage.

5. A differential amplifier has a common-mode input voltage of $+1.75$ V. If -2.2 V is on base 2, what is the voltage on base 1?

6. $+7.2$ V is the common-mode output voltage for a differential amplifier. What is the voltage on collector 2 if the voltage on collector 1 is $+5$ V?

7. With one base of a differential amplifier grounded, the common-mode input voltage is -0.75 V. What is the voltage on the other base?

8. At t_1, $V_{i_1} = -3$ V and $V_{i_2} = +1$ V; at t_2, $V_{i_1} = +2$ V and $V_{i_2} = +0.75$ V.

   a. Find the common-mode input voltage at t_1.

   b. Calculate the common-mode input voltage at t_2.

   c. Find the change in common-mode voltage between t_1 and t_2 by solving for $\Delta V_{i_{CM}}$.

   d. Check your answer by subtracting the common-mode input voltage in part (a). from that in part (b).

9. The ac voltage on collector 1 is $\Delta V_{o_1} = 0.5\sin(100\pi t)$. The ac voltage on collector 2 is $\Delta V_{o_2} = 0.7\sin(100\pi t)$. Find the equation for the ac common-mode output voltage.

10. The ac and dc components of the collector 1 output voltage are

$$V_{o_1} = +3 + 0.85\sin(1000\pi t)$$

and the collector 2 voltage is given by the expression

$$V_{o_2} = +2.3 - 0.45 \sin(1000\pi t)$$

a. Determine the dc component of the common-mode output voltage.

b. Determine the ac component of the common-mode output voltage.

c. Find the equation for the total common-mode output voltage (include both the ac and dc components).

11. A differential amplifier has a difference in common-mode input voltage of -6 V. At the same time the change in common-mode output voltage is 3 V. Calculate the common-mode gain.

12. A differential amplifier has a common-mode gain of -0.4. The dc component of the output voltage when $V_{i_{CM}} = 0$ V is $+6.2$ V. The common-mode input voltage is $\Delta V_{i_{CM}} = 2 \sin[(\pi/3)t] - 0.8 \sin(10\pi t)$.

a. Find the equation for the combined dc and ac common-mode output voltages.

b. Determine the maximum and minimum common-mode output voltage.

c. Using your answer in part b., verify the dc output common-mode voltage given.

13. The input voltages to a bipolar transistor differential amplifier are $V_{i_1} = -6$ V and $V_{i_2} = +4.7$ V. Calculate the input differential voltage.

14. Find the deviation voltage for Problem 13.

15. A bipolar transistor differential amplifier has the following output voltages on its collectors:

$$\text{At } t_1: \quad V_{o_1} = +3.8 \text{ V and } V_{o_2} = -2.3 \text{ V};$$

$$\text{At } t_2: V_{o_1} = -0.7 \text{ V and } V_{o_2} = +1.2 \text{ V}.$$

a. Find the output deviation voltage at t_1.

b. Find the output deviation voltage at t_2.

c. Determine the change in output deviation voltage in the time interval between t_1 and t_2.

d. Verify the answer in part (c). by subtracting the deviation in part (a). from that in part (b).

16. Calculate the change in differential output voltage for Problem 15.

17. A transistor differential amplifier has a change in input deviation voltage over a time interval of ± 3.7 V. The change of voltage in base 1 during this time interval is -6.8 V.

a. Find the change in voltage on base 2 over the time interval.

b. Check your answer.

18. The input deviation voltage on a differential amplifier is $\Delta V_{i_{DEV}} = 0.0125 \sin(100\pi t)$. The output deviation voltage is $\Delta V_{o_{DEV}} = -2.5 \sin(100\pi t)$.

a. Find the deviation gain.

b. Find the differential gain.

19. The input voltages on a differential amplifier are as follows:

At t_1: $V_{i_1} = +2$ V, $V_{i_2} = -4$ V, $V_{o_1} = -6$ V, and $V_{o_2} = +12.2$ V

At t_2: $V_{i_1} = -2.5$ V, $V_{i_2} = -4.6$ V, $V_{o_1} = +3$ V, and $V_{o_2} = +5.0$ V

   a. Find the change in common-mode input voltage over the time interval.
   b. Find the change in common-mode output voltage over the time interval.
   c. Determine the common-mode gain.
   d. Determine the change in the input deviation voltage over the time interval.
   e. Determine the change in the output deviation voltage over the time interval.
   f. Find the deviation gain.
   g. Find the differential gain.

20. Using the results from parts (a), (b), (d), and (e) of Problem 19, show that the output and input terminal voltage changes are as expected.

21. A differential amplifier produces the following oscilloscope readings:

$$V_{i_1} = 3.7 \sin(200\pi t) \qquad V_{o_1} = +8.5 + 6.5 \sin(200\pi t)$$

$$V_{i_2} = 3.8 \sin(200\pi t) \qquad V_{o_2} = +8.4 - 6.6 \sin(200\pi t)$$

   a. Determine the common-mode gain.
   b. Find the deviation gain.
   c. Calculate the differential gain.
      (*Hint:* Let the positive peak voltage of V_{i_1} occur at t_1, and the negative peak voltage of V_{i_1} occur at t_2. Build a table of terminal voltages at t_1 and t_2.)
   d. Calculate the CMRQ.

22. A differential amplifier has a differential gain of 160 and a common-mode gain of 0.02. Find the CMRQ.

23. An operational amplifier uses a differential amplifier with a deviation gain of 400 and a common-mode gain of 2×10^{-3}. Find the CMRQ.

24. A bipolar transistor differential amplifier with a resistor tail has $R_C = 4$ kΩ, $R_E = 4$ kΩ, $R_S = 2$ kΩ $h_{fe} = 100$, $h_{ie} = 2.7$ kΩ, $V_{be} = 0.8$ V, and is biased with ± 15 V supply voltages.

   a. Find the input impedance of each base to ground.
   b. Find the exact common-mode voltage gain.
   c. Determine the approximate A_{CM}.

25. The circuit of Problem 24 has base 1 driven by a signal source with an internal impedance equivalent to a 1.5 kΩ resistor to ground. Find the differential gain A_D.

26. Calculate the CMRQ for the circuit of Problem 25.

   a. Use the solutions from part (b) of Problem 24, and Problem 25.
   b. Use Eq. (1-27).
   c. How can CMRQ be increased?
   d. What problem is encountered by the method of part (c)?

27. Select typical power supply voltages of ± 15 V and then find the resistor values to be used in a transistor differential amplifier with a resistor tail. The amplifier must meet the following specifications: $A_D = -75$, $A_{CM} = -0.6$, CMRQ $= 125$, $R_s = 2$ kΩ, $h_{fe} = 150$, $h_{ie} = 3.3$ kΩ, $V_{be} = 0.7$ V, and $I_{c_1} = I_{c_2} = 1.1$ mA. (*Note:* If the two bases have a O-V dc average, the two emitter voltages are at a constant -0.7 V dc.)

28. A differential amplifier with a constant current tail, shown in Fig. P1-28, uses transistors with the following characteristics: $h_{fe} = 100$, $h_{oe} = 37 \times 10^{-6}$, and $V_{be} = 0.7$ V. The amplifier must perform as follows: $A_D = -75$, $A_{CM} = -2.5 \times 10^{-3}$. If the supply voltages are ± 10 V and the bases have 1.5 kΩ resistors in series:

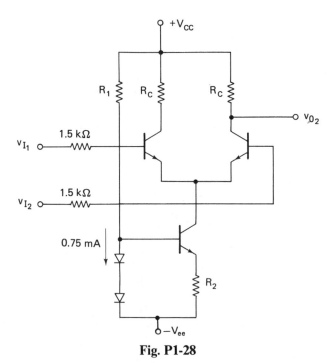

Fig. P1-28

   a. Calculate all resistor values for the circuit.

   b. Calculate the common-mode input voltage range.

   c. When both inputs are connected together and a 2 V peak sine wave is applied, how much ac voltage appears on collector 2? What is its phase relationship to the input voltage?

29. Draw a block diagram of an op amp showing only the basic stages. Describe the purpose of each stage.

30. A differential amplifier has a dc collector voltage of $+8.2$ V when no differential voltage is on the input. The emitter follower output stage must have O V on the emitter. $V_S = \pm 12$ V.

   a. Select the values of R_1 and R_2 if the level shifter draws 1.5 mA collector current.

   b. How much gain does the level shifter have?

31. The same circuit is used, but the level shifter is replaced with a V_{be} multiplier circuit as shown in Fig. P1-31.

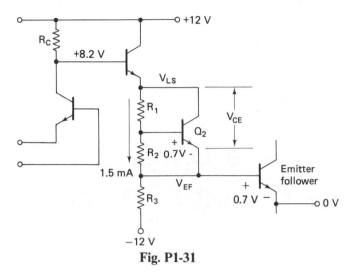

Fig. P1-31

a. Calculate the values of R_1, R_2, and R_3.

b. Determine the loss of gain through the V_{be} multiplier.

32. An op amp has 75 μV rms on its output when the inputs have a common-mode voltage of 5 V p-p. Determine the CMRR if the closed loop gain is unity.

33. Find the CMR(dB) for Problem 32.

34. The CMR(dB) of an operational amplifier is given as 83.5. Find the CMRR.

35. An operational amplifier has a CMRR of 50,000. A common-mode signal voltage of 9 V p-p is on the input terminals. How much voltage appears on the output if the closed loop gain is unity?

36. An operational amplifier has a CMRQ $= 2 \times 10^9$ and a differential gain of 92,500. Determine the CMRR if $A_{\text{VCL}} = 1$.

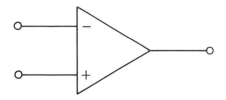

2 Using the Operational Amplifier as a Circuit Element

2.1 Introduction

The operational amplifier, when connected with input and feedback resistors, has a "transfer function" (v_o/v_{in}) that contains only the resistor values in the equation. None of the op amp parameters appear in the transfer-function equation. This phenomenon occurs because of the large differential gain and the feedback between input and output on the operational amplifier.

2.2 Operational Amplifier

The operational amplifier is a device with a differential amplifier at the input and a single-ended output which has a large differential voltage gain. This gain was A_D, as defined in Chapter 1, but will hereafter be called A_{VOL} (voltage-gain open loop). This gain is between the differential inputs and the output of the op amp. The common-mode gain will be presumed to be extremely small; thus only the differential gain is significant. Figure 2-1 illustrates such an amplifier, where it can be seen that

$$v_{Id} = v_B - v_A \qquad (2\text{-}1)$$

and

$$v_o = A_{VOL}(v_B - v_A) = A_{VOL}(v_{Id}) \qquad (2\text{-}2)$$

With v_B grounded (zero volts), $v_{id} = -v_A$ and

$$v_A = -\frac{v_o}{A_{VOL}} \qquad (2\text{-}3a)$$

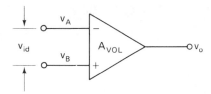

Fig. 2-1 Differential operational
amplifier.

Thus after rearranging Eq. (2-2) with $v_B = 0$ and $v_{Id} = -v_A$, it follows that

$$v_{Id} = \frac{v_o}{A_{VOL}} \qquad (2\text{-}3b)$$

An example will help to clarify the use of Eqs. (2-1) and (2-2).

EXAMPLE 2-1

An operational amplifier has an open loop gain of 100,000. The different input voltage is -10 μV rms; what is the output voltage?

Solution

$$v_o = (A_{VOL}) v_{Id}, \; v_o = (100,000)(-10 \; \mu V) = -1 \text{ V rms}$$

This means that for $v_A = 10 \; \mu V$ and $v_B = 0$ V or for $v_A = 1.000010$ V and $v_B = 1.000000$ or for $v_A = -1.99999$ and $v_B = -2.00000$ V. In all these cases, $v_{Id} = (v_B - v_A) = -10$ μV rms.

Example 2-2 turns the problem around using Eq. (2-3b).

EXAMPLE 2-2

An op amp has an open loop gain of 100,000. Its output voltage is 2 V rms; what is the differential input voltage?

Solution

$$v_{Id} = \frac{v_o}{A_{VOL}} = \frac{2 \text{ V}}{100,000} = 20 \; \mu V$$

Thus it can be seen that even with 2 V on the op amp output, the input has only 20 μV between the two inputs. Thus an extremely small voltage between the two inputs produces a large voltage on the output.

This leads us to Rule 1.

Rule 1: *For any output voltage in the linear operating region of an op amp with negative feedback, the two inputs are <u>virtually</u> at the same potential.*

This rule presumes that the op amp is operating well within its various limitations.

Alternate Rule 1: *An op amp, connected with negative feedback, will make every effort to keep the inverting (−) input potential equal to the noninverting (+) input potential.*

2.3 Inverting Amplifier Gain

This amplifier, as the name implies, inverts the phase of the input signal while amplifying it. Figure 2-2 illustrates the inverting amplifier.

In the inverting amplifier, the input signal is applied through a resistor connected to the inverting input of the op amp "summing junction." A second resistor is connected between the op amp output and the summing junction. Connecting these resistors to the op amp is referred to as "strapping" the op amp with the gain resistors. The gain equation is derived as follows:

$$i_{in} = \frac{v_{in_1} - v_A}{R_I} \qquad v_A = -v_{Id} \qquad \text{since } v_B = 0 \qquad (2\text{-}4)$$

$$i_F = \frac{v_A - v_o}{R_F} \qquad (2\text{-}5)$$

But because of the high input impedance at the inverting input, no appreciable current flows into this input, and thus

$$i_F = i_{in} \qquad (2\text{-}6)$$

and

$$\frac{v_{in_1} - v_A}{R_I} = \frac{v_A - v_o}{R_F} \qquad (\text{for any } v_A) \qquad (2\text{-}7a)$$

or

$$\frac{v_{in_1} + v_{Id}}{R_I} = \frac{-v_{Id} - v_o}{R_F} \qquad (\text{for } v_A \simeq 0) \qquad (2\text{-}7b)$$

Since $v_B = 0$, then $v_{Id} = -v_A$. After rearranging, Eq. (2-7b) becomes

$$v_o(R_I) + v_{Id}(R_F + R_I) = -v_{in_1} R_F \qquad (2\text{-}8)$$

Inserting Eq. (2-3b) into Eq. (2-8) and rearranging yields the closed-loop-gain transfer-function equation,

$$A_{VCL} = \frac{v_o}{v_{in_1}} = -\frac{AR_F}{(A+1)R_I + R_F} \qquad (2\text{-}9)$$

and

$$A = A_{VOL} \qquad \text{and} \qquad \frac{v_o}{v_{in_1}} = A_{VCL}$$

[*Note:* Hereafter, A_{VOL} without any subscripts is A (open loop gain).] If $(A + 1) \gg$

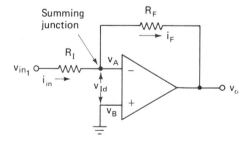

Fig. 2-2 Inverting amplifier.

R_F/R_I, and it usually is, the transfer function for closed loop gain reduces to

$$A_{\text{VCL}} = \frac{v_{\text{out}}}{v_{\text{in}_1}} = -\frac{R_F}{R_I} \qquad (2\text{-}10)$$

An example will help to clarify the point where Eq. (2-10) should be replaced by Eq. (2-9).

EXAMPLE 2-3

An operational amplifier that normally has an open loop gain of 100,000 has failed and now has a gain of 10. Compare the output voltage for a closed loop gain of unity when the input voltage is -1.000 V dc for both cases (i.e., $A = 100,000$ and $A = 10$). For unity closed loop gain, $R_I = R_F = 1$ kΩ.

Solution

For $A = 100,000$,

$$\frac{v_o}{v_{\text{in}_1}} = -\frac{(100,000)(1 \text{ k}\Omega)}{(100,001)(1 \text{ k}\Omega) + 1 \text{ k}\Omega} = -\frac{(100,000)(1 \text{ k}\Omega)}{(100,002)(1 \text{ k}\Omega)}$$

$$= -0.99998 \text{ (actual)}$$

$$\frac{v_o}{v_{\text{in}_1}} = -\frac{R_F}{R_I} = -\frac{1 \text{ k}\Omega}{1 \text{ k}\Omega} = -1 \qquad \text{(ideal)}$$

Thus Eq. (2-9) is only in error by -0.002%. For $A_{\text{VOL}} = 10$,

$$\frac{v_o}{v_{\text{in}_1}} = -\frac{(10)(1 \text{ k}\Omega)}{(11)(1 \text{ k}\Omega) + 1 \text{ k}\Omega} = -\frac{(10)(1 \text{ k}\Omega)}{(12)(1 \text{ k}\Omega)} = -0.8333$$

and Eq. (2-9) is in error by -20.0% given by

$$\left(\frac{1 - 0.8333}{0.8333}\right)(100) = -20.0\%$$

The limiting case is where $A = 1$ and

$$\frac{v_o}{v_{\text{in}_1}} = -\frac{(1)(1 \text{ k}\Omega)}{(2)(1 \text{ k}\Omega + 1 \text{ k}\Omega)} = -\frac{(1)(1 \text{ k}\Omega)}{(3)(1 \text{ k}\Omega)} = -0.333$$

where the ideal value is -1.

A rule of thumb that is a good approximation for the ratio $AR_I/(R_F + R_I)$ greater than 10 is that the actual closed loop gain will be in error from the ideal gain by a percent error between ideal and actual closed loop gains of

$$\% \text{ error} \simeq -\left[\frac{R_F + R_I}{(A)R_I}\right](100)\% \simeq -\frac{1}{A_{\text{VEX}}}(100)\% \qquad (2\text{-}11)$$

(*Note:* See Section 2.5 for A_{VEX}.) The minus sign indicates that the output voltage is lower in absolute magnitude than expected from the actual equation. For percentages less than 2% in magnitude, the equation is quite accurate.

For most operational amplifiers normally encountered in industry, Eq. (2-10) will be an accurate representation of the closed loop gain (A_{VCL}). This leads us to Rule 2.

Inverting Amplifier Gain **39**

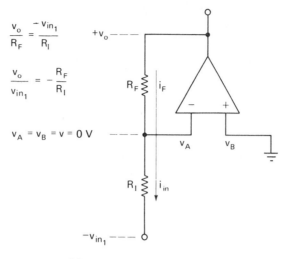

$$\frac{v_o}{R_F} = \frac{-v_{in_1}}{R_I}$$

$$\frac{v_o}{v_{in_1}} = -\frac{R_F}{R_I}$$

$$v_A = v_B = v = 0 \text{ V}$$

(a) Inverting amplifier circuit

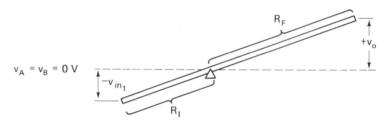

$$v_A = v_B = 0 \text{ V}$$

(b) Teeter-totter representation

Fig. 2-3 Teeter-totter representation of inverting amplifier.

Rule 2: *Any voltage applied to the outer end of a resistor, connected to the inverting input, will be multiplied by the inverting gain as it appears on the amplifier output.*

   Rule 2 is valid even though other voltages are applied to the amplifier circuit at the same time; thus there is no exception when it is not valid as long as the op amp has a large open loop gain and it is operating well within its limitations.

   The inverting amplifier can be thought of as a teeter-totter with a sliding fulcrum point, as depicted in Fig. 2-3. The fulcrum point is at zero volts for the following reasons:

1. The noninverting (+) input is at ground potential (zero volts dc).
2. The differential input voltage (v_{Id}) is *virtually zero volts* dc (by *Rule 1*).
3. Thus the inverting (−) input is at a *virtual ground* potential. The *summing junction*, as noted in Fig. 2-2, is always at nearly zero volts for a normally operating inverting amplifier.

An example will help to clarify the use of *Rule 2*.

EXAMPLE 2-4

An inverting amplifier has an $A_{VOL} = 50,000$. $R_F = 36$ kΩ and $R_I = 9$ kΩ. A dc voltage of -450 mV dc is applied to the circuit input (see Fig. 2-2). What is the magnitude and polarity of the output voltage?

Solution

The closed loop (transfer-function) gain is $-(36$ k$\Omega/9$ k$\Omega) = -4$. The output voltage is the input voltage (magnitude and sign) multiplied by the transfer function. Thus $v_o = (-450$ mV$)(-4) = +1800$ mV or $+1.8$ V dc.

2.4 Noninverting Amplifier Gain

The noninverting amplifier has the same phase on the output as on the input; only the magnitude of the output voltage is different. A noninverting amplifier is illustrated in Fig. 2-4.

In the noninverting amplifier, the signal is applied directly to the noninverting input (v_B) of the operational amplifier. The inverting input (summing junction) to the op amp is again at the connection point of a voltage divider between output and ground. The closed-loop-gain (A_{VCL}) transfer-function equation is derived as follows. Equation (2-2) solved for ($v_B - v_A$), where v_B is v_{in}, becomes $v_{in_2} - v_A = v_{0_2}/A$, and either Eq. (2-7a) solved for v_A with $v_{in_1} = 0$ or by observation of the voltage division from v_o to v_A in Fig. 2-4 yields

$$v_A = \left(\frac{R_I}{R_F + R_I}\right)v_o \tag{2-12}$$

which when inserted into Eq. (2-2) as modified above yields

$$\frac{v_o}{v_{in_2}} = +\frac{A(R_F + R_I)}{(A + 1)R_I + R_F} \tag{2-13}$$

If $(A + 1) \gg (R_F + R_I)/R_I$, Eq. (2-13) reduces to

$$\frac{v_o}{v_{in_2}} = +\frac{R_F + R_I}{R_I} = +\left(\frac{R_F}{R_I} + 1\right) \tag{2-14}$$

An example will help to clarify the error involved through using the approximate solution, Eq. (2-14), with low open loop gains.

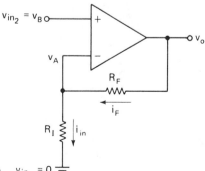

Fig. 2-4 Noninverting amplifier.

EXAMPLE 2-5

An operational amplifier that normally has an open loop gain of 100,000 has failed and now exhibits a gain of 10. Calculate the expected ideal closed loop gain and the actual closed loop gain if $R_I = R_F = 1 \text{ k}\Omega$.

Solution

$$\frac{v_o}{v_{in}} = +\frac{(100,000)(1 \text{ k}\Omega + 1 \text{ k}\Omega)}{(100,001)(1 \text{ k}\Omega) + 1 \text{ k}\Omega} = +\frac{100,000(2 \text{ k}\Omega)}{100,002(1 \text{ k}\Omega)}$$

$$= +1.99996 \quad (\text{actual})$$

$$\frac{v_o}{v_{in}} = +\frac{1 \text{ k}\Omega + 1 \text{ k}\Omega}{1 \text{ k}\Omega} = +2 \quad (\text{ideal})$$

Thus the percent error between the actual and ideal closed loop gains at $A_{\text{VOL}} = 100,000$ and $A_{\text{VCL}} = 2$ is

$$-\left(\frac{2 - 1.99996}{1.99996}\right)(100) = -0.002\%$$

If $A = 10$, then the actual closed loop gain is

$$\frac{v_o}{v_{in}} = +\frac{10(1 \text{ k}\Omega + 1 \text{ k}\Omega)}{(11)(1 \text{ k}\Omega) + 1 \text{ k}\Omega} = +\frac{20(1 \text{ k}\Omega)}{12(1 \text{ k}\Omega)} = +1.667$$

and the percent error between the ideal gain of 2 and the actual gain of 1.667 is

$$-\left(\frac{2 - 1.667}{1.667}\right)(100) = -19.97\% \text{ or } -20\% \text{ error}$$

Thus it can be seen from Example 2-5 that Eq. (2-11) for the percent error between actual and ideal closed loop gains still applies.

The potential on the inverting input is identical to that on the noninverting input (by **Rule 1**). The gain between the noninverting input and the output is the noninverting gain. It follows, then, that the gain between the voltage on the inverting input and the output is also the noninverting gain. This leads us to Rule 3.

Rule 3: *Any voltage appearing directly on <u>either input</u> of the operational amplifier will be multiplied by the <u>noninverting gain.</u>*

Two clarifications of **Rule 3** are warranted here:

1. The case where the voltage is applied to the noninverting ($+$) input of the op amp was derived in this section. The case where the voltage appears on the inverting ($-$) input is covered in Appendix A3. It has the same result.
2. The sign of the output voltage is dependent upon the polarity of a voltage appearing on one of the op amp inputs and to which input lead it is applied. Figure 2-5 will help to clarify the output sign for various applications of **Rule 3**.

Figure 2-5a and b illustrate the output sign for cases where the voltage generator is on the noninverting op amp input. Figure 2-5c and d illustrate the case for the voltage generator on the inverting op amp input. The resulting output sign is the

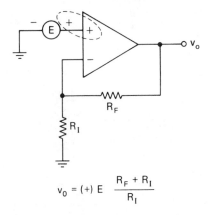

$$v_0 = (+)\ E\ \frac{R_F + R_I}{R_I}$$

(a) Output positive, noninverting

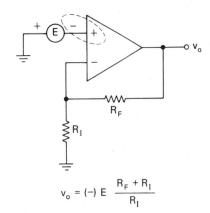

$$v_0 = (-)\ E\ \frac{R_F + R_I}{R_I}$$

(b) Output negative, noninverting

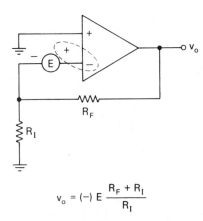

$$v_0 = (-)\ E\ \frac{R_F + R_I}{R_I}$$

(c) Output negative, inverting

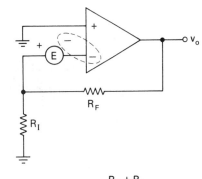

$$v_0 = (+)\ E\ \frac{R_F + R_I}{R_I}$$

(d) Output positive, inverting

Fig. 2-5 Output-voltage sign for various input signs.

product of the generator sign and the op amp input sign, which are adjacent. These signs are multiplied as in basic algebra and produce the following output signs:

$$(+) \times (+) = (+)\quad (+) \times (-) = (-)\quad (-) \times (+) = (-)\quad (-) \times (-) = (+)$$

The noninverting amplifier can be represented by a teeter-totter with the fulcrum point at one end and the input voltage being a displacement somewhere along the teeter-totter, as illustrated in Fig. 2-6.

The fulcrum point in Fig. 2-6 for the noninverting amplifier is at zero volts, because it is at the end of the "summing junction" resistor (Fig. 2-3). The inverting ($-$) input is now elevated above ground potential for the following reasons:

1. The noninverting input is at the input voltage level (v_{in_2}).
2. The differential input voltage (v_{Id}) is *virtually zero volts* (**Rule 1**).

Noninverting Amplifier Gain

43

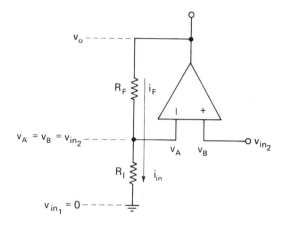

(a) Noninverting amplifier circuit

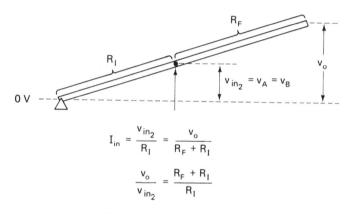

$$I_{in} = \frac{v_{in_2}}{R_I} = \frac{v_o}{R_F + R_I}$$

$$\frac{v_o}{v_{in_2}} = \frac{R_F + R_I}{R_I}$$

(b) Teeter-totter representation

Fig. 2-6 Teeter-totter representation of noninverting amplifier.

3. Thus the inverting ($-$) input is at the same voltage as the noninverting ($+$) input voltage level.

An example will help to clarify use of this reasoning.

EXAMPLE 2-6

An operational amplifier with an open loop gain of 60,000 is connected in the noninverting mode with $R_I = 2$ kΩ and $R_F = 8$ kΩ. An input signal of -1.5 V dc is applied to the noninverting input of the op amp (see Fig. 2-4).

   a. Determine the output voltage using the input voltage and the closed loop gain.
   b. Determine the output voltage using Eq. (2-7a).

                                                 Using the Operational Amplifier as a Circuit Element

Solution

a. The closed loop gain is

$$\frac{2\text{ k}\Omega + 8\text{ k}\Omega}{2\text{ k}\Omega} = \frac{10\text{ k}\Omega}{2\text{ k}\Omega} = +5$$

$$v_o = (-1.5\text{ V})(+5) = -7.5\text{ V dc} \quad (\textbf{\textit{Rule 3}})$$

b. In Fig. 2-4, since v_A must equal v_B and $v_B = -1.5$ V, it follows that $v_A = -1.5$ V also, by **_Rule 1_**. But Eq. (2.7a) is written for the inverting amplifier, not the noninverting amplifier. How, then, can Eq. (2-7a) be applied to the noninverting amplifier? By grounding v_{in} in Fig. 2-2 and applying -1.5 V to v_B (now ungrounded), Eq. (2-7a) becomes

$$\frac{v_{\text{in}} - v_A}{R_I} = \frac{v_A - v_o}{R_F}$$

or

$$\frac{0\text{ V} - (-1.5\text{ V})}{2\text{ k}\Omega} = \frac{(-1.5\text{ V}) - v_o}{8\text{ k}\Omega}$$

and

$$v_o = \frac{(-1.5\text{ V})(2\text{ k}\Omega) - (+1.5\text{ V})(8\text{ k}\Omega)}{2\text{ k}\Omega}$$

$$= \frac{-(1.5\text{ V})(10\text{ k}\Omega)}{2\text{ k}\Omega} = -7.5\text{ V dc}$$

Thus either equation, the fundamental gain equation (2-7a) or the **_Rule 3_** equation (2-14), will yield identical results.

Example 2-6 also used **_Rule 1_**; this demonstrates how **_Rule 1_** can be applied to ease the solution of a problem.

2.4.1 Noninverting Amplifier with Voltage-Divider Input

Figure 2-7 illustrated the circuit where a voltage divider is used to reduce the amplitude to the input of a noninverting amplifier. It has already been established that the voltage gain between v_B and v_o is

$$v_o = +v_B\left(\frac{R_F + R_I}{R_I}\right)$$

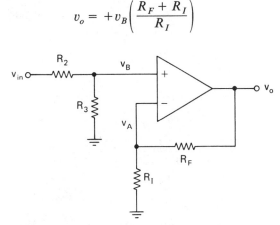

Fig. 2-7 Noninverting amplifier with voltage-divider input.

Since the input impedance to the noninverting (+) input of the op amp is very high [see Eq. (1-23)], the voltage divider is, for all practical purposes, unloaded; thus

$$v_B = v_{in} \frac{R_3}{R_2 + R_3} \qquad (2\text{-}15)$$

and the total circuit transfer function (gain) is the product of these two equations,

$$\frac{v_o}{v_{in}} = + \left(\frac{R_3}{R_2 + R_3} \right) \left(\frac{R_F + R_I}{R_I} \right) \qquad (2\text{-}16)$$

$$\underbrace{\hspace{2cm}}_{\substack{\text{voltage-} \\ \text{divider} \\ \text{gain}}} \quad \underbrace{\hspace{2cm}}_{\substack{\text{closed} \\ \text{loop} \\ \text{gain}}}$$

this leads us to Rule 4.

Rule 4: *Any voltage applied through a voltage divider to the noninverting input of an operational amplifier will be multiplied by the product of the voltage-divider gain (loss) and the noninverting gain.*

Example 2-7 illustrates this application.

EXAMPLE 2-7

A noninverting amplifier with a voltage divider on the input is strapped for a closed loop gain of +6. The voltage-divider gain (loss) is 0.8. The op amp has an open loop gain of 40,000. Find the output voltage for an input voltage of 2.4 V dc.

Solution

Equation (2-16) separates the voltage-divider gain from the closed loop gain and is the one that is appropriate for solution to this problem. Thus

$$\frac{v_o}{v_{in}} = \left(\frac{R_3}{R_2 + R_3} \right) \left(\frac{R_F + R_I}{R_I} \right) = (0.8)(6) = +4.8$$

$$\underbrace{\hspace{2cm}}_{\substack{\text{voltage-} \\ \text{divider} \\ \text{gain}}} \quad \underbrace{\hspace{2cm}}_{\substack{\text{closed} \\ \text{loop} \\ \text{gain}}}$$

Thus the total circuit gain is 4.8 and the output voltage is

$$v_0 = (2.4 \text{ V})(4.8) = 11.52 \text{ V dc} \qquad (\textbf{\textit{Rule 4}})$$

An alternative solution would be one using **_Rule 1_** and Eq. (2-7a), as in Example 2-6. In this case the divided voltage would have to be calculated beforehand.

EXAMPLE 2-8

The circuit illustrated in Fig. 2-7 has $R_I = 5$ kΩ, $R_2 = 2$ kΩ, $R_3 = 6$ kΩ, and $R_F = 9$ kΩ. Determine the peak output voltage for $v_{in} = 3 \sin(2\pi \cdot 0.1t)$.

Solution

$$v_o = \left(\frac{R_3}{R_2 + R_3} \right) \left(\frac{R_F + R_I}{R_I} \right) v_{\text{in}}$$

$$= \left(\frac{6\,k\Omega}{8\,k\Omega} \right) \left(\frac{14\,k\Omega}{5\,k\Omega} \right) 3 \sin(2\pi \cdot 0.1t)$$

$$= (0.75)(2.8)3 \sin(2\pi \cdot 0.1t)$$

$$= 6.3 \sin(2\pi \cdot 0.1t)$$

Thus the peak voltage is 6.3 V.

2.4.2 Rule 5 Combines Rules 2 and 3 or 4

Rule 5: *All input voltages, times their respective closed loop gains, will add algebraically on the amplifier output.*

This rule is valid because of **Rule 1**, which causes all input voltages to be independent of one another as they appear on the circuit output. Thus **Rule 5** for an inverting amplifier with a voltage on the noninverting input causes Eqs. (2-10) and (2-14) to sum algebraically, becoming

$$v_o = \left(-\frac{R_F}{R_I} \right) v_1 + \left(\frac{R_F + R_I}{R_I} \right) v_B \qquad (2\text{-}17)$$

$$\underset{\text{Rule 2 \quad Rule 5} \qquad \text{Rule 3}}{\uparrow}$$

EXAMPLE 2-9

An inverting amplifier with $R_I = 2\,k\Omega$ and $R_F = 10\,k\Omega$ has an input voltage of -3 V. The noninverting input (v_B) is connected to a -5 V power-supply bus. Determine the output voltage using both **Rule 5** and Eq. (2-7a).

Solution

Rule 5 solution:

$$\text{inverting gain} = -\frac{10\,k\Omega}{2\,k\Omega} = -5$$

$$\text{noninverting gain} = \frac{10\,k\Omega + 2\,k\Omega}{2\,k\Omega} = +6$$

$$v_o = (-5)(-3\,V) + (6)(-5\,V) = +15\,V - 30\,V$$

$$= -15\,V$$

Equation (2-7a) solution:

$$\frac{-3\,V - (-5\,V)}{2\,k\Omega} = \frac{(-5\,V) - v_o}{10\,k\Omega}$$

$$(-3\,V + 5\,V)(10\,k\Omega) = (-5\,V)(2\,k\Omega) - (v_o)2\,k\Omega$$

$$v_o = -\frac{(2\,V)(10\,k\Omega) + (5\,V)(2\,k\Omega)}{2\,k\Omega} = -15\,V$$

To bring all the rules together, they will be summarized here and placed in usable equation form. This summary of rules will refer to Fig. 2-8.

Noninverting Amplifier Gain

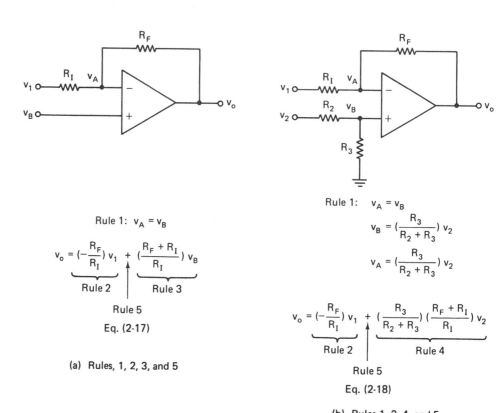

Rule 1: $v_A = v_B$

$$v_o = (-\frac{R_F}{R_I}) v_1 + (\frac{R_F + R_I}{R_I}) v_B$$

Rule 2   Rule 3

Rule 5

Eq. (2-17)

(a) Rules, 1, 2, 3, and 5

Rule 1: $v_A = v_B$

$$v_B = (\frac{R_3}{R_2 + R_3}) v_2$$

$$v_A = (\frac{R_3}{R_2 + R_3}) v_2$$

$$v_o = (-\frac{R_F}{R_I}) v_1 + (\frac{R_3}{R_2 + R_3}) (\frac{R_F + R_I}{R_I}) v_2$$

Rule 2   Rule 4

Rule 5

Eq. (2-18)

(b) Rules 1, 2, 4, and 5

Fig. 2-8 Summary of rules.

Rule 5, summarized in Fig. 2-8b, yields the equation

$$v_o = \left(-\frac{R_F}{R_I}\right)v_1 + \left(\frac{R_3}{R_2 + R_3}\right)\left(\frac{R_F + R_I}{R_I}\right)v_2 \qquad (2\text{-}18)$$

Rule 2   Rule 5   Rule 4

for a circuit (Fig. 2-8b) called a "scaling subtractor." This circuit merely takes v_1 and multiplies it by the inverting gain and takes v_2 and multiplies it by the combination of the voltage-divider gain times the noninverting gain and sums the results. The inverting and noninverting gains are not equal; thus the term "scaling" is used to denote this fact. The differencing, or subtraction, occurs because the inverting gain is opposite in polarity to the noninverting gain.

The gains can be made equal for the inverting and noninverting inputs by selecting a particular resistor relationship. If in Fig. 2-8b $R_F/R_I = R_3/R_2$, Eq. (2-18) reduces to

$$v_o = \left(-\frac{R_F}{R_I}\right)v_1 + \left(+\frac{R_F}{R_I}\right)v_2 = \left(\frac{R_F}{R_I}\right)(v_2 - v_1) \qquad (2\text{-}19)$$

and the circuit becomes a "difference amplifier" which will take the difference between v_2 and v_1 with equal gains. It can be seen that this circuit is a special case of the general circuit

     Using the Operational Amplifier as a Circuit Element

for **Rule 5**. If $R_I = R_2 = R_3 = R_F$, then

$$v_o = v_2 - v_1 \qquad \qquad (2\text{-}20)$$

and the circuit is a "subtractor" which takes the difference between v_2 and v_1 with unity gain.

EXAMPLE 2-10

An operational amplifier is connected as a scaling subtractor. The inverting gain is -5 and the noninverting gain is $+3$. Using resistors between 1 and 10 kΩ, select resistor values that will yield the desired gains.

Solution

Let $R_F = 10$ kΩ and $R_I = 2$ kΩ. The inverting gain is then (-5), thus the noninverting gain is $(+6)$ and

$$\left(\frac{12 \text{ k}\Omega}{2 \text{ k}\Omega} \right)\left(\frac{R_3}{R_2 + R_3} \right) = +3 \quad \text{or} \quad (6)\left(\frac{R_3}{R_2 + R_3} \right) = +3$$

It follows that the voltage divider must have a gain (loss) of 0.5; thus $R_2 = 1$ kΩ and $R_3 = 1$ kΩ. Equation (2-18) now becomes $v_o = 3v_2 - 5v_1$ and the scaling subtractor is finished. Also, through a judicious choice of resistors, the input impedance to both inputs is 2 kΩ.

EXAMPLE 2-11

A difference amplifier has a 10 kΩ feedback resistor. It is to have a differencing gain of 4. Select resistors from the 1% table in Appendix B1 which will yield an input impedance of 2.5 kΩ to each input.

Solution

R_I must equal 2.5 kΩ, so select the 2.49 kΩ from the table. The proportion, $R_F/R_I = R_3/R_2 = 4$, must be satisfied. The input impedance restriction establishes that $R_2 + R_3 = 2.5$ kΩ. Then $R_3 = 4R_2$ and $5R_2 = 2.5$ kΩ; $R_2 = 499$ Ω and $R_3 = 2$ kΩ.

EXAMPLE 2-12

Remove the input impedance restriction and alter the necessary resistors to make the circuit a subtractor (unity gain).

Solution

$$R_I = R_2 = R_3 = R_F = 2.49 \text{ k}\Omega$$

Table 2-1 lists the five rules.

2.5 Excess Loop Gain

Use of the term "excess loop gain" is unique to this author. All other references refer to this quantity as loop gain. It is the opinion of this author that the term "excess" attached to the basic term, loop gain, is useful. It helps remind the reader that the operation of a closed loop system is dependent on the "excess" loop gain for control, and it does not have the full "open" loop gain available, except when the "closed" loop gain is unity.

Table 2-1   Summary of Feedback Rules

Rule 1:   An op amp, connected with feedback, will make every
          effort to keep the inverting (−) input voltage equal to
          the noninverting (+) input voltage.

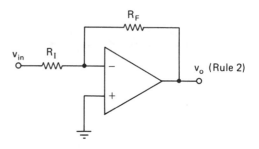

Rule 2: $v_o = (-\dfrac{R_F}{R_I}) v_{in}$

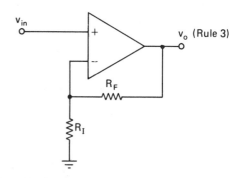

Rule 3: $v_o = (+\dfrac{R_F + R_I}{R_I}) v_{in}$

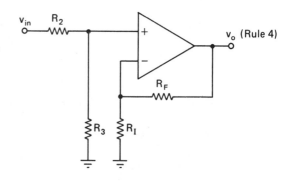

Rule 4: $v_o = (\dfrac{R_3}{R_2 + R_3})(\dfrac{R_F + R_I}{R_I}) v_{in}$

Rule 5:   Rule 2 "and" Rules 3 or 4 sum algebraically on the output:

v_o (Rule 2) + v_o (Rule 3) = v_o (Rule 5)

or

v_o (Rule 2) + v_o (Rule 4) = v_o (Rule 5)

50

Presume that an operational amplifier circuit is strapped for a closed loop gain of 100. The basic operational amplifier has an open loop gain of 100,000. This means that 1/100 of the 100,000 open loop gain is used to overcome signal losses from output back to input due to the ratio of feedback to input components. After the signal loss has been overcome, there is only 100,000/100, or 1000 loop gain left to correct for signal error. The 1000 is the excess loop gain and is represented by the equation

$$A_{VEX} = \frac{A_{VOL}}{A_{VCL}} \qquad (2\text{-}21)$$

where A_{VCL} is *always* the noninverting gain.

EXAMPLE 2-13

An operational amplifier has an open loop gain of 100,000. It is strapped for a closed loop, noninverting gain of 1000. How much excess loop gain is left for signal control?

Solution

$$A_{VEX} = \frac{100,000}{1000} = 100$$

If gains are expressed in (dB), they appear as follows:

$$\text{dB gain} = 20 \log_{10}(\text{gain in ratio}) \qquad (2\text{-}22)$$

To find the excess loop gain in dB, the A_{VCL} (dB) is subtracted from the A_{VOL} (dB). The equation for this appears as

$$A_{VEX}(\text{dB}) = A_{VOL}(\text{dB}) - A_{VCL}(\text{dB}) \qquad (2\text{-}23)$$

Example 2-14 is solved as follows for dB.

EXAMPLE 2-14

An operational amplifier has an open loop gain of 100 dB. It is strapped as an inverting amplifier with a closed loop gain of 60 dB. How much excess loop gain is left for signal control?

Solution

60 dB = 1000 inverting; 1001 (noninverting) = 60.01 dB.

$$A_{VEX}(\text{dB}) = 100 \text{ dB} - 60.01 \text{ dB} = 39.99 \text{ dB}$$

The operational amplifier does not "know" what closed loop gain it is connected to deliver. It only knowns what excess loop gain is available for control. If the closed loop gain is unity, the full open loop gain is available as excess loop gain, but if the amplifier is connected for a large closed loop gain, the excess loop gain is reduced. The amplifier's ability to control is impaired and several performance characteristics, only one of which is output voltage precision, degrade.

EXAMPLE 2-15

In Example 2-14 the excess loop gain was found to be 40 dB. Convert this to gain ratio.

Solution

$$40 \text{ dB} = 20 \log_{10} \text{ (gain ratio)}$$

$$\frac{40}{20} = \log_{10} \text{ (gain ratio)} = 2$$

$$\text{gain ratio} = 10^2 = 100$$

Thus 40 dB is equivalent to a gain ratio of 100.

Since scientific calculators are readily available, this problem requires little or no effort to solve, even with irrational exponents.

An example of a problem with irrational exponents would be:

EXAMPLE 2-16

The excess loop gain was found to be 57 dB. Convert this to gain ratio.

Solution

$$57 \text{ dB} = 20 \log_{10} \text{ (gain ratio)}$$

$$\frac{57}{20} = \log_{10} \text{ (gain ratio)} = 2.85$$

$$\text{gain ratio} = 10^{2.85} = 707.95$$

Since 60 dB is a ratio of 1000, this answer seems reasonable.

2.6 Connecting the Op Amp Circuit Using the Four Methods of Feedback

The operational amplifier circuits just discussed are small control systems. There are four generic types of feedback circuits that describe Type 0 (position) control systems. These are discussed in this section.

2.6.1 Control System Representation

The control system diagram illustrated in Fig. 2-9a contains four elements: the forward gain element (A or G), the feedback element (β or H), the sampler element, and the summing junction (Σ). In mechanical control systems, such as servo systems, the sampler element is a transducer or sensor that converts a physical property to an electrical signal. Examples of these are a thermistor for heat sensing or a tachometer for shaft speed sensing.

In an *all-electronic* control system the sensing signal is *already* a voltage, so no conversion is required. Figure 2-9b illustrates this situation where the feedback element is connected *directly* to the forward gain element in the sensor element, and the sensor element does not *actually* exist. This is the connection that will be expected in the balance of the feedback discussion in this book.

The sign on v_{in} determines whether the control system is noninverting ($+$) or inverting ($-$). The sign on v_f determines whether the feedback is negative ($-$) or positive ($+$). For any feedback control system, the sign on v_f is *always* negative ($-$); for an oscillator or switching (saturated output) mode circuit, this sign is

Using the Operational Amplifier as a Circuit Element

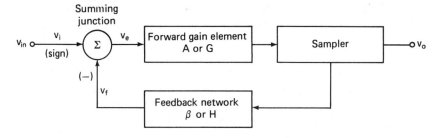

(a) General control system.

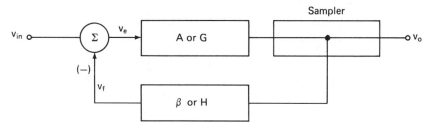

(b) Electronic amplifier control system.

Fig. 2-9 Simple control system.

always positive (+). The amplifiers we will be considering here all possess *negative* feedback.

As the loss through the feedback network becomes greater, the circuit compensates by increasing the closed loop gain. The limiting factor occurs when the closed loop gain equals the open loop gain and the excess loop gain becomes 1; the feedback loop now opens up and the system ceases to control (no feedback exists).

To define the relationship between input and output, a simple derivation will be made. The sign on v_{in} determines whether this is a noninverting (+) or an inverting (−) amplifier.

$$v_e(\text{error voltage}) = v_{\text{in}} - v_f$$

$$v_o = A(v_{\text{in}} - v_f) = Av_{\text{in}} - Av_f$$

But $v_f = \beta v_o$; thus

$$v_o = Av_{\text{in}} - A\beta v_o$$

or

$$v_o(1 + A\beta) = Av_{\text{in}}$$

and

$$\frac{v_o}{v_{\text{in}}} = \frac{A}{1 + A\beta} \tag{2-24}$$

The alternative form of Eq. (2-24) is

$$\frac{v_o}{v_{\text{in}}} = \frac{G}{1 + GH} \tag{2-25}$$

which uses the alternative symbols for the gain and feedback blocks in Fig. 2-9.

This can be stated another way. The excess loop gain (loop gain) is represented by the product of the loop gain (A or G) and the feedback loss factor (β or H). This product is either $A\beta$ or GH, which represents the excess loop gain (loop gain). The amplifier can control only when it has an excess loop gain ($A\beta$ or GH) greater than unity (i.e., the open loop gain is greater than the feedback loss).

The *ideal* closed loop gain is the reciprocal of β or H. This is the circuit gain *expected* by the approximate (ideal) gain equation. See Eq. (2-26) or (2-28) for examples of ideal gain equations; the actual gain equation for a noninverting amplifier is Eq. (2-13).

2.6.2 Voltage Amplifier

The voltage amplifier is shown in Fig. 2-10. Its ideal transfer function (ideal gain equation) is given by

$$\frac{V_o}{V_{\text{in}}} = \frac{R_F + R_I}{R_I} \tag{2-26}$$

while its actual gain equation [Eq. (2-13) rearranged] is

$$\frac{V_o}{V_{\text{in}}} = + \frac{A}{1 + A\left(\dfrac{R_I}{R_F + R_I}\right)} \tag{2-27}$$

A_{OL} for this amplifier is A_V and has the units of v/v; it is thus pure voltage gain.

$$\beta = \frac{R_I}{R_F + R_I} \tag{2-28}$$

which is the reciprocal of the ideal voltage gain.

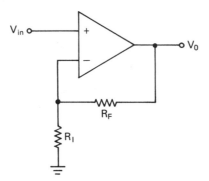

Fig. 2-10 Voltage amplifier.

The actual gain is given by the control function relationship as

$$\frac{V_o}{V_{\text{in}}} = \frac{A(R_F + R_I)}{(A + 1)R_I + R_F} = \frac{A}{1 + A\left(\dfrac{R_I}{R_F + R_I}\right)} \tag{2-29}$$

The input impedance is very high (see Appendix A1) and the output impedance is very low (see Appendix A2); also see Sections 2.8.1 and 2.8.2. This amplifier is designated by other sources as *voltage series* and *series-shunt* [see Appendix A18(I) for the derivation of the actual gain equation].

Using the Operational Amplifier as a Circuit Element

2.6.3 Transresistance Amplifier

This configuration, shown in Fig. 2-11 is the building block used to formulate the inverting amplifier. Its closed loop gain (Appendix A20) has the units of v/i or ohms—thus the name *transresistance*. A_{OL} also has the units of v/i or ohms.

$$\beta = \frac{-1}{R_F} \qquad \text{Siemens} \tag{2-30}$$

$$\frac{V_o}{V_{in}} = -R_F \qquad \text{ohms} \tag{2-31}$$

The actual gain equation is

$$\frac{V_o}{V_{in}} = \frac{-A_{VOL}R_F}{1 + (-A_{VOL}R_F)(-1/R_F)} \tag{2-32}$$

where $A_{OL} = -A_{VOL}R_F$ and $\beta = -1/R_F$. The input impedance is very low (see Appendix A19) and the output impedance is very low (see Appendix A2). This amplifier is designated by other sources as *voltage-shunt* or *shunt-shunt* (see Appendix A20 for the derivation of the actual gain equation).

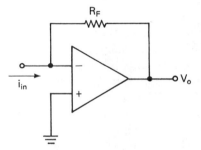

Fig. 2-11 Transresistance amplifier.

The *inverting amplifier* is formed, as shown in Fig. 2-2, by inserting a series resistor (R_I) between the *circuit input* and the low-impedance *summing junction*. The purpose of this resistor is to convert the input voltage on the inverting amplifier input to a current for the transresistance amplifier.

2.6.4 Transconductance Amplifier

The transconductance amplifier is a *constant current* (*current-series or series-series*) amplifier, as shown in Figs. 2-12 and 6-9 and shown in black box form in Appendix A18(III). Its transfer function (closed loop gain) has the units i/v or Siemens. The actual gain equation is derived in Appendix A21 and is composed of:

$$A_{OL} = \frac{A_{VOL}}{R_{CS}} \qquad \text{Siemens} \tag{2-33}$$

$$\beta = R_{CS} \qquad \text{ohms} \tag{2-34}$$

$$\text{ideal gain} = \frac{I_o}{V_{in}} = \frac{1}{R_{CS}} \qquad \text{Siemens} \tag{2-35}$$

$$\text{actual gain} = \frac{I_o}{V_{in}} = \frac{A_{VOL}/R_{CS}}{1 + (A_{VOL}/R_{CS})(R_{CS})} \tag{2-36}$$

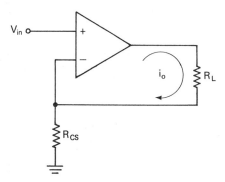

Fig. 2-12 Transconductance amplifier.

The input impedance is very high (see Section 2.8.1) and the output impedance is very high (see Appendix A11). For operational characteristics and typical circuits, see Section 6.5.

2.6.5 Current Amplifier

The current amplifier, illustrated in Fig. 2-13 is the only configuration that cannot be represented using a single op amp. It also does not, in the two-op-amp form, lend itself to being placed in the black-box form (Appendix A18(IV)); an extra connection not conforming to the black-box connections must be used.

At this point it is well to say that the black-box form of feedback representation are two-port networks with no leads on either port necessarily being a common ground. The op amp configuration forces a common ground to occur and this violates the representation in its purest form. This violation causes the extra lead in the current amplifier to be necessary.

The current amplifier has the following parameters:

$$A_{\text{OL}} = (A_{\text{VOL}1}R_F)\frac{A_{\text{VOL}2}}{R_{\text{CS}}} \qquad i/i \text{ unitless} \tag{2-37}$$

$$\beta = -\frac{R_{\text{CS}}}{R_F} \tag{2-38}$$

$$\frac{I_o}{I_{\text{in}}} = \left(\frac{-A_{\text{VOL}}R_F}{1 + (-A_{\text{VOL}}R_F)(-1/R_F)}\right)\left(\frac{A_{\text{VOL}}/R_{\text{CS}}}{1 + (A_{\text{VOL}}/R_{\text{CS}})(R_{\text{CS}})}\right) \tag{2-39}$$

$$\frac{I_o}{I_{\text{in}}} = \frac{-A_{V_1}A_{V_2}R_F/R_{\text{CS}}}{1 + (-A_{V_1}A_{V_2}R_F/R_{\text{CS}})(-R_{\text{CS}}/R_F)} \cong -\frac{R_F}{R_{\text{CS}}} \tag{2-40}$$

As can be seen by Eq. (2-39), the actual gain equation is the product of that for the transresistance and transconductance amplifiers; if the voltage gains A_{V_1} and A_{V_2} are very large, Eq. (2-40) is the result. The amplifier is designated as *current-shunt* or *shunt-series*. The input impedance (see Appendix A18(IV) and A19) is very low and output impedance (see Appendix A11) is very high.

All feedback circuits used in electronics can be categorized into one of the four types shown in this section. They are used in both discrete and monolithic circuits. They are not only used for gain stabilization, but are also used for impedance matching in monolithic circuits.

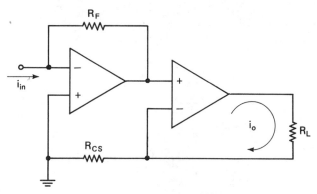

Fig. 2-13 Current amplifier.

2.7 Special Cases of Op Amp Amplifiers

Several special cases of the standard circuits just discussed can be implemented through slight circuit changes. These will be analyzed individually in this section.

2.7.1 Noninverting Amplifier with Unity Gain (Voltage Follower)

The gain equation for the noninverting amplifier is

$$\frac{v_o}{v_{\text{in}}} = \frac{R_F + R_I}{R_I}$$

and referring to Fig. 2-4, it can be seen that R_I is the shunt resistor between the inverting input and common.

When R_I becomes very large with respect to R_F, the numerator tends to become R_I alone, and since the denominator is R_I the gain tends toward unity. If R_I is removed, leaving only R_F in the circuit, the gain equation becomes

$$\frac{v_o}{v_{\text{in}}} = \frac{R_F + \infty}{\infty} \simeq 1 \tag{2-41}$$

which becomes unity since R_F is insignificant with respect to infinity (∞).* It makes no difference what value of R_F is used; the gain remains at unity.

This circuit is extremely useful as an impedance transformer. The input impedance is nearly infinite, the output impedance is nearly zero (Section 2.8.2), and the voltage gain is unity.

Figure 2-14 illustrates a voltage follower. The feedback resistor (R_F) can be of any reasonable value (from 1 to 50 kΩ); the input resistor, R_I, is for current compensation purposes (Chapter 3) and is always equal to R_F. Neither resistor has any effect on Eq. (2-41) for voltage gain; it is still unity.

---

*It must be noted here that the term infinity (∞) represents a very large but still finite number. This assertion eliminates the need for a formal proof that v_o/v_{in} is indeed unity.

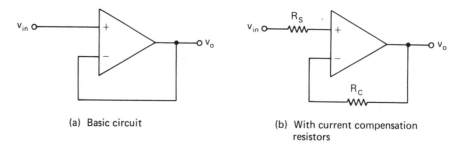

| (a) Basic circuit | (b) With current compensation resistors |

Fig. 2-14 Voltage follower.

2.7.2 Inverting Summer

A special case of the inverting amplifier is created when multiple inputs are connected at the summing junction (op amp inverting input), as illustrated in Fig. 2-15a.

Since by **Rule 1** the summing junction has nearly zero volts on it, the resistors and their respective voltages are isolated from one another, as the summing junction appears to be grounded as shown in Fig. (2-15b). Thus each resistor acts as though

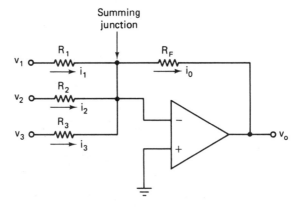

(a) Circuit diagram

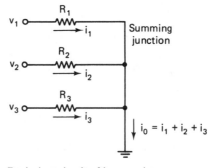

(b) Equivalent circuit of input resistors

Fig. 2-15 Inverting summer.

it is the only one in the circuit and the effect of the three is only felt on the amplifier output where the individual voltages are summed through **Rule 5**. Thus

$$v_o = \left(-\frac{R_F}{R_1}\right)v_1 + \left(-\frac{R_F}{R_2}\right)v_2 + \left(-\frac{R_F}{R_3}\right)v_3 \qquad (2\text{-}42)$$

This circuit operates as it does because each current, i_1, i_2, and i_3, must each pass through R_F to the output of the op amp; each current adds its own component to the total current which produces the output voltage.

One of the problems encountered when using this circuit with many inputs is the loss of excess loop gain. The multiple input resistors appear to the amplifier to be in parallel, which causes the apparent closed loop gain to appear very high (because the apparent input resistance is very low). This reduces the excess loop gain and the amplifier has more amplifying error than is at first expected. This circuit is used in digital-to-analog (D/A) converters where many (8 to 16) summing resistors are used; here the problem must be considered carefully. Operational amplifiers with large open loop gains should be chosen for this application.

EXAMPLE 2-17

A 4-bit D/A converter (see Chapter 8) uses an inverting summer as its output stage. The input voltages can be either 0 or 5 V in a binary sequence. The input resistors are as follows: $R_1 = 1$ kΩ, $R_2 = 2$ kΩ, $R_3 = 4$ kΩ, $R_4 = 8$ kΩ, and $R_F = 1$ kΩ. Determine the output voltage for the case where the $v_1 = 0$ V, $v_2 = 5$ V, $v_3 = 5$ V, and $v_4 = 5$ V. Also, determine the apparent closed loop gain and the excess loop gain if $A = 50{,}000$.

Solution

$$v_o = \left(-\frac{1\text{ k}\Omega}{1\text{ k}\Omega}\right)0\text{ V} + \left(-\frac{1\text{ k}\Omega}{2\text{ k}\Omega}\right)5\text{ V} + \left(-\frac{1\text{ k}\Omega}{4\text{ k}\Omega}\right)5\text{ V} + \left(-\frac{1\text{ k}\Omega}{8\text{ k}\Omega}\right)5\text{ V}$$

$$= 0 - 2.5\text{ V} - 1.25\text{ V} - 0.625\text{ V} = -4.375\text{ V}$$

$$\text{apparent } R_I = 1\text{ k}\Omega\|2\text{ k}\Omega\|4\text{ k}\Omega\|8\text{ k}\Omega = 0.533\text{ k}\Omega$$

The apparent noninverting closed loop gain is

$$A_{\text{VCL}} = 1 + \frac{R_F}{0.533\text{ k}\Omega} = 1 + \frac{1\text{ k}\Omega}{0.533\text{ k}\Omega} = 2.875$$

The apparent excess loop gain is

$$\frac{A_{\text{VOL}}}{A_{\text{VCL}}} = \frac{50{,}000}{2.875} = 17{,}391$$

or reduced to about one-third of the original open loop gain.

To make this circuit sum with a positive sign instead of negative, one needs only to follow this circuit with a unity-gain inverter.

2.7.3 Noninverting Averager

This circuit appears at first to be like the one in Section 2.7.2; *it is not*. The averager circuit shown in Fig. 2-16a has the resistors connected to an impedance of nearly infinity (noninverting op amp input); thus there is *total interaction* between all input voltages.

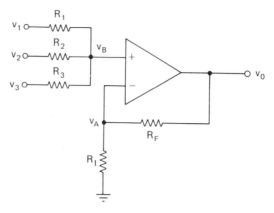

(a) Noninverting averager circuit

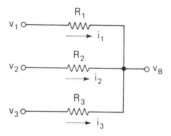

(b) Nodal analysis circuit

Fig. 2-16 Noninverting averager.

The voltage that appears at v_B can be found by solving a nodal network (Fig. 2-16b) and is

$$v_B = \frac{\dfrac{v_1}{R_1} + \dfrac{v_2}{R_2} + \dfrac{v_3}{R_3}}{\dfrac{1}{R_1} + \dfrac{1}{R_2} + \dfrac{1}{R_3}} \qquad (2\text{-}43)$$

and since v_o is the noninverting gain times v_B, it follows that

$$v_o = \left(\frac{R_F + R_I}{R_I}\right)\left(\frac{\dfrac{v_1}{R_1} + \dfrac{v_2}{R_2} + \dfrac{v_3}{R_3}}{\dfrac{1}{R_1} + \dfrac{1}{R_2} + \dfrac{1}{R_3}}\right) \qquad (2\text{-}44)$$

EXAMPLE 2-18

A noninverting averager circuit is used to provide a weighted sum of four voltages.

$$v_1 = 1 \text{ V} \qquad v_2 = 2 \text{ V} \qquad v_3 = 4 \text{ V} \qquad v_4 = 8 \text{ V}$$

$$R_1 = 1 \text{ k}\Omega \qquad R_2 = 2 \text{ k}\Omega \qquad R_3 = 4 \text{ k}\Omega \qquad R_4 = 8 \text{ k}\Omega$$

Determine the output voltage of a voltage follower being driven by this averager.

Solution

The gain of the voltage follower is unity.

$$v_o = \frac{\dfrac{1\text{ V}}{1\text{ k}\Omega} + \dfrac{2\text{ V}}{2\text{ k}\Omega} + \dfrac{4\text{ V}}{4\text{ k}\Omega} + \dfrac{8\text{ V}}{8\text{ k}\Omega}}{\dfrac{1}{1\text{ k}\Omega} + \dfrac{1}{2\text{ k}\Omega} + \dfrac{1}{4\text{ k}\Omega} + \dfrac{1}{8\text{ k}\Omega}} = \frac{4\text{ m}A}{1.875\text{ mS}} = 2.13\text{ V}$$

If R_1, R_2, and R_3 are all equal, Eq. (2-44) reduces to

$$v_B = v_1\left(\tfrac{1}{3}\right) + v_2\left(\tfrac{1}{3}\right) + v_3\left(\tfrac{1}{3}\right) = (v_1 + v_2 + v_3)\left(\tfrac{1}{3}\right)$$

and Eq. (2-44) reduces to

$$v_o = \left(\frac{v_1 + v_2 + v_3}{3}\right)\left(\frac{R_F + R_I}{R_I}\right) \quad \text{when } R_1 = R_2 = R_3 \qquad (2\text{-}45)$$

and the output voltage is the average of the three input voltages.

EXAMPLE 2-19

A noninverting averager circuit is averaging the voltages on three lines. $v_1 = 3$ V, $v_2 = -7$ V, and $v_3 = -2$ V. $R_1 = R_2 = R_3 = 2.4$ kΩ. The noninverting amplifier has a gain of $+2$. Determine the output voltage.

Solution

$$v_o = \left(\frac{3\text{ V} - 7\text{ V} - 2\text{ V}}{3}\right)(2) = -4\text{ V}$$

The reasoning can also be extended to many inputs. Thus for four inputs the output would be the average of the four:

$$v_o = \left(\frac{v_1 + v_2 + v_3 + v_4}{4}\right)\left(\frac{R_F + R_I}{R_I}\right) \qquad (2\text{-}46)$$

when $R_1 = R_2 = R_3 = R_4$.

2.8 Op-Amp-Circuit Impedance Modification Through Feedback

Both the input and output impedances are modified through the use of negative feedback. The input impedance of the noninverting amplifier is increased while the output impedance of all negative feedback circuits is decreased.

2.8.1 Input Impedance of the Noninverting Amplifier

Since the differential amplifier is the first stage of the operational amplifier, it is in this stage that the impedance increase occurs. Figure 2-17 illustrates the equivalent circuit of an operational amplifier connected as a voltage follower (unity gain). Open loop gain is introduced in the output voltage generator.

R_{in}, from each base to common, is the common-mode input impedance given by Eq. (1-23). A typical value for this impedance is 400 MΩ. Since the two emitters are connected together, there is a path between them which contains a resistance

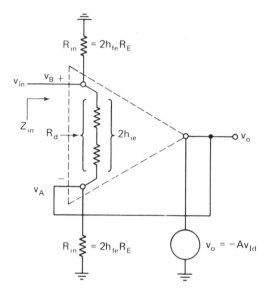

(a) Equivalent circuit of voltage follower

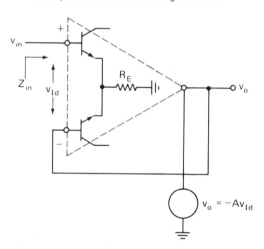

(b) Transistor representation

Fig. 2-17 Equivalent-circuit representation of voltage follower.

equivalent to h_{ie} for each transistor. Thus the impedance between the two transistor bases is $2h_{ie}$.

The approximate relationship for h_{ie} [1] in terms of I_B is

$$h_{ie} \cong \frac{V_T{}^*}{I_B(\text{dc})} \qquad (2\text{-}47)$$

If I_B is 100 nA (0.1 μA), then $h_{ie} = 260$ kΩ. This is an extreme difference from the

---

*$V_T = 26$ mV $= kT/q$ at $T = 301°$ Kelvin or $28°$C (see Section 5.4 for constants), where 26 mV is the electron-volt equivalent of temperature (or work function) from the Einstein equations.

1 to 4 kΩ normally encountered in transistor amplifiers, but the differential amplifier is normally biased to conduct very small base currents. Typical values for h_{ie} range from 50 to 500 kΩ for op amps.

For Fig. 2-17, $2h_{ie} = R_d$, which has a value of 520 kΩ. Thus without feedback the input impedance to each base appears to be 400 MΩ to common and 520 kΩ to the other input (which is connected to the op amp output—low impedance). This appears to be a quite low impedance when it was indicated that a follower has a very high input impedance. The equivalent input impedance between the two bases with feedback becomes

$$R_{df} = \left(\frac{A_{VOL}}{A_{VCL}}\right)2h_{ie} = (A_{VEX})2h_{ie} \tag{2-48}$$

which, if A_{VEX} is large, is very large.

The total input impedance to each base is then the common-mode input impedance, R_{in}, in parallel with R_{df} (differential mode, with feedback). Thus

$$Z_{in} = \frac{(R_{df})(R_{in})}{R_{df} + R_{in}} \quad \text{(derivation appears in Appendix A1)} \tag{2-49}$$

An example of how the input impedance is calculated and varies with closed loop gain is given next.

EXAMPLE 2-20

An operational amplifier, connected as a voltage follower (unity gain), has an open loop gain of 100,000. $I_B = 110$ nA and $R_{in} = 420$ MΩ. Find the input impedance to the follower.

Solution

$$h_{ie} = \frac{26 \text{ mV}}{110 \text{ nA}} = 236 \text{ k}\Omega \qquad R_d = 2h_{ie} = 472 \text{ k}\Omega$$

$$A_{VEX} = \frac{100,000}{1} = 100,000$$

Thus

$$R_{df} = (472 \text{ k}\Omega)(100,000) = 47,200 \text{ M}\Omega$$

$$Z_{in} = \frac{(420 \text{ M}\Omega)(47,200 \text{ M}\Omega)}{420 \text{ M}\Omega + 47,200 \text{ M}\Omega} \approx 420 \text{ M}\Omega$$

As can be seen from Example 2-20, the input impedance was very large, owing to the large excess loop gain present for multiplying $2h_{ie}$.

Just as this phenomenon can be used to an advantage, it can also work to trap the unsuspecting user. Example 2-21 will serve to illustrate this case.

EXAMPLE 2-21

An operational amplifier with an open loop gain of 10,000 is strapped in the noninverting mode for a closed loop gain of 1000. A voltage divider, using two 500 kΩ resistors, precedes the noninverting amplifier. Because of the expected high input impedance, no problem is presumed to exist from the high-value voltage-divider resistors.

Solution

$$R_{df} = (472 \text{ k}\Omega)(A_{VEX}) = (472 \text{ k}\Omega)(10) = 4.7 \text{ M}\Omega$$

$$Z_{in} = \frac{(420 \text{ M}\Omega)(4.7 \text{ M}\Omega)}{420 \text{ M}\Omega + 4.7 \text{ M}\Omega} = 4.65 \text{ M}\Omega$$

Thus the shunt leg of the voltage divider (500 kΩ) is loaded by a 4.6 MΩ load, and a 5% error results in the actual gain of the circuit.

Thus one should not take for granted that the input impedance of a noninverting amplifier is nearly infinite.

2.8.2 Output Impedance of the Op Amp Circuit

The output stage of most operational amplifiers is some form of emitter follower. Thus the open loop output impedance of this amplifier ranges from 50 to 100 Ω. But someone experienced in op amp circuits knows that the output impedance of an operational amplifier circuit is less than 1 Ω for most circuits. How can this apparent enigma exist?

For the inverting amplifier, the output impedance is reduced by an amount given by

$$R_o = \frac{r_o}{1 + \dfrac{A_{VOL}}{\left(\dfrac{R_F}{R_I} + 1\right)}} = \frac{r_o}{1 + A_{VEX}} \tag{2-50}$$

where R_o is the closed loop output impedance and r_o is the open loop impedance of the op amp output stage. The derivation for Eq. (2-50) can be found in Appendix A2. Equation (2-50) applies equally well for the noninverting amplifier.

An example will help to clarify the use of Eq. (2-50).

EXAMPLE 2-22

An operational amplifier with an open loop gain of 80,000 is strapped in the inverting mode for a closed loop gain of 10. The output impedance of the output stage in the op amp is 80 Ω. Determine the circuit output impedance.

Solution

$$A_{VEX} = \frac{80,000}{11} = 7272.7$$

Remember that A_{VEX}, even for the inverting amplifier, is that of the noninverting amplifier.

$$R_o = \frac{80}{7272.7} = 0.011 \ \Omega$$

Thus even though the op amp has an output impedance of 80 Ω, the circuit has an output impedance of 11 mΩ.

This example serves to illustrate the meaning of the expression "an op amp circuit has zero ohms output impedance;" it is nearly zero for all practical purposes.

Output impedance is an extremely useful tool for measuring the excess loop gain of any position-type feedback system. The system need not be an op amp with

Using the Operational Amplifier as a Circuit Element

feedback resistors; it can be a large control system with many amplifiers and other components which tend to mask the true nature of the system. It must be remembered that the system performance depends on the excess loop gain for control; thus a measurement of excess loop gain provides a deep insight into the performance characteristics of a feedback control system. An example will help to clarify these assertions:

EXAMPLE 2-23

A large system with many amplifiers and high power stages is a driver to supply electric current to a power-line simulator. The power output stage of the feedback control system (driver) has an open loop output impedance of 3 Ω. The power-line simulator draws 9 A from driver at 11.00 V. When the power-line simulator is disconnected from the driver, the driver's output voltage rises to 11.05 V. Determine the closed loop output impedance and the excess loop gain of the driver.

Solution

11.05 V is the unloaded output voltage (V_T) of a Thévenin equivalent circuit. 11.00 is the loaded output voltage at a 9 A load current. The output impedance (R_T) is

$$R_T = \frac{11.05 \text{ V} - 11.00 \text{ V}}{9 \text{ A}} = \frac{0.05 \text{ V}}{9 \text{ A}} = 5.5 \text{ m}\Omega = R_o$$

$r_o = 3 \ \Omega$; thus the excess loop gain, from Eq. (2-50),

$$A_{\text{VEX}} = \frac{r_o}{R_o} = \frac{3 \ \Omega}{5.5 \text{ m}\Omega} = 54$$

If tighter control over output voltage is needed, more excess loop gain is required.

EXAMPLE 2-24

An inverting amplifier circuit is "strapped" for a closed loop gain of 40. It is known that the output stage of the op amp is an emitter follower which has an approximate output impedance of 75 Ω. When the amplifier is loaded and draws 20 mA, the output voltage drops by 1 mV. Determine the op amp's open loop gain in dB.

Solution

$$R_o = \frac{\Delta v_o}{\Delta i_L} = \frac{1 \text{ mV}}{20 \text{ mA}} = 50 \text{ m}\Omega \qquad r_o = 75 \ \Omega$$

The output impedance formula is a function of the noninverting gain, so $40 + 1 = 41$ (noninverting gain). Then

$$A_{\text{VEX}} = \frac{75 \ \Omega}{50 \text{ m}\Omega} = 1500$$

and the open loop gain is $41(1500) = 61{,}500$.

$$A_{\text{VOL}} \text{ in dB} = 20 \log_{10}(61{,}500) = 20(4.79) = 95.8 \text{ dB}$$

REFERENCE

1. **Jacob Millman and Christos C. Halkias**, *Integrated Electronics: Analog and Digital Circuits and Systems* (New York: McGraw-Hill Book Company, 1972), p. 352.

PROBLEMS

1. An operational amplifier has an open loop gain of 75 dB. The positive input is connected to ground, and the output has -5.5 V on it. Determine the dc voltage on the inverting input lead.

2. Determine the output voltage and polarity of an operational amplifier running open loop (no feedback resistor). It has -1.00005 V dc on the noninverting input and -1.00010 V dc on the inverting input. A_{vol} is 75,500.

3. An op amp with an open loop gain of 94 dB has 7.99997 V on the noninverting input and 8.00002 V on the inverting input. Determine the magnitude and polarity of the output voltage.

4. An inverting amplifier is connected with an input resistor of 1 kΩ and a feedback resistor of 10 kΩ. A dc voltage of -0.32 V is applied to the circuit input. Calculate the output voltage and polarity.

5. An inverting amplifier with $R_I = 2$ kΩ and $R_F = 5$ kΩ has an applied input voltage of 3.2 V dc.

   a. Determine the magnitude and direction of the current through the feedback resistor.

   b. Find the magnitude and polarity of the output voltage using the current found in part (a).

6. An inverting amplifier has an input resistance of 2 kΩ and a feedback resistor of 8 kΩ. A 10-V dc voltage is applied to its input. The noninverting input is removed from ground and connected to 5 V dc.

   a. Find the output voltage using Eq. (2-7a).

   b. What voltage appears at the amplifier's summing junction?

   c. Which rule was required to solve this problem?

7. An operational amplifier with an open loop gain of 5000 is connected as an inverting amplifier with an R_I of 2 kΩ and a R_F of 150 kΩ.

   a. What is the ideal closed loop gain?

   b. Find the actual closed loop gain.

   c. What is the percent error in actual closed loop gain?

8. Using Eq. (2-11), determine the percent error in actual closed loop gain for Problem 7. Compare your answer with that for part (c) of Problem 7.

9. The op amp of Problem 7 now has an open loop gain of 100,000; $R_I = 1$ kΩ and $R_F = 75$ kΩ.

   a. What is the new ideal closed loop gain?

   b. Find the new actual closed loop gain.

   c. Compare the actual closed loop gain in part (b) with that in part (b) of Problem 7. Draw a conclusion about the relationship between open loop gain and closed loop gain error.

10. A dc voltage of 3 V is applied to a noninverting amplifier with an R_I of 1 kΩ and a R_F of 5 kΩ.

    a. Calculate the closed loop gain.

    b. Find the output voltage magnitude and polarity.

11. A noninverting amplifier is "strapped" with gain resistors as follows: $R_I = 3$ kΩ and $R_F = 6$ kΩ. -2 V is applied to the circuit input.

    a. Find the output voltage and polarity.

    b. Identify the rules used to solve this problem.

12. An op amp with an open loop gain of 2000 is connected as a noninverting amplifier. R_I is 3 kΩ and R_F is 21 kΩ.

    a. Determine the ideal output voltage if $Vin_2 = +0.01$ V.

    b. Find the actual output voltage.

    c. What is the percent error in output voltage?

13. A noninverting amplifier must amplify a -200 mV signal to -5 V with a gain error not exceeding 1%.

    a. Select a resistor combination from Appendix B1 (1%) in the range 1 to 50 kΩ.

    b. Determine the excess loop gain required for the amplifier. [*Hint*: use Eq. (2-11).]

    c. Find the minimum open loop gain required by the op amp.

14. A scaling subtractor (Fig. 2-8b) has $R_1 = 1$ kΩ, $R_2 = 3.3$ kΩ, $R_3 = 6$ kΩ, and $R_F = 4$ kΩ. The applied voltages are $V_1 = -2.2$ V dc and $V_2 = +1.5 \sin(3\pi t)$.

    a. Find the gain between V_1 and the output.

    b. Find the gain between V_2 and the output.

    c. Find the equation for the output voltage.

    d. Which resistor should be changed in order to change the scaling subtractor to a difference amplifier? What will its new value be?

    e. Make the circuit a subtractor by changing one or more resistors while leaving $R_F = 4$ kΩ. What are their values?

15. Repeat Problem 6 using only the rules for the solution. Identify the rules.

16. A noninverting amplifier with an open loop gain of 80 dB is strapped for a closed loop gain of 90. Determine the values of A and β in Eq. (2-24). Express the equation with numerical values in place of all constants.

17. A feedback control system (not using op amps) has an actual closed loop gain of $+20$. The feedback network has an overall voltage loss of 26.5 dB.

    a. Determine the ideal closed loop gain of the system.

    b. Find the open loop gain.

    c. What is the excess loop gain?

18. An inverting summer has the following closed loop gains: $A_{VCL_1} = -10$, $A_{VCL_2} = -7$, $A_{VCL_3} = -5$. If R_F is 100 kΩ, find the values of R_1, R_2, and R_3.

19. An inverting summer has three input voltages:

$$V_1 = -0.5 + 1.5\sin(2\pi t) \text{ V}$$
$$V_2 = -1.3 \text{ V}$$
$$V_3 = +2.0 - 3.2\sin(2\pi t) \text{ V}$$

The biasing resistors are: $R_1 = 2 \text{ k}\Omega$, $R_2 = 4 \text{ k}\Omega$, $R_3 = 6 \text{ k}\Omega$, and $R_f = 5R_3$.

a. Determine the equation of the output voltage.

b. Draw the output waveform to scale.

20. The inverting summer of Problem 19 uses an op amp with an open loop gain of 20,500. Calculate the apparent closed loop gain and the excess loop gain for this summer circuit.

21. An averager circuit with a voltage follower as its amplifier has six inputs with 16 kΩ resistors on each input. The input voltages are: $V_1 = 2 \text{ V}$, $V_2 = +6 \text{ V}$, $V_3 = +0.5 \text{ V}$, $V_4 = -3.5 \text{ V}$, $V_5 = -5.0 \text{ V}$, and $V_6 = +10.2 \text{ V}$. Determine the output voltage of the voltage follower.

22. The open loop gain of an amplifier with $h_{ie} = 300 \text{ k}\Omega$ and $R_{I_{CM}} = 100 \text{ M}\Omega$ is 10,000. Calculate the input impedance of the system if the op amp is used in a noninverting amplifier and is strapped for a gain of 2.5.

23. A μA 741 op amp is strapped for a noninverting gain of 5000. Determine the approximate input impedance of the amplifier using the specified values.

24. An inverting amplifier uses an op amp with an emitter-follower output stage with an impedance of 50 Ω. The open loop gain is 75,000 and is strapped for a closed loop gain of -375. Determine the circuit's output impedance.

25. An op amp with an output impedance of 75 Ω is strapped for an inverting gain of 200. When the amplifier circuit is loaded to 20 mA, the output voltage drops by 15 mV.

a. Determine the excess loop gain.

b. Calculate the open loop gain.

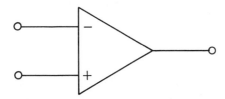

3 Offsets and Offset Compensation

3.1 Introduction

Chapter 1 introduced the differential amplifier and the statement was made that this stage, alone, contributes most of the good features and many of the bad features to the operational amplifier. Chapter 3 deals with voltage offset, current offset, bias currents, and offset drift, all of which are bad features of the operational amplifier and occur in the differential amplifier stage. Offset compensation, including bias-current compensation, is discussed and the merits of the various forms are evaluated.

3.2 Output-Voltage Offset Due to Input Voltage Offset

Offsets appear as *changes* from the expected output voltage on an operational amplifier circuit; these are due to internal imbalances in the input differential amplifier of the op amp.

The type of operational amplifiers being discussed in this text are for the most part of the monolithic type (i.e., all components are deposited on a single silicon substrate). This means that all transistors on a particular integrated circuit "chip" should be identical engineering designs; the transistors are "identical" only within a statistical deviation around some mean. This means that there are minor variations between transistors, even though they are identical in design. This phenomenon produces slight variations in the nominal value of the collector voltages on the differential amplifier when both inputs are grounded; as a result, the op amp has an output voltage deviation from zero. This unbalance is known as an input voltage offset.

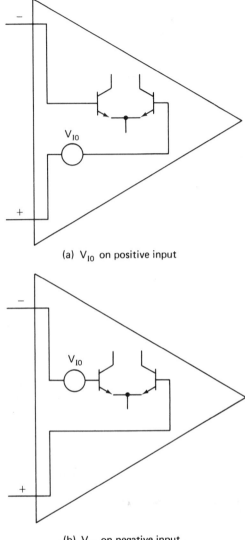

(a) V_{IO} on positive input

(b) V_{IO} on negative input

Fig. 3-1 Offset-voltage generators in operational amplifier.

Input voltage offset can be represented by a voltage generator in series with one of the input leads of the operational amplifier, as illustrated in Fig. 3-1.

3.2.1 Inverting Amplifier

The operational amplifier is useful for linear applications when used as the active element of an amplifier circuit employing input and feedback resistors. Let us now construct an inverting amplifier using the op amp of Fig. 3-1a.

Figure 3-2 illustrates this amplifier circuit. Figure 3-2b is the circuit of greatest interest where the offset voltage generator has been moved outside the op amp. The

Offsets and Offset Compensation

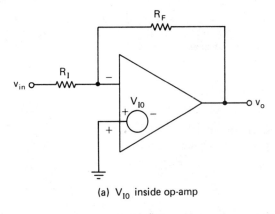

(a) V_{IO} inside op-amp

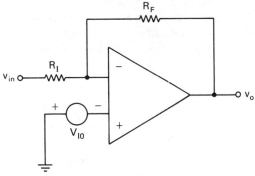

(b) V_{IO} outside op-amp

Fig. 3-2 Inverting amplifier using op amp with voltage-offset generator.

input voltage offset (V_{IO}) is now just a voltage source attached to the positive input of the op amp. The output voltage is the sum of two components (Appendix A3):

$$v_o = \left(-\frac{R_F}{R_I}\right)v_{in} \pm \left(\frac{R_F}{R_I} + 1\right)V_{IO} \qquad (3\text{-}1)$$

expected output voltage	output voltage change

One, the input signal voltage, is multiplied by the inverting gain *(Rule 2)* and the second, the input offset voltage (V_{IO}), is multiplied by the noninverting gain *(Rule 3)*; the two components sum on the amplifier's output *(Rule 5)*.

An example will help to illustrate this phenomenon.

EXAMPLE 3-1

An inverting amplifier, employing an op amp with a voltage offset, is connected for a closed loop gain of -5. The input signal is -1.000 V dc. The output voltage is $+5.120$ V dc. How much offset voltage does the op amp have?

Solution

First, calculate the expected value of the output voltage using Eq. (2-10).

$$v_o = -1.000(-5) = +5.000 \text{ V dc}$$

Second, subtract the expected output voltage from the actual output voltage.

$$+5.120 - (+5.000) = +0.120 \text{ V dc}$$

Thus the actual output voltage is 120 mV higher than expected. Since the second component on the output of the amplifier circuit is found using **Rule 3**, $V_{IO}[(R_F/R_I) + 1]$, it follows that

$$+120 \text{ mV dc} = V_{IO}(5 + 1) = 6(V_{IO})$$

$$V_{IO} = +20 \text{ mV dc}$$

Example 3-1 shows that the op amp has a $+20$ mV dc offset, which causes the output to be higher than expected. If the offset generator had its signs reversed, the output voltage would be 120 mV dc lower than expected, or $+4.880$ V dc. Thus it can be seen that not only the magnitude of the offset generator, but also the sign, contributes to the error in output voltage.

Equation (3-2) relates the output voltage change (V_{OV}) to the input offset voltage:

$$V_{OV} = \left(\frac{R_F}{R_I} + 1 \right) V_{IO} \qquad (3\text{-}2)$$

where the term $[(R_F/R_I) + 1]$ is *always* the *noninverting gain*.

It should be noted here that when an op amp is purchased for use, it does not have its input offset voltage stamped on the case along with the type number. The input offset voltage is created by small differences in the two transistors of the differential amplifier, and neither the magnitude nor the sign is known at the time of purchase. The manufacturer usually selects devices from a yield, grades them according to the desired parameters, and marks them with an appropriate type number. Examples of this grading process are the National Semiconductor LM 108, 208, and 308 integrated-circuit operational amplifiers. The 108 is a device meeting full military specifications and has 3 mV of maximum input offset voltage; the 208 is an industrial grade and also has 3 mV of maximum input offset voltage; the 308 is a commercial grade and has the worst of the guaranteed parameters (10 mV of maximum input offset). There is a lower grade than the 308, which is sold in bulk to the large hobby distributors; parameters on these devices are usually not guaranteed.

Thus the user can buy devices that are guaranteed to have input offsets falling within a prescribed tolerance range, but cannot buy a device with a particular offset voltage (including zero voltage offset).

In Fig. 3-2 the op amp of Fig. 3-1a (input voltage offset generator on positive input) was used to construct the inverting amplifier. It makes no difference to which input the offset voltage generator is attached; the output voltage is related by the same equation (i.e., the V_I generator voltage is multiplied by the noninverting gain). See **Rule 3**, which states that a voltage on *either input* will be multiplied by the noninverting gain. The output sign is explained in the discussion for **Rule 3** (Section 2.4). The derivation of this phenomenon is performed in Appendix A3.

Offsets and Offset Compensation

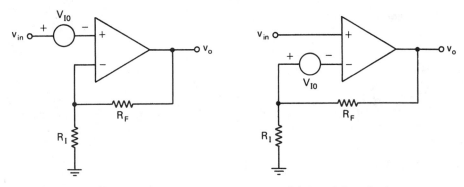

(a) V_{IO} in series with signal input   (b) V_{IO} on inverting input

Fig. 3-3 Noninverting amplifier using op amp with offset.

3.2.2 Noninverting Amplifier

The noninverting amplifier is also plagued by input offset voltage. Figure 3-3 illustrates the noninverting amplifier employing an op amp with internal offsets.

In Fig. 3-3a it can be seen that the input voltage (V_I) is in series with the signal voltage. Thus the two voltages sum:

$$v_o = \left(\frac{R_F}{R_I} + 1 \right) v_{in} \pm \left(\frac{R_F}{R_I} + 1 \right) V_{IO} \tag{3-3}$$

expected   output
output   voltage
voltage   change

and are multiplied by the same gain. Since the input signal is multiplied by the noninverting gain, it follows that the input offset voltage is also multiplied by the same noninverting gain and the change, alone, in output voltage due to an input offset voltage is

$$V_{OV} = \pm \left(\frac{R_F}{R_I} + 1 \right) V_{IO} \tag{3-4}$$

This can be clearly seen to be the same as the output-voltage change for the inverting amplifier case.

It makes no difference which amplifier configuration is used; the input offset voltage is *always* multiplied by the *noninverting gain*. It appears on the output as a change from the expected voltage value.

EXAMPLE 3-2

An operational amplifier has among its listed parameters: offset voltage ± 15 mV maximum. This op amp is connected as a noninverting amplifier with a gain of 8. The output voltage is -5.85 V when the input voltage is -0.75 V dc. Is this op amp within its rated maximum for offset voltage?

The expected output voltage is found using Eq. (2-14) as -6.00 V dc. The actual output voltage is -5.85 V dc. The difference (or voltage change) is 0.15 V dc on the output. This voltage referred to the input is (150 mV)/8, or 18.75 mV. Thus the op amp voltage offset is in excess of its rated maximum.

Again, it makes no difference to which input the V_{IO} generator is assigned; the ratio is still the noninverting gain as it appears on the output. The sign for the output change direction can be determined by comparing the signs on the V_{IO} generator symbol (see Fig. 2-5).

3.3 Output-Voltage Offset Due to Input Current

The input differential amplifier transistors of an operational amplifier are biased to conduct extremely small currents on the inputs. This current is so small as not to affect the circuit gain equations adversely, as illustrated in Chapter 2. As small as these currents are, they still contribute to the output voltage under certain circumstances.

3.3.1 Output-Voltage Offset Due to Bias Current

The typical input bias current (I_B) into each base of the differential-stage transistors is approximately 100 nA for a bipolar-transistor operational amplifier. All differential amplifiers in op amps will be presumed to be of the *NPN* transistor variety; this is not necessarily true in the industry, but it makes the material in this chapter more uniform.

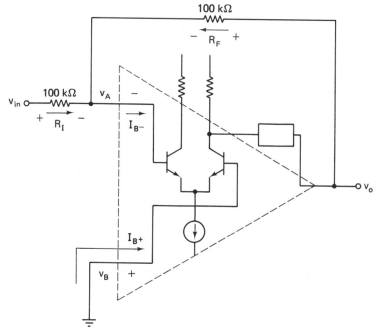

Fig. 3-4 Inverting amplifier showing differential input pair.

Since the transistors are *NPN*, both bias currents flow *into* the bases of each transistor of the differential pair. Figure 3-4 illustrates a typical case. The current flowing into the (+) base is labeled I_B^+, while that flowing into the (−) base is labeled I_B^-, where the superscripts, (+) and (−), on the I_B terms indicate which base is being referenced in a discussion. This bias current causes a voltage offset (current induced) to appear on the amplifier output when the op amp is connected with feedback. The reason for this offset can be seen from the following discussion.

Let v_{in} in Fig. 3-4 be zero volts. The voltage drop across R_I is essentially zero, since v_A is nearly zero volts by **Rule 1**. Essentially all of the base bias current (I_B^-) must pass through R_F, as the current through R_I is zero. The voltage drop across R_F is $(I_B^-)R_F$, which all appears at the circuit output, since the summing junction end of R_F is at zero volts. The sign of the output voltage is (+), since the bias current flows through R_F toward the base. The output voltage is (Appendix A4)

$$V_{OI} = +(I_B^-)R_F \qquad (3\text{-}5)$$

A bias current also flows into the (+) base, but it does not cause an output offset voltage to occur, because the impedance between base and ground is zero (direct wire).

EXAMPLE 3-3

An operational amplifier has rated bias currents of 110 nA maximum. This op amp is connected as an inverting amplifier with $R_I = 1 \ k\Omega$ and a closed loop gain of -100. Determine the output voltage offset due to bias currents alone.

Solution

Using Eq. (2-10), we obtain

$$-\frac{R_F}{R_I} = -100 \quad R_F = 100(1 \ k\Omega) = 100 \ k\Omega$$

Using Eq. (3-5) because it is an inverting amplifier, we obtain

$$V_{OI} = +(110 \ nA)(100 \ k\Omega) = +11 \ mV$$

An 11 mV offset, due to bias current alone, is an extremely large change voltage, and a method has been devised to reduce the effects of bias currents by canceling them at the op amp input.

3.3.2 Output-Voltage Offset Due to Offset Current

Bias currents flow into each base of the differential amplifier. These two currents are of the same order of magnitude and are nearly equal but are almost never exactly equal. The difference between the two bias currents, I_B^- and I_B^+, is the *offset current*. The offset current is

$$I_{IO} = \pm(I_B^+ - I_B^-) \qquad (3\text{-}6)$$

and is usually less than 10% of the average of the two bias currents. Thus the output change voltage could be reduced to 10% of its value, due to bias currents, if the change voltage were related to offset current. This can be accomplished by placing a

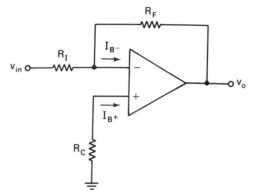

Fig. 3-5 Inverting amplifier with bias-current compensation resistor.

resistor in the noninverting input to ground, as illustrated in Fig. 3-5. This resistor is called a bias-current compensation resistor (R_C). The bias current (I_B^+) flowing through this resistor would then produce a negative voltage on the noninverting input, which, when multiplied by the noninverting gain, would partially cancel the change voltage due to I_B^- on the inverting input. But what value of resistor should be used?

A voltage appearing on the noninverting input will be multiplied by the noninverting gain as it appears on the op amp output. This voltage must equal the change voltage already on the output due to the inverting input bias current; it is opposite because of the noninverting gain. Thus

$$(I_B^-)R_F = (I_B^+)R_C\left(\frac{R_F + R_I}{R_I}\right)$$

where I_B^- and I_B^+ will be presumed to be equal. Solving for R_C, we get

$$R_C = \frac{R_F R_I}{R_F + R_I} \tag{3-7}$$

and it can be seen that R_C has a value equal to the parallel combination of R_I and R_F.

An example will help to illustrate the use of Eq. (3-7).

EXAMPLE 3-4

An operational amplifier is connected as an inverting amplifier with $R_I = 10$ kΩ and $R_F = 100$ kΩ. Determine the proper value for R_C.

Solution

Using Eq. (3-7), we obtain

$$R_C = \frac{(10\text{ k}\Omega)(100\text{ k}\Omega)}{10\text{ k}\Omega + 100\text{ k}\Omega} = 9.1\text{ k}\Omega$$

The output change voltage due to a bias current on the noninverting input with R_C in place is

$$V_{OIB} = -(I_B^+)R_F \tag{3-8}$$

and the output change voltage due to offset current (I_{IO}) is then

$$V_{OIO} = \pm(I_B^+ - I_B^-)R_F$$

and

$$V_{OIO} = \pm(I_{IO})R_F \tag{3-9}$$

where R_C is defined by Eq. (3-7) and the $(\pm)$ signs are used because it is never known which bias current is larger; thus the polarity of the offset current is never known, although it can be determined.

EXAMPLE 3-5

An operational amplifier has the following maximum specifications:

$$\text{Maximum average bias current:} \quad \pm 115 \text{ nA}$$
$$\text{Maximum offset current:} \quad \pm 12 \text{ nA}$$

Calculate the output offset voltage with and without a bias-current compensation resistor for an inverting amplifier with $R_I = 10 \text{ k}\Omega$ and $R_F = 100 \text{ k}\Omega$.

Solution

Assuming that $V_{in} = 0$, without R_C, the output offset change voltage is given by Eq. (3-5) as

$$V_{OI} = +(115 \text{ nA})(100 \text{ k}\Omega) = +11.5 \text{ mV}$$

With R_C,

$$R_C = \frac{(10 \text{ k}\Omega)(100 \text{ k}\Omega)}{10 \text{ k}\Omega + 100 \text{ k}\Omega} = 9.1 \text{ k}\Omega$$

The output change voltage is given by Eq. (3-9) as

$$V_{OIO} = \pm(12 \text{ nA})(100 \text{ k}\Omega) = \pm 1.2 \text{ mV}$$

It can be seen from Example 3-5 that the output offset voltage can be reduced by 10 times through the use of a current compensation resistor in the noninverting input of an inverting amplifier. High values for the gain-selecting resistors were deliberately chosen to illustrate the large output offset created. If lower values of R_I and R_F were chosen, the output offset voltage would be lower.

EXAMPLE 3-6

An operational amplifier has maximum bias currents of 115 nA and a maximum offset current of 12 nA. The op amp is connected as an inverting amplifier with a gain of -10, utilizing an input resistor of 1 kΩ. What typical output offset is created by:

a. Grounding the noninverting input?
b. Inserting a current compensation resistor in the noninverting input?

Solution

a. The feedback resistor (R_F) is found using Eq. (2-10) as 10 kΩ. Equation (3-5) yields

$$V_{OI} = 115 \text{ nA} (10 \text{ k}\Omega) = +1.15 \text{ mV maximum}$$

b. Equation (3-9) gives

$$V_{OIO} = \pm 12 \text{ nA} (10 \text{ k}\Omega) = \pm 120 \text{ } \mu\text{V maximum}$$

Output-Voltage Offset Due to Input Current

77

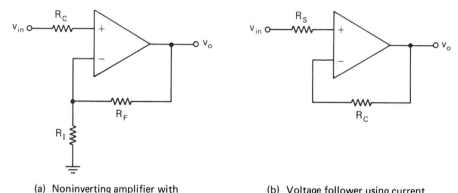

(a) Noninverting amplifier with current compensation resistor

(b) Voltage follower using current compensation resistors

Fig. 3-6 Current compensation in noninverting amplifiers.

In comparing the output offset voltage with the small resistors versus that with the large resistors, it can be seen that there is a significant advantage to using a small-value feedback resistor. Thus as a general rule, it can be said: *Use the smallest-value feedback resistor permissible* consistent with the op amp drive capability and other circuit considerations.

3.3.3 Current Compensation in Noninverting Amplifiers

The noninverting amplifier can be thought of as being an inverting amplifier upside down for current-offset-compensation purposes. Figure 3-6a illustrates the noninverting amplifier with a current compensation resistor.

It may seem strange at first to see a resistor in series with an input that is known to be nearly infinite in impedance. The reason, as is now known, is for current compensation purposes. The voltage follower would then be connected as shown in Fig. 3-6, which explains why Fig. 2-14b contains an apparent redundancy of resistors.

3.4 Output Changes Due to Input Voltage and Current Offsets

It is now apparent that the operational amplifier possesses imbalances that produce a voltage offset as well as a current offset. These two effects combine on the amplifier output. They are completely random and statistical as to their magnitude and polarity. Thus they could aid or cancel each other and each op amp, even of the same type, is an individual case.

The total maximum output offset is then the sum of the voltage and current offset effects. Thus the total output change due to offsets is

$$V_{OC} = \pm V_{OV} \pm V_{OI} \qquad (3\text{-}10)$$

where V_{OI} can be V_{OIB} or V_{OIO}.

We are now prepared to analyze an actual case of output offset in an operational amplifier circuit.

Offsets and Offset Compensation

EXAMPLE 3-7

The specifications for an LM 308 are as follows:

	Typical	Maximum	Units
Voltage offset	± 12	± 18	mV
Bias currents	105	145	nA
Offset current	± 10	± 15	nA

An inverting amplifier circuit, current-compensated, is being used where no offset adjust is possible. What is the largest feedback resistor that can be used in the inverting amplifier for a gain of -5 if the output offset is never to exceed 110 mV in either polarity?

Solution

The inverting gain is -5; the noninverting gain, $(R_F/R_I) + 1$, is $+6$. Thus the output offset due to voltage offset is given by Eq. (3-4):

$$V_{OV} = \pm(18 \text{ mV})(+6) = \pm 108 \text{ mV}$$

Solving Eq. (3-10) for V_{OIO} yields

$$\pm 110 \text{ mV} = \pm 108 \text{ mV} \pm V_{OIO}$$

Thus $V_{OIO} = \pm 2$ mV. Now using Eq. (3-9), 2 mV $= R_F(15 \text{ nA})$ and $R_F = 133 \text{ k}\Omega$ maximum. Thus any value of R_F less than 133 kΩ is permissible.

Operational amplifier circuits seldom use input or feedback resistors less than 1 kΩ because of the current drive problems from this or the previous op amp stage. Thus, the range of feedback resistors for the circuit in Example 3-7 would be 5 kΩ to 133 kΩ, as neither R_I nor R_F should go below 1 kΩ.

3.5   Offset Drifts

Offset drifts with temperature fall into two categories.

1. The current offset drift is labeled "coefficient of input offset current" and is measured in pA/°C.
2. The voltage offset drift is labeled "coefficient of input voltage offset" and is measured in μV/°C. These two drifts can be seen on the electrical characteristics table for the LM 308 in Fig. 3-7.

The user is a victim of these two effects and can take measures only to *reduce* the effects due to each drift. Two situations usually exist:

1. The user can select an operational amplifier with drift characteristics within an acceptable maximum.
2. The user is faced with an existing op amp, with no possibility of change, and must make the best of a poor situation. In this case, the user has two

PARAMETER	CONDITIONS	MIN	TYP	MAX	UNITS
Input Offset Voltage (Note 5)	$T_A = 25°C$		0.7	2.0	mV
Input Offset Current	$T_A = 25°C$		0.05	0.2	nA
Input Bias Current	$T_A = 25°C$		0.8	2.0	nA
Input Resistance	$T_A = 25°C$	30	70		MΩ
Supply Current	$T_A = 25°C$		0.3	0.6	mA
Large Signal Voltage Gain	$T_A = 25°C$, $V_S = ±15V$ $V_{OUT} = ±10V$, $R_L \geq 10\ k\Omega$	50	300		V/mV
Input Offset Voltage (Note 5)				3.0	mV
Average Temperature Coefficient of Input Offset Voltage (Note 5)			3.0	15	μV/°C
Input Offset Current				0.4	nA
Average Temperature Coefficient of Input Offset Current			0.5	2.5	pA/°C
Input Bias Current				3.0	nA
Supply Current	$T_A = +125°C$		0.15	0.4	mA
Large Signal Voltage Gain	$V_S = ±15V$, $V_{OUT} = ±10V$ $R_L \geq 10\ k\Omega$	25			V/mV
Output Voltage Swing	$V_S = ±15V$, $R_L = 10\ k\Omega$	±13	±14		V
Input Voltage Range	$V_S = ±15V$	±13.5			V
Common Mode Rejection Ratio		85	100		dB
Supply Voltage Rejection Ratio		80	96		dB

typical performance characteristics

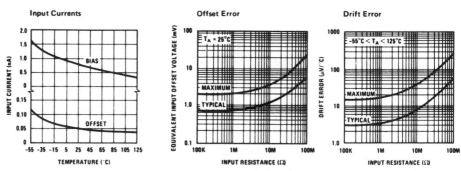

Fig. 3-7 Offset drifts with temperature change. (Courtesy of National Semiconductor Corp., Santa Clara, Calif.)

Offsets and Offset Compensation

alternatives to reduce the output effects of the existing drift to an acceptable minimum.

a. To reduce current offset drift to its minimum, use a bias-current compensation resistor (R_C) and make the feedback resistor (R_F) as small as possible.

b. To reduce voltage offset drift to its minimum, make the noninverting gain as low as possible.

EXAMPLE 3-8

An inverting amplifier employing a LM308 op amp with $A_{VCL} = -50$ has $R_I = R_C = 1\,k\Omega$ and $R_F = 49.9\,k\Omega$. The amplifier has a combined voltage and current *output* offset voltage of $+5$ mV at 25°C. Determine the output offset voltage at $+65$°C if the input offset current drift is 10 pA/°C and the input offset voltage drift is 5 μV/°C and they both cause the output to go more positive.

Solution

The temperature change is 65°C $-$ 25°C = 40°C. The total change in input current offset is (10 pA/°C)(40°C) = 0.4 nA. The output voltage will increase by

$$\Delta V_{OIO} = (0.4\ nA)(49.9\ k\Omega) = 0.02\ mV$$

The total change in input offset voltage is (5 μV/°C)(40°C) = 0.2 mV, which is multiplied by the noninverting gain or (R_F/R_I) + 1, which yields

$$\Delta V_{OV} = (0.2\ mV)(50.9) = 10.2\ mV$$

Thus the total change in output offset voltage can be found using Eq. (3-10) as

$$\Delta V_{OC} = \pm 0.02\ mV + 10.2\ mV = +10.22\ mV$$

and the circuit output voltage moves to $+10.22$ mV $+$ 5 mV $=$ $+15.22$ mV.

Differentiators with long time constants are particularly subject to offset drift.

3.6 Offset Compensation

Since offsets exist and are undesirable effects, they must be removed for certain circuit applications. Not all applications are affected by offsets. Examples of applications that *are not* affected by offsets include:

1. Circuits that have large dc output voltages and low closed loop gains. A circuit whose output voltage is 5 V is not normally affected by an error of 5 mV or even 20 mV.
2. Circuits that are ac-coupled on the output.
3. Circuits that are only ac-amplitude-sensitive; an example is ac filter circuits.

Circuits that *are* affected by offsets include:

1. Circuits that amplify small dc voltages; for example, output is 43 mV dc.
2. Circuits having large dc output voltages accurate to two or more decimal places; for example, a 20 mV output offset would be intolerable for an amplifier that was required to yield $+5.258$ V dc.

3. Amplifiers that have large closed loop gains where a small input offset would be multiplied many times on the output; for example, an op amp with a 15 mV input offset used in a circuit strapped for a gain of 200 will have a 3 V change in output voltage due to offset.

Thus it can be seen that judgment must be used in determining the need for offset compensation. This section presumes that the need is present.

Care should be taken first to reduce the offsets to an absolute minimum through the techniques described in Section 3.3, such as inserting a current compensation resistor and making the feedback resistor as small as possible. After these precautions have been observed and the output offset is still intolerable, additional offset compensation is required.

It must be understood that there is no way to separate voltage and current offsets on the output of an operational amplifier circuit. A voltage change is present and it must be compensated away; part of it is voltage and part of it is current, but without measurement, the contribution for each is unknown.

The most obvious method for offset-compensating an operational amplifier circuit is the attachment of a potentiometer on the two "offset null" terminals of the op amp. A variety of suggested locations, one for each op amp type, exists for connecting the potentiometer wiper terminal. This method for offset compensation is preferred for most circumstances, but for a variety of reasons it is not always used.

When internal compensation is deemed not appropriate, external methods for offset compensation are available.

3.6.1   Inverting Amplifiers

The inverting amplifier is the configuration most suitable for external offset compensation. Offset may be compensated through the use of an additional summing resistor, as shown in Fig. 3-8. The value of R_O is usually at least 100 times R_I and is preferably 1000 times R_I. R_O is attached to the wiper of a potentiometer connected between the + and − power supply voltages in order to permit offset compensation for either polarity. The wiper voltage (V_n) is the "null" voltage. A 10

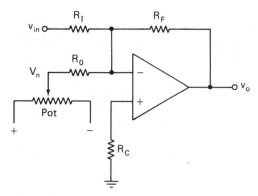

Fig. 3-8 Offset compensation by summing a voltage.

kΩ potentiometer is usually selected, as it draws little dividing current while still being small in comparison to a large R_O impedance (including the potentiometer), essentially independent of the pot-wiper position.

The output voltage is now the algebraic *sum* of the two input voltages times their respective inverting gains, or

$$v_o = \left(-\frac{R_F}{R_I} \right) v_{\text{in}} + \left(-\frac{R_F}{R_O} \right) V_n \qquad (3\text{-}11)$$

which is **Rule 2** twice and **Rule 5** once.

The compensation (second) term has $1/100$ the effect of the signal (first) term if $R_O = 100 R_I$.

EXAMPLE 3-9

An inverting amplifier with a 12 kΩ feedback resistor uses ± 10 V power-supply voltages. The output offset can be as much as ± 8 mV. Select an R_O to offset the amplifier through the use of an additional summing resistor.

Solution

Figure 3-8 illustrates the circuit for use with the additional summing resistor. The offset pot is selected to be 10 kΩ. The upper limit for R_O is determined by rearranging Eq. (3-11) with $v_{\text{in}} = 0$; then Eq. (3-11) becomes

$$\pm 8 \text{ mV} = \left(-\frac{12 \text{ k}\Omega}{R_O} \right)(\pm 10 \text{ V}).$$

Then

$$R_O(\text{max}) = (12 \text{ k}\Omega)\frac{10 \text{ V}}{8 \text{ mV}} = 15 \text{ M}\Omega.$$

Thus R_O can have any value between 1.2 and 15 MΩ and satisfy the requirements. But as R_O becomes smaller, the number of degrees of pot rotation to move ± 8 mV becomes smaller as well. It is therefore wise to select R_O near the high limit of the usable range so that the pot has more resolution. Make $R_O = 12$ MΩ.

Two methods are available for introducing the offset compensation on the noninverting input to the op amp.

1. A current may be passed through R_C by making it the shunt leg of a voltage divider with R_O being the series leg, as illustrated in Fig. 3-9a. The voltage change on the output, due to the compensation, follows **Rule 4**, so that the total output is

$$v_o = \left(-\frac{R_F}{R_I} \right) v_{\text{in}} + \left(\frac{R_C}{R_O + R_C} \right)\left(\frac{R_F + R_I}{R_I} \right) V_n \qquad (3\text{-}12)$$

where the signal gain is governed by **Rule 2** and the compensation gain is governed by **Rule 4**, which sums on the output according to **Rule 5**.

EXAMPLE 3-10

The circuit in Example 3-9 is to be offset-nulled using the circuit in Fig. 3-9a. The circuit gain is -10 and $R_C = 1.1$ kΩ. Determine the value of R_O to use to be able to null the output offset by ± 8 mV.

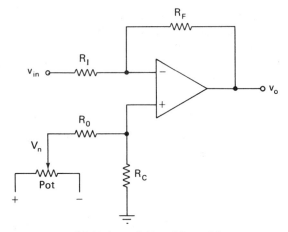

(a) Voltage divider of R_0 and R_C

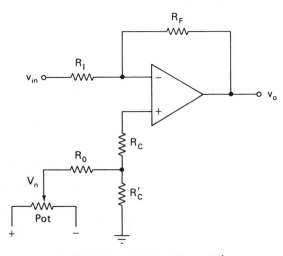

(b) Voltage divider of R_0 and R_C'

Fig. 3-9 Compensation voltage drop across R_C.

Solution

$R_O(\text{min}) = 10R_C = 11 \text{ k}\Omega$; from Example 3-9, $R_F = 12 \text{ k}\Omega$. The gain is -10; thus $R_I = 1.2$ kΩ. The upper limit for R_O is found by rearranging Eq. (3-12) for the case where $v_i = 0$ as

$$\pm 8 \text{ mV} = \left(\frac{1.1 \text{ k}\Omega}{R_O + 1.1 \text{ k}\Omega} \right)(10 + 1)(\pm 10 \text{ V})$$

$$R_O(\text{max}) = (1.1 \text{ k}\Omega)\frac{(11)(10 \text{ V}) - 8 \text{ mV}}{8 \text{ mV}} = 15 \text{ M}\Omega$$

Again, it is better to select R_O to be near the upper limit for greatest pot resolution: $R_O = 12$ MΩ.

2. A current may be passed through a small resistor appended to the ground side of R_C as illustrated in Fig. 3-9b. The small resistor is labeled R_C' as it is part of the

     Offsets and Offset Compensation

total R_C for offset current compensation purposes; thus $R_C + R'_C$ should be equal to the parallel combination of R_F and R_I. R'_C should be at least $\frac{1}{10}$ the value of R_C and possibly smaller.

The voltage drop across R'_C, created by the voltage division of R_O and R'_C, is passed through to the circuit output according to **Rule 4** as the voltage drop across R_C is zero; thus for Fig. 3-9b,

$$v_o = \left(-\frac{R_F}{R_I} \right) v_{\text{in}} + \left(\frac{R'_C}{R_O + R'_C} \right) \left(\frac{R_F + R_I}{R_I} \right) V_n \qquad (3\text{-}13)$$

where the signal gain is governed by **Rule 2** and the compensation gain is governed by **Rule 4**, which sums on the output according to **Rule 5**.

EXAMPLE 3-11

The circuit of Example 3-10 is to be nulled by splitting R_C into two parts $R_C = 1$ kΩ and $R'_C = 100$ Ω. The circuit is now illustrated in Fig. 3-9b. Select an R_O that will permit the ± 8 mV output offset to be nulled in 60% of the pot rotation.

Solution

The offset pot is connected to ± 10 V; 60% of the voltage at the end points will make V_n be ± 6 V. Thus rearranging Eq. (3-13), where $v_{\text{in}} = 0$, we get

$$\pm 8 \text{ mV} = \left(\frac{100}{R_O + 100} \right)(10 + 1)(\pm 6 \text{ V})$$

$$R_O = (100 \ \Omega) \left[\frac{(11)(6 \text{ V}) - 8 \text{ mV}}{8 \text{ mV}} \right] = 825 \text{ k}\Omega$$

Select R_O to be 820 kΩ, as that is a standard value and the slightly smaller resistor will assure that the 60% rotation can be met.

The one common factor in all the situations concerning the inverting amplifier is the *independence* between the signal and the compensation components at the circuit output. This one factor makes the inverting amplifier quite conducive to a variety of compensation techniques.

3.6.2 Noninverting Amplifiers

The noninverting amplifier always creates a compromise situation when it is being externally offset-compensated. This occurs because the offset voltage must usually be introduced somewhere in the gain-division resistors between the inverting $(-)$ input (summing junction) and ground. As a result, the circuit gain is modified whenever the offset potentiometer is rotated. This gain modification can be made small and most of the time it is insignificant, but it is always present. Therefore, it is strongly recommended to offset-compensate the noninverting amplifier through the op amp "offset null" terminals whenever possible. When this is not possible, the following methods are available:

1. A resistor (R_O) is connected to the junction of R_I and R_F at the inverting input to the op amp with a value of at least 10 times, and preferably 100 times, the

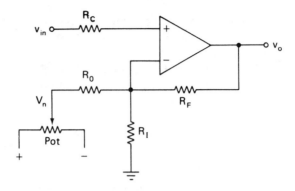

(a) R_O to inverting input

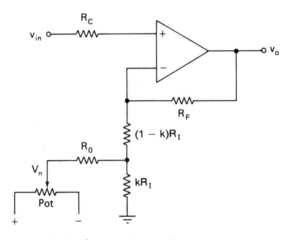

(b) $R_O R_I'$ divider between R_I and ground

Fig. 3-10 External offset compensation of noninverting amplifier.

value of R_I. The other end of this resistor is connected to a potentiometer, as shown in Fig. 3-10a. The equation for the output voltage is

$$ v_o = \left[\frac{R_F + R_I \| (R_O + R_P)}{R_I \| (R_O + R_P)} \right] v_{in} \mp \left(\frac{R_F}{R_O + R_P} \right) V_n \qquad (3\text{-}14) $$

where R_P is the parallel combination of the two legs of the potentiometer (one leg on each side of the wiper). R_P is largest when the wiper is in the center and is zero when the wiper is located at either end of the pot. R_P is usually small in comparison to R_O. The offset potentiometer can be seen as one term in the gain equation for the signal (v_{in}) position of Eq. (3-14). The potentiometer is again usually 10 kΩ. As can be seen in Fig. 3-10a, R_O plus the potentiometer is in parallel with R_I and

R_O plus R_P and kR_I are in series and modify the signal-gain equation. This effect is minor and can usually be ignored; the exception is for signals requiring a high degree of signal accuracy (5 decimal places).

Equation (3-14) for Fig. 3-10a can be approximated by

$$V_o = \left(\frac{R_F}{R_I} + 1 \right) V_{in} \pm \left(\frac{R_F}{R_O} \right) V_N \qquad (3\text{-}15)$$

where R_O is large in comparison to both R_P and kR_I. Calculations for output offset voltages are done in the same manner, as illustrated in Example 3-9.

2. Another method for offset-compensating the noninverting amplifier involves splitting R_I into large and small proportions; the smaller value (kR_I) is the one on the ground end where $k \le 0.1$. R_O is then connected between the junction of the large and small components of the R_I resistor and R_P (10 KΩ pot) as illustrated in Fig. 3-10b. The exact equation for the output voltage is

$$V_o = \left[\frac{R_F + R_I - \dfrac{k^2 R_I^2}{R_O + R_P + kR_I}}{R_I - \dfrac{k^2 R_I^2}{R_O + R_P + kR_I}} \right] V_{in} \pm \left[\frac{kR_F}{R_O + kR_I} \right] V_N \qquad (3\text{-}16)$$

It can be seen again that the potentiometer modifies the signal gain as it is rotated for adjustment of offset. Indeed, the modification is small, but ever-present.

EXAMPLE 3-12

The circuit of Fig. 3-10b is implemented using the following resistor values: $R_F = 10$ KΩ, $R_I = 1$ KΩ, $k = 0.1$, $R_O = 1$ MΩ, $R_c = 1$ KΩ and the pot has a total resistance of 10 KΩ. Determine the circuit gain for input signals when the pot wiper is in three positions: (a) $-V_{CC}$, (b) center, and (c) $+V_{CC}$. Find the percent change in gain.

Solution

Only two solutions are required as R_P is zero ohms when the pot wiper is at either end.

$$V_O = \left[\frac{10\text{ K} + 1\text{ K} - \dfrac{0.01(1\text{ K})^2}{1\text{ M} + 0\text{ K} + 100}}{1\text{ K} - \dfrac{0.01(1\text{ K})^2}{1\text{ M} + 0\text{ K} + 100}} \right] V_{in} = \frac{11{,}000 - 0.01}{1{,}000 - 0.01} = 11.0001\, V_{in}$$

$$V_O = \left[\frac{10\text{ K} + 1\text{ K} - \dfrac{0.01(1\text{ K})^2}{1\text{ M} + 2.5\text{ K} + 100}}{1\text{ K} - \dfrac{0.01(1\text{ K})^2}{1\text{ M} + 2.5\text{ K} + 100}} \right] V_{in} = \frac{10{,}999.99}{999.99} = 11.0001\, V_{in}$$

Any change in gain occurs in the 7th and 8th decimal place.

When R_O is large in comparison to both R_P and kR_I, the approximation to Eq. (3-16) for Fig. 3-10b is

$$V_o = \left[\frac{R_F}{R_I} + 1 \right] V_{in} \pm \left[k \frac{R_F}{R_O} \right] V_N \qquad (3\text{-}17)$$

It can be seen again that the potentiometer modifies the signal gain as it is rotated

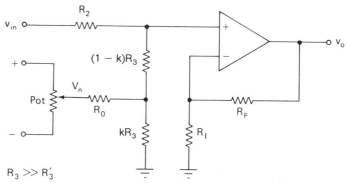

Fig. 3-11 Offset introduced at input voltage divider.

3. The third method for nulling the noninverting amplifier is implemented by inserting a voltage divider on the input, as illustrated in Fig. 3-11. The closed loop gain is the product of the input divider ratio and the noninverting gain. Thus

$$V_O = \frac{R_3 - \dfrac{k^2 R_3^2}{(R_O + R_P + kR_3)}}{R_2 + R_3 - \dfrac{k^2 R_3^2}{(R_O + R_P + kR_3)}} \left(\frac{R_F}{R_I} + 1\right)$$

(3-18)

$$\pm V_N \frac{kR_2 R_3}{R_O(R_2 + R_3) + R_2 R_3 + kR_3^2 - k^2 R_3^2} \left(\frac{R_F}{R_{I_I}} + 1\right)$$

The change of circuit gain as the pot wiper is rotated can be seen. When $k \leq 0.1$, Eq. (3-18) reduces to the approximation

$$V_O = V_{in}\left(\frac{R_3}{R_2 + R_3}\right)\left(\frac{R_F}{R_{I_I}} + 1\right) \pm V_N \frac{k}{R_O}\left(\frac{R_2 R_3}{R_2 R_3}\right)\left(\frac{R_F}{R_{I_I}} + 1\right)$$

(3-19)

As R_O becomes larger, the gain change becomes smaller.

Methods 2 and 3 are probably the most desirable for externally nulling the output offset voltage in a noninverting amplifier.

PROBLEMS

1. A noninverting amplifier is strapped for a gain of 5. The operational amplifier used has an internal voltage offset of 1.5 mV. How much error will be introduced to the output of the amplifier due to the voltage offset?

2. An op amp is used as an inverting amplifier with a voltage offset of 2.5 mV. For this circuit, $R_I = 3.3$ kΩ and $R_F = 10$ kΩ. The input signal is -1.500 V dc. Determine the output voltage to three decimal places.

3. An inverting amplifier has 8.020 V on the output when the input voltage is equal to -2.000 V. The circuit is strapped for a closed loop gain of -4. What is the input voltage

offset to the operational amplifier? Draw the voltage offset generator on the inverting input, and show its polarity.

4. 80 nA of current is drawn into each base of an input differential amplifier. The noninverting amplifier is connected to the common. There is no input offset voltage. R_F is 150 kΩ and R_I is 7.5 kΩ. What voltage appears on the circuit output if the input signal is at zero volts dc? Give both the magnitude and the polarity.

5. Repeat Problem 4 for $R_I = 7.5$ kΩ and $R_F = 1.5$ MΩ.

6. Repeat Problem 4 for $R_I = 500$ Ω and $R_F = 15$ kΩ; draw a conclusion about the relationship between R_F and the effects of bias currents on the amplifier output.

7. An input offset voltage of 2.5 mV causes the output of an operational amplifier with bias currents of 90 nA to go more negative. With an R_I of 25 kΩ, the amplifier is strapped for a gain of -5. No current compensation resistor (R_C) is in place. If the input voltage is $+2.250$ V, what is the output voltage to three decimal places?

8. An op amp has an output offset current of 10 nA, and I_B^+ is greater than I_B^-. The amplifier is connected in the noninverting mode with a gain of 45. With $R_I = 5$ kΩ and negligible offset voltage, determine the error in output voltage.

9. Calculate the value of the current compensation resistor (R_C) in Problem 2-21.

10. Compensate an inverting amplifier with an output offset of 30 mV by using the "inverting summer" method illustrated in Fig. 3-8. What should be the value of R_O be with the following parameters: $R_I = 1$ kΩ, $R_F = 5$ kΩ, offset potentiometer $= 10$ kΩ with ± 10 V applied voltage? The pot wiper setting is to be set at two-thirds of its total range. Select a standard value resistor for R_O from the 5% table in Appendix B1.

11. A μA 741A is specified to have 15 μV/°C and 500 pA/°C offset drift. If the op amp is used in an inverting amplifier with a gain of 50 and $R_F = 150$ kΩ, determine the output change as the circuit changes from 25°C to 60°C.

12. An inverting amplifier has an output voltage of ± 20 mV, $R_I = 2$ kΩ, and $R_F = 12$ kΩ. The offset pot is 15 kΩ and has a ± 15 V voltage applied to it. If R_O is attached between the pot wiper and the summing junction, determine the value of R_O for the offset to be nulled out using not more than 85% of the pot rotation. Select a standard value for R_O from the 5% table in Appendix B1.

13. The output of an inverting amplifier should be -5.295 V, but it is actually -5.372 V. The op amp is one section of a quad op amp where the offset terminals are not brought out. Determine the value of R_O to be used if it is to be connected between the pot and the noninverting op amp input. The gain is -25.25, and the input resistor has a value of 2 kΩ. The voltage on the 10 kΩ pot is ± 10 V, and the pot must be able to exceed the required output offset by 22 mV.

14. A quad op amp is known to have $I_B = 100$ nA average, $I_{IO} = \pm 10$ nA maximum, and $V_{IO} = \pm 2$ mV maximum. The op amp is used in an inverting amplifier with $R_I = 20$ kΩ and a gain of -12. The offset is nulled using a split current-compensation resistor. R_C' is $\frac{1}{10} R_C$. Select a R_O that will just suffice if the pot value is 10 kΩ and its ends are connected to ± 12 V (use Appendix B1, 5% values).

15. Five 20 kΩ input resistors are on the noninverting input of an op amp used as a noninverting averager (see Section 2.7.3). The noninverting amplifier gain is 10. The circuit is to be offset by connecting the offset resistor between a pot and the amplifier

summing junction (**inverting input**). Calculate the value of R_O to compensate for an output offset of ± 25 mV. The pot is 10 kΩ and is connected to ± 12 V.
(*Hint:* R_C is the parallel combination of the averaging resistors.)

16. The circuit of Problem 15 yields a value for R_O that is larger than desired. To reduce it, split R_I of the noninverting amplifier into two parts. Let R'_I equal $\frac{1}{10}R_I$. Recalculate the value of R_O.

17. The circuit for a scaling subtractor (see Fig. 2-8b) is already laid out on a printed circuit board when it is discovered that offset nulling will be necessary. The board layout only permits splitting R_3. R_O will then be connected between R_3 and R'_3. Calculate the value of R_O for $R_2 = 5$ kΩ, $R_3 = 20$ kΩ, $R_I = 2$ kΩ, $R_F = 20$ kΩ, and $R'_3 = \frac{1}{10}R_3$. The output offset voltage is ± 12 mV and the 10kΩ pot is connected to ± 10 V.

18. A noninverting amplifier is offset-compensated through the method illustrated in Fig. 3-10a. $R_I = 2$ kΩ, $R_F = 10$ kΩ, $R_C = 1.667$ kΩ, $R_O = 90$ kΩ, and the pot has a 10 kΩ total resistance. How much total percent signal-gain variation will occur when the pot is rotated over its whole range?

19. Calculate the percent change in gain as the offset potentiometer is rotated from its center position to one end for:

   a. Problem 16.
   b. Problem 17.

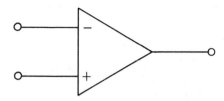

4 Power Supplies for Op Amp Circuits

4.1 Introduction

The design of power supplies, including transformers, rectifiers, filters, and regulators is discussed. Both design and analysis philosophies are given. The comparison between CMRR and PSRR is made and both are explained.

4.2 Dual-Polarity Power Supply

Operational amplifier circuits almost always require two power supplies: a positive supply and a negative supply. The reason for this can be seen in Chapter 1, where the differential amplifier and output emitter-follower amplifier were discussed. For the output complementary-symmetry emitter-follower stage to amplify both positive and negative voltages on its output, both positive and negative supply voltages must be applied. If a supply voltage of only one polarity is applied, the amplifier whose collector is grounded will saturate when a voltage of opposite polarity appears on its base. Thus this section describes two power supplies, one positive and one negative, which are identical in operation except for polarity.

A power supply is composed of a string of sections (blocks) that are typical of most supplies. Figure 4-1 illustrates a typical power supply for use with operational amplifiers.

4.3 Filter Capacitor

What is filtering, and how much filtering does one get from a particular size of filter capacitor?

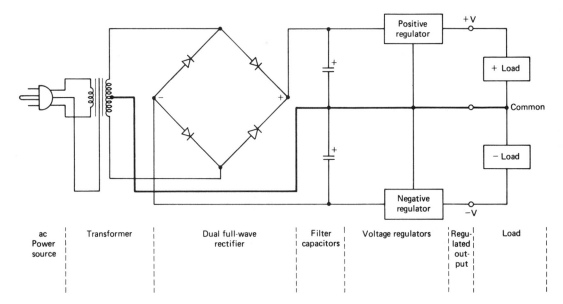

Fig. 4-1 Dual polarity-regulated power-supply block diagram.

The waveform entering the filter is a full-wave rectified ac voltage, as illustrated by the dashed lines in Fig. 4-2a. The waveform leaving the filter is a pulsating dc voltage, as shown by the solid lines in Fig. 4-2a.

A capacitive filter section functions as follows:

1. The incoming rectified voltage produces current that *pumps* the capacitor up with charge to V_{max} during the low-impedance, forward-biased-diode part of the cycle (see Fig. 4-2a, region A).

2. The incoming rectified voltage then diminishes to zero and back toward the capacitor voltage; during this time the diodes are reverse-biased. The only path for current to flow is into the regulator and then the load (see Fig. 4-2a, region B). During the time in region B, the capacitor is discharging as though a resistive load were across the capacitor; it discharges by Δv_C and stops at V_{min}.

The peak-to-peak ripple is caused by the capacitor discharging during the time Δt when the diodes are reverse-biased and is given by

$$\Delta v_C = \left(\frac{I_o}{C}\right)\Delta t \qquad (4\text{-}1)$$

where the actual *amount* of ripple voltage is *independent* of the initial voltage on the capacitor but the *percentage* of ripple voltage is *dependent* on the initial capacitor voltage. The percentage of ripple is given by

$$\% \text{ ripple} = \left(\frac{\Delta v_c}{V_{max}}\right)100 \qquad (4\text{-}2)$$

Let us now bring this information together for a useful relationship.

Power Supplies for Op Amp Circuits

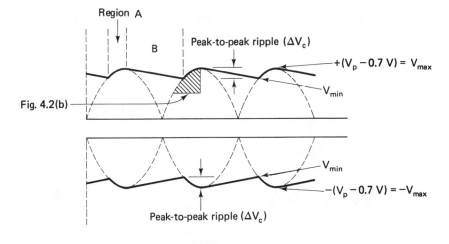

Fig. 4.2(b)

(a) Showing voltages

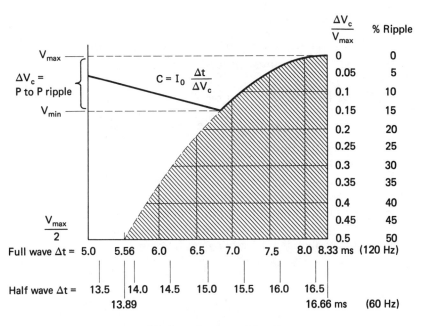

(b) Curve for determining filter
    capacitor size

Fig. 4-2 Filtering waveforms.

Figure 4-2b expands the shaded section of the positive waveform in Fig. 4-2a and yields the significant voltages and times involved. For the full-wave rectified supply, Δt ranges from 5.56 ms to 8.33 ms for corresponding Δv_c's of 0.5 V_{max} to 0 (zero). For the half-wave rectified supply, Δt ranges from 13.89 ms to 16.67 ms for the same change in Δv_c. Notice, that if there is no filter output current there is no capacitor voltage decay and no ripple voltage. Likewise, as the load current increases, the ripple increases as well. The filter capacitor size can be determined by rearranging Eq. (4-1) as

$$C = I_O\left(\frac{\Delta t}{\Delta v_c}\right) \tag{4-3}$$

and using the curve in Fig. 4-2b to determine the Δt value. The use of Fig. 4-2b is best illustrated by example.

EXAMPLE 4-1

A power supply whose full-wave rectifier delivers 12.1 V peak is to have a ripple voltage of no greater than 1 V p-p when 100 mA of current is drawn by the regulator. Find the minimum filter capacitor size.

Solution

$$\frac{\Delta v_c}{V_{max}} = \frac{1\ V}{12.1\ V} = 0.083$$

1. Find 0.083 on the $\dfrac{\Delta v_c}{V_{max}}$ scale.
2. Extend a line across to the left until it intersects the curve.
3. Extend a line down from the curve until it intersects a Δt value for full-wave.

$$\Delta t \cong 7.2\ ms$$

Use Eq. (4-3) to find the filter capacitor size as follows:

$$C = I_O\left(\frac{\Delta t}{\Delta v_c}\right) = (100\ mA)\left(\frac{7.2\ ms}{1\ V}\right) = 720\ \mu F$$

Choose the 1200 μF, 15 V or the 1000 μF, 25 V capacitor from Appendix B3.

Figure 4.2b and Eq. (4-3) are cumbersome to use. For a "quick and dirty" filter capacitor selection, a series of three specific points have been plotted and values determined for the approximate capacitor sizes. They are: 5%, for an exceedingly good supply (probably over designed); 10%, for an average supply; and 20%, for a rather poor supply; all are for full-wave filtering.

Ripple %

$$5 \qquad\qquad C = I_O\left(\frac{7.5\ ms}{\text{p-p ripple}}\right) \tag{4-4}$$

$$10 \qquad\qquad C = I_O\left(\frac{7.2\ ms}{\text{p-p ripple}}\right) \tag{4-5}$$

$$20 \qquad\qquad C = I_O\left(\frac{6.6\ ms}{\text{p-p ripple}}\right) \tag{4-6}$$

The use of these equations is given in Example 4-2.

EXAMPLE 4-2

A power supply for powering a series of operational amplifier circuits is being constructed. $V_{max} = 13$ V and the load draws an essentially constant load current of 1 A. The ripple voltage should be less than 1.5 V p-p on each polarity. Select a filter capacitor size from the following list:

2,500 μF	15 V
4,000 μF	15 V
8,000 μF	15 V
25,000 μF	15 V

Solution

Ripple (%)	Δv_c (p-p V)
5	0.65
10	1.3
20	2.6

The 20% ripple is greater than the specified 1.5 V p-p, so only the 5% and 10% ripple solutions can be selected.

$$C(5\%) = (1 \text{ A})\left(\frac{7.5 \text{ ms}}{0.65 \text{ V}}\right) = 11,538 \ \mu\text{F}$$

$$C(10\%) = (1 \text{ A})\left(\frac{7.1 \text{ ms}}{1.3 \text{ V}}\right) = 5461 \ \mu\text{F}$$

Choose 10%, as it meets the criteria and uses the smaller capacitor; select the 8000 μF capacitor.

The op amp user is sometimes faced with the problems of determining the ripple voltage for an already existing power supply. This is a more difficult task than finding the capacitor size, for it requires knowledge of three factors:

1. The filter capacitor size (C).
2. The load current from the filter (I_o).
3. The value of the peak filter voltage (V_{max}).

Having the value of the three parameters, the peak-to-peak ripple voltage may be calculated through

$$\Delta v_C = \frac{0.95}{118.25(C/I_o) + 1/V_{max}} \text{ V p-p} \qquad (4\text{-}7)$$

This equation is a first-order approximation to the exact solution, but is still reasonably accurate. The equation becomes increasingly in error as the ripple increases to greater than 30% ($\Delta v_C/V_{max} > 0.3$).

Example 4-3 explains the use of Eq. (4-7).

EXAMPLE 4-3

Find the peak-to-peak ripple for a power supply having a 200 μF filter capacitor, a peak filter voltage of 15 V, and an output current (current into the regulator) of 100 mA.

Solution

$$\Delta v_C = \frac{0.95}{118.25(200 \times 10^{-6}/100 \times 10^{-3}) + 1/15} = 3.13 \text{ V p-p}$$

4.4 Power-Supply Rejection Ratio

In Section 1.8, common-mode-rejection ratio was discussed. This related the amount of output voltage change that appeared when the two input voltages were moving in unison.

Let us now compare power-supply rejection ratio (PSRR) with CMRR. If the two input voltages are held constant and the plus and minus power-supply voltages are moved up and down in unison, from the op amp's vantage point it is as though the two power-supply voltages are held constant and the two input voltages are moved up and down together (see Fig. 4-3).

CMRR and PSRR can be related to the joke about painting battleships in which one person holds the paint brush while 10,000 other persons move the battleship. This joke is an almost perfect analogy of the distinction between CMRR and PSRR. For CMRR, the op amp inputs (paintbrush) move. For PSRR, the two power supplies [thus the op amp (battleship)] move.

Thus PSRR and CMRR are both caused by a common-mode voltage shift and are therefore essentially the same phenomenon. The real difference lies in the fact that the CMRR affects only the differential amplifier, whereas the PSRR affects the output stage as well. Thus there is a slight difference in numerical value between PSRR and CMRR.

The same relationship that exists between CMRR and CMR(dB) also exists for PSRR and PSR(dB):

$$\text{PSR(dB)} = 20 \log_{10}(\text{PSRR}) \tag{4-8}$$

Let us now visualize a voltage follower, having a noninverting gain (A_{VCL}) of unity (1), with plus and minus power supplies attached. For a moment presume that the plus supply has a 1 V peak-to-peak ripple while the minus supply has zero volts ripple. The op amp is now raising and lowering in voltage by the average of the 1 V ripple and the zero ripple voltage. The apparent input voltage appears to be changing by 0.5 V peak to peak and the op amp responds as though the ripple were a common-mode voltage on the op amp inputs; the ripple voltage appears on the op amp output reduced by the PSRR as given by Eq. (4-10). It now is evident that the op amp is sensitive to differences between the two power supplies that move in the *same* direction, but insensitive to power-supply movements in equal but opposite directions.

It is now time to review ripple again. Observe Fig. 4-2a. The ripple on the plus supply is going positive while the ripple on the minus supply is moving negative. If the two ripples are equal, there is *no average level shift* on the op amp. Thus if the ripples on the plus and minus supplies are equal, no ripple appears on the op amp output. If the ripples on the plus and minus power supplies have different voltage levels, one-half the difference (average) appears on the op amp as an effective level

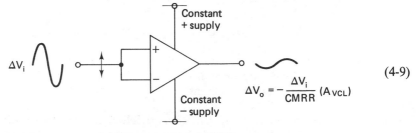

$$\Delta V_o = -\frac{\Delta V_i}{CMRR}(A_{VCL}) \qquad (4\text{-}9)$$

(a) Representation of CMRR

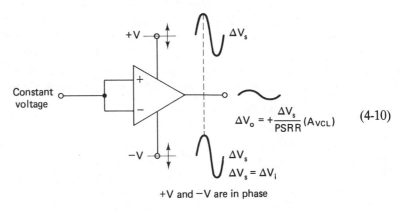

$$\Delta V_o = +\frac{\Delta V_s}{PSRR}(A_{VCL}) \qquad (4\text{-}10)$$

+V and −V are in phase

(b) Representation of PSRR

Fig. 4-3 Comparison between CMRR and PSRR.

shift which is reduced by the PSR ratio as it appears on the op amp output. When the amplifier has a noninverting gain (A_{VCL}), we are led to the relationship

$$\Delta V_{oPS} = \frac{|+V_{ripple}| - |-V_{ripple}|}{2(PSRR)}(A_{VCL}) \qquad (4\text{-}11)$$

In this equation the term V_{ripple} can be used to represent any change in voltage on one supply voltage that does not occur in an equal but opposite manner on the other supply voltage.

An example will help to clarify the use of Eq. (4-11).

EXAMPLE 4-4

An industrial and control mechanism uses op amp circuits, discrete transistors, and logic devices. The load on the two (plus and minus) supplies is different, such that the ripple on the plus supply is 1.5 V p-p and that on the minus supply is 0.9 V p-p. The op amps used have a PSR of 85 dB. The noninverting gain is +5. How much ripple appears on the op amp output?

Solution

PSR(dB) is 85 dB; thus PSRR is

$$\text{antilog}\left(\frac{85}{20}\right) = 10^{4.25} = 17{,}783 = PSRR$$

Then

$$\Delta V_{o\text{PS}} = \frac{|1.5| - |-0.9|}{2(17,783)}(5) = 84.35 \ \mu\text{V p-p}$$

This ac voltage will cascade through the system of op amp circuits, being amplified by the closed loop gain of each circuit it passes through. If the gains are not large, as is usually the case, the effect will be insignificant.

The effects due to both CMRR and PSRR can be represented by small voltage generators inserted in series with one of the op amp input leads, as illustrated in Fig. 4-4. This is purely a hypothetical representation, as they are still created as defined previously. Representing these as generators on the input allows all disturbances, offsets, CMRR, PSRR, and noise (Chapter 5) to be connected in series on the input to the op amp; thus their respective effects can be compared.

The value of the voltage generator is less than the output voltage by A_{VCL}. The method of calculating the equivalent input generator voltage value (V_{CMRR}) is as

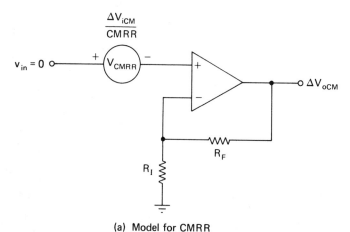

(a) Model for CMRR

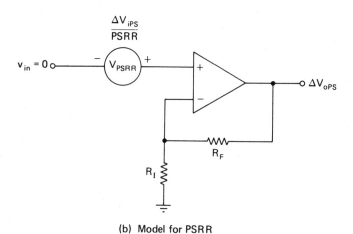

(b) Model for PSRR

Fig. 4-4 CMRR and PSRR equivalent generators.

follows:

1. Determine the output voltage change ($\Delta V_{o\text{CM}}$) by dividing the average common-mode input voltage by CMRR.

2. Divide this output voltage change by the noninverting gain (A_{VCL}) and assign this numerical voltage value to the input voltage generator (V_{CMRR}).

3. Multiply the numerical value of V_{CMRR} by CMRR and assign this numerical value to $\Delta V_{i\text{CM}}$.

4. The numerical value of the V_{CMRR} generator is then $\Delta V_{o\text{CM}}/A_{\text{VCL}}$ and the numerical value of $\Delta V_{i\text{CM}}$ is $(V_{\text{CMRR}})(\text{CMRR})$.

5. This gives $V_{o\text{CM}}$ the numerical value, $-(\Delta V_{i\text{CM}}/\text{CMRR})(A_{\text{VCL}})$, which satisfies Eq. (1-35); it is *out of phase* with V_{CMRR} or $\Delta V_{i\text{CM}}$.

The same sequence of steps yields the various voltage values for PSRR and satisfies Eq. (4-11). $\Delta V_{o\text{PS}}$ is *in phase* with V_{PSRR} or $[| + V_r| - | - V_r|]$.

It can be seen by these procedures that the CMRR or PSRR generator voltage (V_{CMRR} or V_{PSRR}) value is given by

$$V_{\text{CMRR}} = -\frac{\Delta V_{o\text{CM}}}{\dfrac{R_F + R_I}{R_I}} = \frac{\Delta V_{i\text{CM}}}{\text{CMRR}}. \qquad (4\text{-}12)$$

where $\Delta V_{o\text{CM}}$ is defined in Eq. (1-35):

$$V_{\text{PSRR}} = \frac{\Delta V_{o\text{PS}}}{\dfrac{R_F + R_I}{R_I}} = \frac{| + V_r| - | - V_r|}{2(\text{PSRR})} \qquad (4\text{-}13)$$

4.5   Voltage Regulator

A voltage regulator is a circuit that tends to have a constant output voltage regardless of the input voltage variations, which include ripple. For regulation to occur, the output voltage must always be less than the input voltage. The degree of ripple reduction and the consistency of output voltage depend on the type and complexity of the regulator.

4.5.1   Zener Diode Regulator

The most simple regulator is composed of a zener diode and a current-limiting resistor R_D to maintain a "keep alive" current in the zener diode at full load current. Figure 4-5 shows a zener-diode regulator and the zener characteristics.

1. The zener diode voltage follows the slope ($\Delta v_z/\Delta i_z$) on Fig. 4-5b. It can be seen that the zener voltage varies as the load current changes. This slight variation in voltage is specified in the zener parameters as a zener impedance

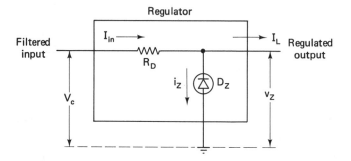

(a) Zener diode regulator circuit

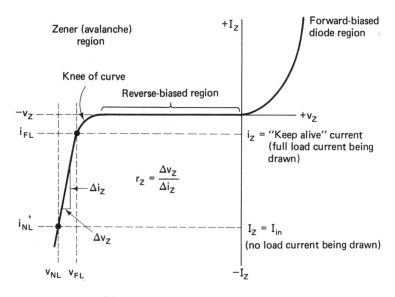

(b) Zener diode characteristics

Fig. 4-5 Zener-diode voltage regulator.

given by

$$r_Z = \frac{\Delta v_Z}{\Delta i_Z} \tag{4-14}$$

which ranges between 5 and 30 Ω for 400 mW zeners and 0.2 to 2 Ω for large (10 A) current zeners for voltages below 20 V dc. The manufacturers specify two zener impedances: one at a specific current on the slope and one at the knee of the curve. The one at the knee is the larger of the two and is not useful for zener operation; the one on the slope is the smaller and is the one to use.

2. V_{min} of the filtered supply must *never* go below the zener-diode voltage (v_Z) or regulation will cease and ripple will be passed directly through the regulator,

as the zener will be in the reverse-biased region (see Fig. 4-5b) rather than in the regulation region of zener avalanche.

3. The amount of ripple is reduced as it passes through the regulator and the output ripple is given by

$$v_{o\text{-rip}} = \frac{r_Z}{R_D + r_Z}(v_{\text{in-rip}}) \qquad (4\text{-}15)$$

The current drawn by the zener-diode regulator from the filtered power supply is *constant* regardless of load current as long as the zener is in the avalanche region. This makes this regulator quite inefficient for varying load currents.

An example will best illustrate the use of a zener-diode regulator.

EXAMPLE 4-5

A zener-diode regulator is to yield 10 V dc from a filtered supply whose average dc voltage (V_C) is 13 V with a 1 V peak-to-peak ripple. $R_D = 22$ Ω and $r_Z = 3$ Ω. Determine the amount of output ripple.

Solution

$V_C = 13$ V; the p-p ripple is 1 V; thus the amount of ripple appearing at the regulator output is given by Eq. (4-15).

$$v_{o\text{-rip}} = \left(\frac{3}{3 + 22}\right)(1 \text{ V p-p}) = 120 \text{ mV p-p}$$

Thus the ripple has been reduced to 12% of its input value.

4.5.2   Feedback Voltage Regulator

This voltage-regulator circuit appears as illustrated in Fig. 4-6. The circuit has three major components, characteristic of most feedback-type voltage regulators:

1. A stable voltage reference.
2. A noninverting amplifier.
3. A "pass transistor" or emitter follower (current-gain stage).

Since the current amplifier (emitter follower) is *in the feedback loop*, it has no effect on the circuit gain. The output voltage is given by *Rule 3* as

$$v_{\text{out}} = \left(\frac{R_F + R_I}{R_I}\right)V_{\text{REF}} \qquad (4\text{-}16)$$

It is important to recognize that any ripple appearing on V_{REF} will not be removed by PSRR (Section 4.4), as it is an *input signal* and it will thus appear on the output multiplied by the closed loop gain.

EXAMPLE 4-6

A zener-diode regulator is used as the voltage reference for a feedback-type voltage regulator. The reference has an output voltage of $+6.2$ V and a peak-to-peak ripple of 35 mV. Select

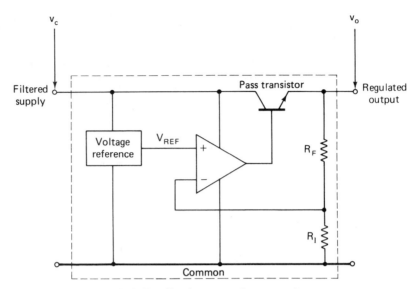

v_c ... v_o

Filtered supply ... Pass transistor ... Regulated output

Voltage reference ... V_{REF} ... + ... R_F ... − ... R_I

Common

Fig. 4-6 Feedback-type voltage regulator.

values for R_F and R_I in Fig. 4-6 which will yield a $+10$ V output voltage. Use resistor values from the 1% table in Appendix B1 between 1.00 and 10.0 kΩ. Find the peak-to-peak output ripple voltage.

Solution

The closed loop gain must be

$$\frac{10}{6.2} = 1.61 = \frac{R_F + R_I}{R_I}$$

In searching for a set of resistor ratios that fits the requirements, it is found that when $R_I = 1.64 R_F$, the gain requirements are met. Select $R_F = 1.69$ kΩ, and $R_I = 2.74$ kΩ, the ratio is then 1.62, which is 1.2% in error. The input ripple is multiplied by the closed loop gain; thus $v_{o\text{-rip}} = (1.61)(35$ mV$) = 56.4$ mV p-p.

The voltage reference can be a zener reference diode with a current-limiting resistor. This voltage reference has problems, in that it allows some ripple to pass through (see Example 4-5).

Three solutions to this problem are available:

1. The zener-diode voltage reference can derive its input voltage from the regulated output voltage. This occasionally creates a problem whereby the regulator will not "come up" when power is first applied and the regulator has to be "started" manually.

2. Two zener regulators can be cascaded where the first has a larger output voltage than the second and drives it. This usually results in a large ripple reduction.

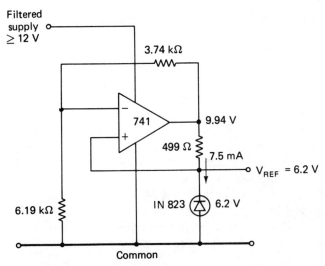

Filtered supply ≥ 12 V

3.74 kΩ

741

9.94 V

499 Ω

7.5 mA

V_{REF} = 6.2 V

IN 823   6.2 V

6.19 kΩ

Common

Fig. 4-7 Preregulator type of voltage reference.

3. A preregulator circuit can be used as the voltage reference. This is usually an op amp type of circuit employing feedback to hold the zener reference current constant. Figure 4-7 illustrates this circuit.

The ripple on the output (V_{REF}) of the preregulator in Fig. 4-7 is reduced from that on the filtered supply by the PSRR (about 50,000).

The noninverting amplifier seldom uses a single op amp, as shown in Fig. 4-6. The current required from a typical regulator circuit usually requires several stages of current drivers between the actual op amp and the pass transistor. These additional stages in drawing the circuit can be assigned to either the op amp or the pass transistor in Fig. 4-6. An example of this situation occurs when a darlington pair is used for the pass transistor. The driver transistor of the darlington pair is assigned to the pass transistor in the regulator drawing of Fig. 4-6.

The actual voltage regulation occurs in the pass transistor. It is this stage that adjusts its collector-to-emitter voltage to keep the emitter (output) voltage constant. The voltage from the feedback voltage reference (Fig. 4-7) is ripple-free. The regulator, including the pass transistor, appears to be a noninverting amplifier with its closed loop gain set by R_F and R_I. If the reference is 6.2 V and the regulated output is to be 6.2 V, the circuit takes the form of a voltage follower with unity gain. In this form it has the full open loop gain (A_{VOL}) for A_{VEX}. Thus the output ripple is reduced by the full PSR ratio. Also, the output voltage is held equal to the reference voltage by a feedback loop with about 95 dB of excess loop gain. It is thus constant, regardless of load current, to a high degree. These regulators are virtually perfect.

4.5.3 Three Terminal Regulators

Three-terminal regulators are in wide use today. Many models are made by various manufacturers. One of the widely used three-terminal regulators is the 78xx series.

The last two digits specifies the regulated output voltage which ranges from 5 V (xx = 05) to 24 V (xx = 24); xx = 05, 06, 08, 12, 15, 18, and 24. The minimum input voltage is 2.5 V above the regulated output voltage.

In order to analyze the operation of a voltage regulator, certain parameters must be known; one of these is the reference voltage inside the regulator. The manufacturer does not publish this value so a devious method must be used to find it.

Observe the schematic diagram in Appendix B-14 for a 78xx series regulator. All component values are specified except the top feedback, gain setting, resistor. By reading the specifications, the resistor is said to vary from 1 KΩ to 19 KΩ; thus at 5 V, this resistor is 1 KΩ and the voltage division to the reference point (same as reference voltage) is 5 V[4/(1 + 4)] = 4 V. Thus the reference voltage is 4 V; it is 4 V for all regulators in the series. When the feedback resistor is 19 KΩ, the output voltage is 4 V[(4 + 19)/4] = 23 V; the maximum voltage for the regulator series is 24 V so this author believes that either the feedback resistor actually goes to 20 KΩ rather than the 19 KΩ specified or that the reference voltage is 4.17 V.

4.6   Line and Load Regulation

The manufacturer specifies both line and load regulation for regulators. These parameters can also be found from the op amp parameters in the regulator.

4.6.1   Line Regulation

Line regulation is specified in one of two ways: either as a percentage of output voltage change or as an absolute output voltage change. Various manufacturers choose one or the other method.

Line regulation is specified as a change in regulated output voltage for a specified change in input (filtered) voltage. If the regulator is being driven from a transformer and filter, changes in line voltage are converted to corresponding changes in filtered input voltage.

As explained in Section 4.4, changes in the output voltage caused by changes in the input voltage are due to PSRR of the op amp in the regulator. Thus the methods of Section 4.4 can be used to determine the changes in output voltage for a given change in input (filtered) voltage; then if the line regulation is to be specified as a percentage of output voltage, this calculation can be done independent of the line regulation calculation. An example will serve to illustrate this process.

EXAMPLE 4-7

A 12 volt regulator is driven from a transformer, rectifier, and filter from a 120 VAC line. The op amp of the regulator has a PSRR of 2,000 and the reference voltage is 6 V. The filtered input voltage is 16 V and this voltage drops by 1 V when the line voltage drops 120 V to 114 V.

    a. Find the line regulation in volts using Eq. (4-11)

    b. Find the line regulation in percent.

Solution

    a. Using Eq. (4-11), $\Delta V_O = (1 \text{ V})(12/6)/2(2{,}000) = 500 \; \mu\text{V}$.

    b. Percent line regulation = $[(500 \; \mu\text{V})/12 \text{ V}] \, 100 = 0.004167\%$.

4.6.2 Load Regulation

Like line regulation, load regulation is specified either as a voltage change or as a percentage of regulated output voltage. Load regulation occurs as a result of the output impedance of the feedback amplifier as explained in Section 2.8.2. As the regulator is loaded and delivers more load current, its output voltage drops as a result of the low output impedance. The r_0 (without feedback) is usually the output impedance of the pass transistor (emitter follower) which can vary from 1 Ω to 20 Ω for typical three-terminal regulators. The output impedance of a regulator is "guessed" and a calculation of ΔV_o for a specified load current range (ΔI_O) is performed using Eq. (2-50) where $R_O = \Delta V_O/\Delta I_O$. An example will serve to illustrate this process.

EXAMPLE 4-8

An MC 7812 three terminal regulator has a pass transistor output impedance of 1.5 Ω. The regulator uses a 6 V internal reference and has an open loop gain of 1500. The regulator will deliver 0–1 amp to the load. The output voltage at no load is 12.065 V and that at 1 amp load current is 12.063 V. Two methods can be used, given this information

    a. Find the load regulation in voltage using Eq. (2-50).

    b. Find the load regulation in percent.

    c. Find the load regulation using the ΔV_o given.

    d. Find the load regulation in percent.

Solution

    a. Using Eq. (2-50),

$$1 + A_{\text{VEX}} = 1 + \left[1500/(12 \text{ V}/6 \text{ V}) \right] = 751,$$

then

$$R_o = 1.5 \; \Omega/751 = 0.002 \; \Omega \text{ and } \Delta V_o = 1 \text{ A}(0.002 \; \Omega) = 0.002 \text{ V}.$$

    b. Percent load regulation = $(0.002 \text{ V}/12 \text{ V}) \, 100 = 0.0167\%$.

    c. Using data given, $\Delta V_o = (12.067 - 12.065) = 0.002 \text{ V}$

    d. Percent load regulation is as in part (b), 0.0167%.

4.7 Slaving Plus and Minus Voltage Regulators

Operational amplifier circuits are subject to common-mode problems, as the CMRR is not infinite, which include power-supply problems. It is desirable to have the dc average of the plus and minus power supplies remain at zero volts even though the actual power-supply voltage may be changing. The common-mode problems can then be essentially eliminated. This is accomplished through a technique called "slaving" or "tracking." The negative supply is slaved to the positive supply so that when the positive supply goes more positive, the negative supply goes more negative. Through this means, the dc average is maintained at zero volts even though the two supplies vary slightly in voltage. This aids in keeping changes in output offset voltage to the absolute minimum.

Figure 4-8 illustrates the circuit connection for slaving the negative supply to the positive supply. A single voltage reference is used only on the positive supply. A voltage divider permits adjusting the positive supply voltage to the desired value. The negative regulator is referenced to the common bus line and its feedback is acquired from a voltage divider connected between the two regulated outputs. The resistors of this divider are of equal value. Thus the regulation point is at zero volts when the two supply voltages are equal but opposite. A potentiometer is provided for trimming the negative supply voltage to be the same value as on the positive-supply voltage.

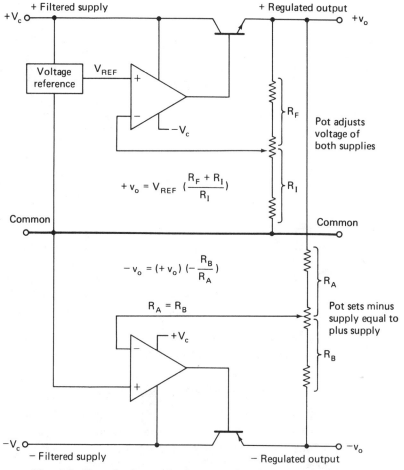

$$+v_o = V_{REF} \left(\frac{R_F + R_I}{R_I} \right)$$

$$-v_o = (+v_o) \left(-\frac{R_B}{R_A} \right)$$

$$R_A = R_B$$

Fig. 4-8 Slaved, plus and minus regulated power supplies.

This technique is employed in a great number of commercial power supplies for use with analog equipment.

An example will help to illustrate the way a slaved power supply is developed (see also Fig. 4-9).

PS 503A
TRIPLE POWER SUPPLY

Independent + and − Controls

Dual Tracking Voltage Control

0 to ±20 V at 1 A (in high-power compartment)

Fixed Output + 5 V @ 1 A

Remote Resistance Programming

Over-Voltage Protection Standard

The PS 503A features superior dual tracking performance, over-voltage protection, and remote resistance programming of voltage. When operated in the high-power compartment of a TM 504 or TM 506 Mainframe, the PS 503A provides up to 1 amp from both 0 to 20 volt supplies.

±20 V FLOATING SUPPLIES

Outputs — 0 to ±20 V dc with respect to the common terminal or 0 to 40 V dc across the + and − terminals. Outputs can be varied independently or at a constant ratio.

Tracking Mode Offset Error — If the two supplies are set independently to any given voltage ratio and then varied by use of the VOLTS DUAL TRACKING control, the two supplies will maintain the same voltage ratio as initially set within ±50 mV.

Current Limit — Adjustable from less than 100 mA to 1 A (high-power compartment) or less than 40 mA to 400 mA (standard compartment) on each supply.

Load Regulation — Within 3 mV for 1 A change (high-power compartment) or 1 mV for 400 mA change (standard compartment).

Ripple and Noise — 3 mV p-p or less at 1 A load (high-power compartment). 0.5 mV p-p or less at 400 mA load (standard compartment).

Indicators — Individual voltage indicators and current limiting indicators for both + and − supplies. Standard compartment (400 mA) indicator.

PS 503A Power Supplies . . . .

Fig. 4-9 Slaved (dual tracking) power supply. (Courtesy of Tektronix Inc., Beaverton, Oreg.)

EXAMPLE 4-9

Design a slaved ±(0 to 15 V) power supply which can deliver 500 mA from each side. The op amps are μA 741's. The transformer is 38 V rms center-tapped, 2 A full load; the reference is that shown in Fig. 4-7. 5% ripple is required on the + supply. Determine the output ripple if the + load is 500 mA and the − load is 250 mA; make both filter capacitors the same value.

Solution

The peak voltage on the filter capacitors is

$$\left(\frac{38 \text{ V rms}}{2}\right)\sqrt{2} - 0.7 \text{ V} = 26.87 \text{ V} = V_{\max}$$

From Eq. (4-2), 5% ripple is (0.05)(26.86 V) = 1.34 V p-p ripple. From Eq. (4-4),

$$C = (500 \text{ mA})\left(\frac{7.5 \text{ ms}}{1.34 \text{ V}}\right) = 2798 \ \mu\text{F}$$

at 26 V; from Appendix B3, choose the 3000 μF at 30 V from the miniature computer-grade column. The gain from V_{REF} to v_o must be 15 V/6.2 V = 2.42. Thus from Eq. (2-14), $R_F/R_I = 1.42$. $R_A = R_B$; choose 10 kΩ. Let $R_I = 10$ kΩ; then $R_F = 14.2$ kΩ. Choose 14.3 kΩ from Appendix B1(1%). From μA 741 specifications, PSRR = 15,000, typical. Thus, from Eq. (4-11),

$$\Delta V_{o\text{PS}} = \frac{|1.34| - |-0.67|}{(2)(15,000)}(2.42) = 52.8 \ \mu\text{V ripple}$$

The no-load to full-load regulation can be determined using Eq. (2-50); $A_{\text{VEX}} = 50,000/2.42 = 20,660$ and the output impedance if the pass transistors have an $r_o = 50 \ \Omega$ of $R_o = 50/20,660 = 0.0024 \ \Omega$. Thus $\Delta v_o/\Delta I_L = 0.0024$ and $\Delta v_o = 0.0024(500 \text{ mA}) = 1.2$ mV; the load regulation is then (1.2 mV/15 V)(100) = 0.008%.

Thus it can be seen from Example 4-9 that the *line regulation* is determined by PSRR while the *load regulation* is determined by the pass transistor output impedance and the *excess loop gain* (A_{VEX}).

PROBLEMS

1. A filtered supply using a dual full-wave rectifier draws a 1.5 A load from each polarity. The peak-to-peak ripple is not to exceed 1.8 V p-p. $V_{\text{sec}} = 12.50$ V rms on each side of the center tap.

   a. What is V_{max}?

   b. What is the percentage of ripple?

   c. What value of filter capacitor will be required?

2. The power supply in Problem 1 is to have 7% ripple voltage.

   a. What is ΔV_c?

   b. What filter capacitor size should be used?

   c. Select the capacitor value from Appendix B3.

3. A 10 V unregulated power supply uses a 4700 μF filter capacitor and draws 1.8 A. V_{max} is approximately 12 V.

   a. Calculate the peak-to-peak ripple voltage.

   b. What is the value of V_{min}?

   c. Is this a good 10 V supply? Why?

   d. What could be done to improve it?

4. A filtered power supply must have the following characteristics: $V_{\text{max}} = 15$ V, $V_{\text{min}} = 13.5$ V, and $I_{\text{out}} = 250$ mA.

   a. Using Eq. (4-4), (4-5), or (4-6), calculate the size of the filter capacitor for this supply.

   b. Pick a capacitor from Appendix B3.

5. A 15 V regulator has been designed to only operate at input voltages that are at least 20 V. The regulator must supply 5 A constantly, and requires 250 mA from the unregulated source for its own operation. What size of filter capacititor is required if a 16 V rms transformer is to be used? Use a full-wave bridge.

6. A 10.2 V, 725 mW zener diode is being used as a voltage regulator. The filtered supply has a $V_{max} = 15$ V and a 1.5 V p-p ripple. The load current is 75 mA, and the diode requires a keep-alive current of 20 mA.

   a. Determine the value and required power rating of R_D.

   b. Find the amount of ripple appearing on the regulator output if $r_z = 6.75$ Ω.

7. Calculate the no-load to full-load regulation for the regulator in Problem 6. The current ranges from 0 to 75 mA, and $r_z = 6.75$ Ω.

8. The load current in Problem 6 suddenly increases to 100 mA due to a component failure in the load circuit.

   a. Calculate the load voltage at V_{min}.

   b. Calculate the load voltage at V_{max}.

   c. How much output ripple is on the regulator?

   d. What is the wave shape of the regulator output voltage?

   e. What happens to the regulated output voltage when the input voltage reaches V_{min}?

9. A feedback-type voltage regulator has a closed loop gain of 5. The positive supply has a ripple voltage of 1.5 V p-p, and the negative supply has a ripple voltage of 0.7 V p-p. The op amp in the regulator has a power supply rejection of 83 dB. Find the output ripple voltage on the power supply.

10. An input signal of 1.5 V p-p is applied to the input of a noninverting amplifier with a gain of 2. The CMRR of the op amp is 15,000.

    a. Calculate the value of $\Delta V_{o_{CM}}$.

    b. Figure 4-3a shows that the output common-mode voltage is out of phase with the input common-mode voltage. What is the actual output voltage of the circuit to five decimal places?

    c. Calculate the numerical value of the CMRR voltage generator.

    d. Put the generator from part (c) in series with the input voltage, and verify that the output voltage is the same as found in part (b).

    e. Verify that the two input leads of the op amp are moving the same amount with V_{CMRR} in place.

11. A feedback-type voltage regulator has a reference voltage of 5.7 V. The output voltage must be 12 V.

    a. Determine the ratio of R_I to R_F.

    b. Select values of R_I and R_F from the 1% table of Appendix B1. Keep the ratio within 1% accuracy.

12. A voltage regulator with a reference of 10.25 V has an output voltage of 21 V. The voltage reference experiences 75 μV of ripple. Calculate the amount of ripple on the regulator output.

13. A 12 V voltage regulator uses a feedback-type voltage reference where both regulator and reference use μA 741 op amps with a PSRR = 15,000. The voltage reference is shown in Fig. 4-7, where the reference output voltage is inadvertently taken from the op amp output rather than from the zener diode. If the filtered supply voltage has 2 V rms ripple and supplies both the regulator and reference, calculate the amount of ripple on the output of the regulator due to both the PSRR of the regulator and the ripple on the output of the reference.

14. A voltage regulator uses an op amp with an open loop gain of 25,000. The reference voltage is 8 V, and the output voltage is 15 V. The output impedance of the pass transistor in the regulator is 35 Ω.

   a. Calculate the output impedance of the regulator.

   b. How much does the output voltage of the regulator drop if the load current changes from 0 to 75 mA?

   c. Calculate the load regulation percentage over the load range in part (b).

   d. What internal factors determine the load regulation?

   e. What component would you change to improve the load regulation?

15. An op amp with a PSRR of 12,000 is used in a 5 V voltage regulator. Make the following calculations considering the voltage reference as a constant +3 V.

   a. Calculate the change in output voltage if the filtered supply voltage drops by 2 V.

   b. Calculate the line regulation percentage for part a.

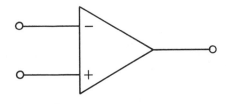

5 Frequency-Related Characteristics of Operational Amplifiers

5.1 Introduction

Operational amplifiers are composed of transistors, FETs, diodes, resistors, and capacitors. The capacitors are parasitic (not intended to be there) for the most part, but they do exist and cause the active devices to exhibit less gain at higher frequencies than at lower frequencies. In this chapter these effects are analyzed and explained in relation to the parameters found for operational amplifiers.

5.2 Open-Loop-Gain Characteristics

In Chapter 2 the topic of open loop gain was introduced, but care was taken to keep the discussion to operational characteristics near zero frequency (dc). The reason for this will become evident as the characteristics for frequencies in excess of a few hertz are explored.

5.2.1 The AC Equivalent Op Amp

An operational amplifier is composed of essentially three separate stages: the differential amplifier; the level shifter, which usually has additional voltage gain; and the output-current-gain stage. Each of these has separate frequency rolloff characteristics which combine to produce the composite frequency rolloff seen for the whole op amp. Each stage of the op amp appears to be an ideal amplifier with an associated low-pass *RC* filter section as illustrated in Fig. 5-1. The op amp is actually a complex device with many active devices, each of which contributes to the frequency rolloff. The simplified analysis that follows is adequate for a basic understanding of the rolloff phenomenon.

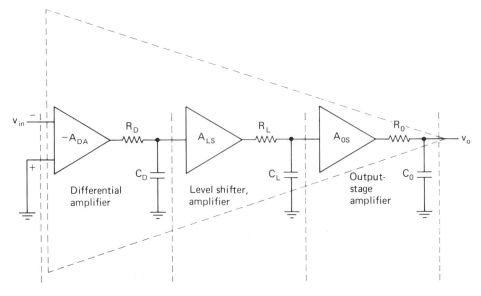

Fig. 5-1 Ac equivalent of an operational amplifier.

5.2.2 The Low-Pass *RC* Filter Section

Each of the individual stages of the equivalent op amp contains its own low-pass *RC* section, having an individual corner (pole) frequency.

It is well at this time to review the low-pass *RC* section so that a better understanding of its operation is acquired. The low-pass *RC* filter has a constant gain (attenuation) out to some pole frequency where the gain begins to roll off at a prescribed rate of decrease with frequency; in this region it has the characteristics of an integrator. Figure 5-2 shows a low-pass *RC* filter section with its gain (Bode plot) and phase shift curves.

The transfer-function equation for frequency response is (see Appendix A5)

$$\frac{v_o}{v_{in}} = \frac{1/2\pi RC}{(1/2\pi RC) + jf} \tag{5-1}$$

This transfer function has three distinct regions of interest:

1. Where $f \ll 1/2\pi RC$, $v_o/v_{in} = 1$ or 0 dB and the phase approaches $0°$.
2. Where $f = 1/2\pi RC$, $v_o/v_{in} = 0.707 \, \underline{/-45°}$ or -3 dB.
3. Where $f \gg 1/2\pi RC$,

$$\frac{v_o}{v_{in}} = \left(\frac{1}{2\pi RC}\right)\left(\frac{1}{f}\right)\underline{/-90°}$$

where the amplitude is decreasing inversely to f as f becomes larger; this is the region where it responds as an integrator.

The three regions are readily identifiable in Fig. 5-2b and c.

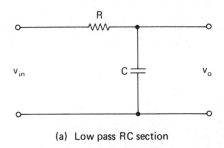

(a) Low pass RC section

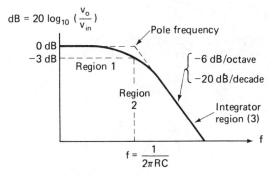

(b) Frequency response (Bode) plot

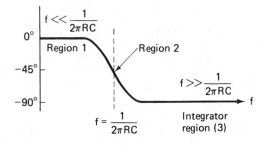

(c) Phase shift plot

Fig. 5-2 Low-pass *RC* filter section: frequency and phase response.

The rate of decline in the integrator region (3) can be reasoned out as follows. A decrease in v_o/v_{in} by $\frac{1}{2}$ is represented in dB as

$$\text{dB change} = 20 \log_{10}\left(\frac{1}{2}\right) = -6 \text{ dB}$$

When the frequency (f) doubles it is an *octave* change; thus the rate of decline is -6 dB/octave.

Similarly, a decrease in v_o/v_{in} by $\frac{1}{10}$ is represented in dB as

$$\text{dB change} = 20 \log_{10}\left(\frac{1}{10}\right) = -20 \text{ dB}$$

When the frequency increases by 10 times, it is a *decade* change; thus the rate of decline is -20 dB/decade.

The two designations represent the *same* rate of decline of v_o/v_{in} as the frequency increases; either can be used to designate the slope of the integrator region of the low-pass RC filter section.

The use of low-pass RC section equations is shown in Example 5-1.

EXAMPLE 5-1

A low-pass RC filter section has $R = 159$ kΩ and $C = 0.1$ μF. Insert $f = 0.1, 10, 1$ kHz, and 10 kHz into Eq. (5-1) and show that interest areas 1, 2, and 3 are true. Verify that between 1 and 10 kHz the rate of decline in amplitude is -20 dB/decade.

Solution

$RC = (159$ k$\Omega)(0.1$ μF$) = 15.9$ ms, which equals the time constant of the RC section. The pole frequency is $1/2\pi RC$ or $1/0.1$ or 10 Hz. Equation (5-1) is

$$\frac{v_o}{v_{in}} = \frac{10}{10 + jf}$$

$f = 0.1$ Hz:

$$\frac{v_o}{v_{in}} = \frac{10}{10 + j(0.1)} \approx \frac{10}{10} = \angle{-0.57°}$$

$f = 10$ Hz:

$$\frac{v_o}{v_{in}} = \frac{10}{10 + j10} = \frac{10}{14.141\angle{+45°}} = 0.707\angle{-45°}$$

$f = 1$ kHz:

$$\frac{v_o}{v_{in}} = \frac{10}{10 + j1000} = \frac{10}{1000.05\angle{+89.43°}} = 0.01\angle{-89.43°}$$

$f = 10$ kHz:

$$\frac{v_o}{v_{in}} = \frac{10}{10 + j10,000} = 0.001\angle{-89.94°}$$

Thus when f has increased by 10 times, the amplitude of v_o/v_{in} has decreased to $\frac{1}{10}$ the previous amplitude; and the -20 dB/decade decay rate is demonstrated.

5.2.3 Internal Response of the Op Amp

Figure 5-1 shows three distinct circuits, each with a low-pass RC section. Each RC section follows the concepts explained in the preceding section with one major exception: the three poles occur at different frequencies.

The differential amplifier pole is probably the lowest in frequency, the level-shifter pole is the next highest, while the output-stage pole is the highest in frequency. Each has an associated ideal amplifier which, for the op amp illustrated in Fig. 5-1, combine to form the open loop gain.

Figure 5-3 illustrates the relationship between the three frequency responses and their corresponding phase shift as they are combined. It will be presumed that the noninverting (+) input of the op amp is grounded while the signal enters the inverting (−) input; a phase inversion must occur. Thus the first stage (differential amplifier) will have a gain with 180° phase reversal (−) associated with it. The open

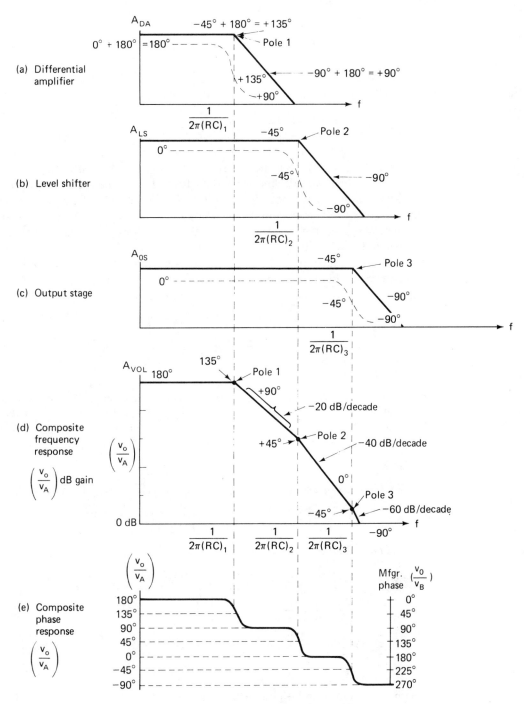

Fig. 5-3 Plot of op amp open loop gain.

loop gain (A_{VOL}) is the product of the individual stage gains, in Fig. 5-3, and is

$$A_{\text{VOL}} = (-A_{\text{DA}})(A_{\text{LS}})(A_{\text{OS}}) \qquad (5\text{-}2)$$

where the $(-)$ sign of A_{VOL} has been accounted for in the phase part of subsequent equations. Each stage gain is a transfer function Eq. (5-1), rearranged, in the form

$$\frac{v_o}{v_{\text{in}}} = (A_{\text{STAGE}}) \frac{1}{1 + j\dfrac{f}{1/2\pi RC}} \qquad (5\text{-}3)$$

where $1/2\pi RC$ is the pole frequency.

In Fig. 5-3d, the composite frequency-response plot shows that at dc (beyond the ordinate of the graph, to the left) the open loop gain is indeed A_{VOL}, as asserted in Chapter 2. But, as the frequency increases, the open loop gain begins to decrease and the phase begins to decline from $180°$ toward $0°$. It passes through the $+90°$ integrator region and reaches $0°$ somewhere within the -40 dB/decade slope. The gain continues to decrease and changes slope to -60 dB/decade until it reaches 0 dB, where the open loop gain is unity; it then decreases to a value less than unity gain, where there is actually a loss through the op amp. By this time the phase has continued to advance to $-90°$ and sometimes beyond if a fourth pole exists where the slope would become -80 dB/decade. The exact frequency of the $0°$ point is given by (where f_1 is the lowest frequency pole)

$$f_{0°} = \sqrt{f_1 f_2 + f_1 f_3 + f_2 f_3} \qquad (5\text{-}4a)$$

$$\cong \sqrt{f_2 f_3} \qquad (5\text{-}4b)$$

EXAMPLE 5-2

An operational amplifier has the following internal characteristics: The pole (1) for the differential amplifier is at 300 Hz and the gain (A_{DA}) is -400. The pole (2) for the level shifter is at 30 kHz and the gain (A_{LS}) is 263. The pole (3) for the output stage is at 460 kHz and the gain (A_{OS}) is 0.95. Calculate the magnitude and phase of A_{VOL} at $f = 118$ kHz.

Solution

Using Eq. (5-3) as a model of the gain equation for each stage, A_{VOL} from Eq. (5-2) is

$$A_{\text{VOL}} = (-400)\left[\frac{1}{1 + j(f/300)}\right](263)\left[\frac{1}{1 + j(f/3 \times 10^4)}\right](0.95)\left[\frac{1}{1 + j(f/4.6 \times 10^5)}\right]$$

and when $f = 118$ kHz,

$$A_{\text{VOL}} = (99{,}940)\left[\frac{-1}{1 + j(118\text{ kHz}/300\text{ Hz})}\right]\left[\frac{1}{1 + j(118\text{ kHz}/30\text{ kHz})}\right]$$

$$\times \left[\frac{1}{1 + j(118\text{ kHz}/460\text{ kHz})}\right]$$

$$= (99{,}940)(-0.00254\underline{/-89.85°})(0.246\underline{/-75.7°})(0.968\underline{/-14.38°})$$

$$= (-60.45)\underline{/-179.93°} = (60.45)\underline{/+0.07°}$$

By placing the inversion in the phase term, we get $(+60.45)\underline{/+0.07°}$. Thus it is evident that if a frequency slightly higher than 118 Hz were chosen, the phase would have progressed to the 0° point; Eq. (5-4b) can be used as a good estimate of the 0° frequency. This would have yeilded $+63.76 \underline{/+0.81°}$. The open loop gain ($A_{\text{VOL}}$) in dB is

$$A_{\text{VOL}} = 20 \log_{10}(60.5) = 35.6 \text{ dB}$$

5.2.4   Op Amp Oscillation

Let us define an oscillator: An oscillator is any device with an excess loop gain where the total loop phase shift, including amplifier and feedback is 0°. Both the excess loop gain and the 0° loop phase shift must be present at the same time for oscillation to occur. The feedback is now positive.

Presume that in inverting amplifier with unity gain is connected using an op amp with the open loop gain characteristic shown in Fig. 5-3d. The closed loop gain is along the 0 dB (unity-gain) line (abscissa), while the excess loop gain equals the open loop gain (see Section 2.5). This means that when the open loop gain has reached the −40 dB/decade slope, where the amplifier phase is 0°, the excess loop gain is still greater than unity (0 dB). All of the necessary conditions for an oscillator exist in this unity-gain amplifier and *it oscillates* at the exact frequency where the phase becomes 0°. The oscillation usually rides on top of the amplified signal waveform.

There is one sure way to prevent an amplifier from becoming an oscillator: *Be sure that the loop phase cannot become 0° while any excess loop gain (greater than unity) exists*. This method is fail-safe for assuring an amplifier rather than an amplifier that oscillates. An op amp circuit that does not oscillate is said to possess "compensated open loop gain."

5.2.5   Frequency Compensation for the Op Amp Circuit

Figure 5-4 illustrates an amplifier with a closed loop gain of 10 (20 dB) employing an op amp with the open loop gain characteristic shown. The difficulty with this

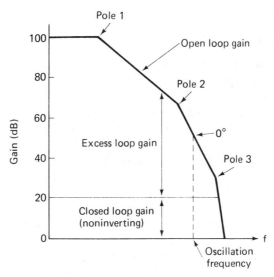

Fig. 5-4 Open and closed loop gain for amplifier.

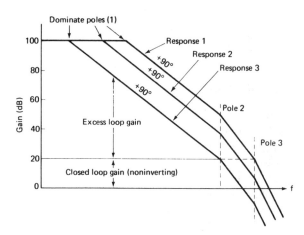

Fig. 5-5 Effect of dominate-pole frequency compensation.

circuit is that the conditions necessary for oscillation exist. Something must be done to eliminate them. The way to prevent oscillation from occurring is accomplished by moving the pole of the differential amplifier stage (dominate pole) to a lower frequency. This shifts the whole −20 dB/decade slope to a lower frequency. All other slopes remain as they were. This effect is shown in Fig. 5-5 for several tries at frequency-compensating the open loop gain. The curve labeled response (1) is the original open loop gain. The curve labeled response (2) is the first try for frequency compensation (lowering of the dominate pole frequency of the differential amplifier stage). The curve labeled response (3) is the second try for frequency compensation.

On curves 1 and 2, excess loop gain still exists where the phase could become 0°. Remember that 0° phase shift occurs within the −40 dB/decade slope. But on compensation curve 3, the excess loop gain has already become unity and less when the phase reaches 0°; this compensation curve will produce an amplifier that will not oscillate.

5.2.6 Gain and Phase Margin

Figure 5-6a illustrates the three-pole op amp circuit with a reference gain drawn at the 0° phase point. Any closed loop gains *above* this reference gain have excess loop gains of less than 1 when the frequency reaches the 0° point. Thus not enough excess loop gain exists to produce oscillation.

The *gain margin* is the ratio (difference in dB) between the reference gain (at 0°) and the closed loop gain; this gain margin should be at least 20 dB for critical damping of any step voltages or sudden changes in sine wave amplitudes.

The *phase margin* is the difference in phase between the reference gain (at 0°) and the closed loop gain where they each strike the open loop gain line. This phase margin should be at least 50° for critical damping to occur on step voltages.

On most *internally* compensated op amps, the dominate pole is placed such that the reference gain line is 20 dB *below* the unity-gain line as shown in Fig. 5.6b. This assures that the amplifier has at least 20 dB of gain margin and 50° of phase margin at unity closed loop gain and more for larger gains. Thus they are unconditionally stable for all gains.

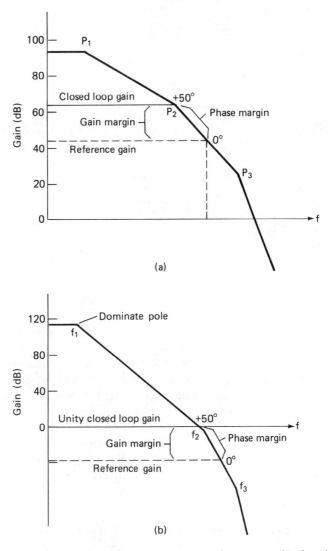

Fig. 5-6 (a) Gain and phase margin for uncompensated op amp; (b) showing how a dominate-pole compensated op amp has 30 dB of gain margin and 50° of phase margin at unity closed loop gain.

Let us see now how this frequency rolloff is accomplished. The frequency rolloff is represented by a capacitor across the two collectors of the differential amplifier, as illustrated in Fig. 5-7. There are several methods actually used for frequency rolloff, and this is only one of them. But all of the methods produce results equivalent to those of the method shown in Fig. 5-7. Notice in Fig. 5-5 that the phase of the compensated open loop gain remains at +90° all along curve 3 until it breaks because of the second pole, where the phase shifts from +90° through +45° (at the second pole) to 0° (near the midpoint of the −40 dB/decade slope). This causes no harm because the excess loop gain has already become less than unity when this

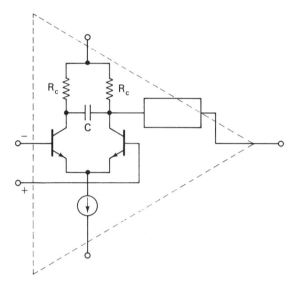

Fig. 5-7 Location of frequency rolloff capacitor.

phase change occurs; it can be seen that this compensation curve has prevented the conditions for oscillation to occur.

The excess loop gain becomes unity where the ideal closed loop gain line (20 dB) intersects the open loop gain response curve (3). Beyond this point (higher in frequency), the closed loop gain equals the open loop gain and they both reduce together; the feedback loop is now open.

Most of today's integrated-circuit operational amplifiers have two terminals for attaching the external compensated capacitor. Specific values of capacitance are recommended and compensation open loop gain curves for each are usually drawn on the response curves (see Appendix B6).

EXAMPLE 5-3

An op amp has an open loop gain of 95,000 and has poles at 10 Hz, 31.623 kHz, and 500 kHz.

a. Find the 0° frequency.

b. Find the reference gain.

c. Find the closed loop gain for a gain margin of 20 dB.

d. What is the phase margin for this gain margin?

Solution

a. $f_{0°} = \sqrt{f_1 f_2 + f_1 f_3 + f_2 f_3} = 125.765$ kHz

b. Gain drops at -20 dB/decade from 10 Hz to 31.623 kHz, which is a drop to a value of 95 k(10/31.623 kHz) = 30.
From 31.623 to 125.765 kHz in the -40 dB/decade region, the gain drops to $30(31.623/125.765)^2 = 1.896$. This is the *reference gain* at $f_{0°}$.

Frequency-Related Characteristics of Operational Amplifiers

c. The closed loop gain for a gain margin of 20 dB is $10(1.896) = 18.96$.

d. The phase margin for a gain of 18.96 is found by first finding the frequency at this gain in the -40 dB/decade region. $f = 125.765$ kHz$/\sqrt{10} = 39.77$ kHz. Now the phase is found from $-95k/(10 + j39.77k)(31.623k + j39.77)(500k + j39.77k) = 180° - 89° - 50° - 5° = 56°$ phase margin.

5.2.7   AC Closed-Loop-Gain Equations

The amplitude of the open loop gain along a -20 dB/decade slope compensation curve varies inversely as the frequency, and the phase remains constant at $+90°$ (see region 3 of the low-pass RC filter section and Fig. 5-5). Thus the open loop gain at any single frequency on a -20 dB/decade slope can be represented by

$$A_{\text{VOL}} = +jA = A\underline{/+90°} \tag{5-5}$$

When the closed loop gain is not unity the excess loop gain at any single frequency along a -20 dB/decade slope (for A_{VOL}) is

$$A_{\text{VEX}} = +jA_{\text{VEX}} = A_{\text{VEX}}\underline{/+90°} \tag{5-6}$$

This is the expression for the excess loop gain at frequencies above the first (dominate) pole.

Returning now to Eq. (2-9) for closed loop gain in the inverting amplifier, we insert $+jA$ as it appears in Eq. (5-5), which yields the ac closed loop gains for the inverting amplifier:

$$A_{\text{VCL}}(\text{ac}) = -\frac{+jAR_F}{(1 + jA)R_I + R_F} \qquad \text{inverting} \tag{5-7}$$

EXAMPLE 5-4

In Example 2-3, a closed loop gain of -0.8333 was calculated for the inverting amplifier when the open loop gain was $10\underline{/180°}$; this was at dc. At a particular frequency where the open loop gain is $+j10$, calculate the closed loop gain using an open loop gain of $+j10$ and compare your answer to the closed loop gain calculated in Example 2-3 for $A_{\text{VOL}} = 10$.

Solution

Inserting Eq. (5-5) into Eq. (5-7) as the desired closed loop gain is unity, we get

$$\frac{v_o}{v_{\text{in}}} = -\frac{(1\ \text{k}\Omega)10\underline{/+90°}}{(1 + j10)(1\ \text{k}\Omega) + 1\ \text{k}\Omega}$$

which computes to

$$\frac{v_o}{v_{\text{in}}} = -\frac{10\ \text{k}\Omega\underline{/+90°}}{2\ \text{k}\Omega + j10\ \text{k}\Omega} = -\frac{10\ \text{k}\Omega\underline{/+90°}}{10.2\ \text{k}\Omega\underline{/+78.7°}}$$

$$= -0.980\underline{/+11.3°}$$

which is in error by only $(1 - 0.98)100 = -2\%$, where at dc the error was -20.0%. The $90°$ phase shift aided the gain accuracy. The penalty paid is the closed loop phase shift of $11.3°$, which in certain circumstances would be intolerable.

Open-Loop-Gain Characteristics **121**

EXAMPLE 5-5

An inverting amplifier is strapped for a closed loop gain of -10; $R_F = 10$ kΩ and $R_I = 1$ kΩ. It has an open loop gain of $+j100$ at a signal frequency of 50 kHz. Determine the actual closed loop gain at 50 kHz.

Solution

The open loop gain from Eq. (5-5) is

$$A_{VOL} = 100\underline{/+90°} = +j100$$

Inserting $+j100$ into Eq. (5-6), we get

$$\frac{v_o}{v_{in}} = -\frac{(10\text{ k}\Omega)100\underline{/+90°}}{(1+j100)(1\text{ k}\Omega) + 10\text{ k}\Omega} = -\frac{1000\text{ k}\Omega\underline{/+90°}}{11\text{ k}\Omega + j100\text{ k}\Omega}$$

$$= -\frac{1000\text{ k}\Omega\underline{/+90°}}{100.6\text{ k}\Omega\underline{/+83.72}} = -9.94\underline{/+6.27°}$$

Thus the actual circuit gain is -9.94 instead of the desired -10, and the phase shift is $+6.27°$ instead of $0°$. The effect of a low excess loop gain is becoming evident.

Equation (2-13) for the closed loop gain in the noninverting amplifier can be altered in a like manner. Equation (5-5) for open loop gain is inserted into Eq. (2-13) to yield the ac equation for the noninverting amplifier:

$$A_{VCL}(ac) = +\frac{+jA(R_F + R_I)}{(1+jA)R_I + R_F} \qquad \text{noninverting} \qquad (5\text{-}8)$$

while the percent error is given by

$$\% \text{ error} = -0.5\left[\frac{R_F + R_I}{AR_I}\right]^2(100)\underline{\bigg/ \tan^{-1}\left(\frac{R_F + R_I}{AR_I}\right)} \qquad (5\text{-}9)$$

EXAMPLE 5-6

A noninverting amplifier is used where the open loop gain is $+j15$, the signal frequency is 60 kHz, and the amplifier has a closed loop gain of 5; $R_F = 8$ kΩ and $R_I = 2$ kΩ. The phase shift through the amplifier must not exceed $+20°$.

    a. Will this circuit meet the specifications?

    b. Verify the gain using Eq. (5-9).

Solution

    a. Equation (5-5) is used for the open loop gain as $+jA$ was given . Then Eq. (5-8) yields

$$\frac{v_o}{v_{in}} = +\frac{+j15(k\text{ k}\Omega + 2\text{ k}\Omega)}{(1+j15)2\text{ k}\Omega + 8\text{ k}\Omega} = +\frac{+j150\text{ k}\Omega}{10\text{ k}\Omega + j30\text{ k}\Omega}$$

$$= +\frac{150\text{ k}\Omega\underline{/+90°}}{31.6\text{ k}\Omega\underline{/71.6°}} = +4.75\underline{/+18.4°}$$

so the circuit barely meets the phase specification, and the closed loop gain is 4.7 instead of the desired 5.

b.

$$\% \text{ error} = -0.5 \times \left[\frac{8 \text{ k}\Omega + 2 \text{ k}\Omega}{(15)2 \text{ k}\Omega} \right]^2 \times 100 = -5.5\%, 5(-0.055) = -0.277 \text{ gain error}$$

$$5 - 0.277 = 4.72 \underline{/\tan^{-1}(0.333)} = 4.72 \underline{/18.43°}$$

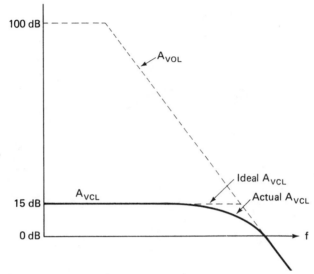

Fig. 5-8 Ideal versus actual closed loop gain.

Figure 5-8 illustrates the loss of actual closed loop gain as the excess loop gain diminishes with increasing frequency. It can be seen that as the excess loop gain becomes less at the higher signal frequencies, it becomes increasingly more difficult for an amplifier to maintain the ideal closed loop gain. When the excess loop gain has finally diminished to unity, the closed loop gain then follows the open loop gain curve. At this point the amplifier has *no* feedback for control and is operating "open loop."

As the excess loop gain becomes larger, the phase effects will diminish and the actual closed loop gain will again approach the ideal gain given by Eq. (2-10). The same effect is true for the noninverting amplifier.

5.2.8 Feedforward Compensation

Often, an operational amplifier manufacturer will compose an applications note where a method of compensation known as feedforward compensation is proposed. This method of compensation takes a portion of the output signal and feeds it in a positive feedback manner, through a network, back to the input with a frequency rolloff that extends the compensation curve out in frequency at lower open loop gains. The compensation curve appears as shown in Fig. 5-9. It can be seen that if

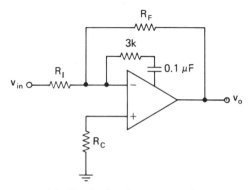

(a) Circuit, showing compensation

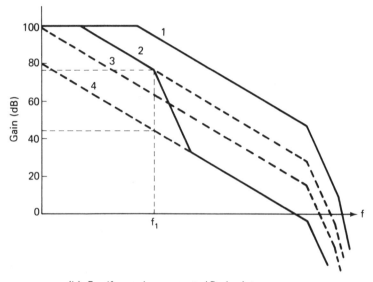

(b) Feedforward compensated Bode plot

Fig. 5-9 Feedforward compensation.

standard compensation were to be used for a unit-gain amplifier operating at a signal frequency of f_1, the compensation curve would have to be 4. The excess loop gain is approximately 50 dB.

It can also be seen that with the feedforward compensation (solid line), the excess loop gain at f_1 is approximately 80 dB, which creates a 30 dB advantage in excess loop gain. The lead-lag network that produces this feedforward compensation does not create a -40 dB/decade slope long enough to permit the phase to reach $0°$. Thus stable operation can be attained while increasing the overall bandwidth.

Feedforward compensation is not without fault. Any time that positive feedback is used around an amplifier, it becomes a potential oscillator and is only conditionally stable rather than being unconditionally stable, as with curve 4.

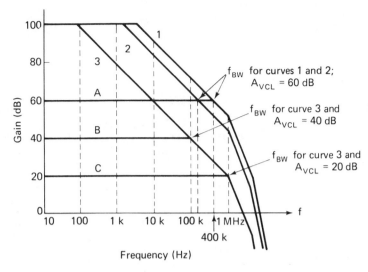

Fig. 5-10 Closed loop gains, showing frequency response.

5.2.9 Frequency Response

The frequency response of an operational amplifier circuit is defined as extending from dc through the frequency where the ideal closed loop gain line strikes the appropriate open loop gain line (excess loop gain = 1). Figure 5-10 redraws a familiar open loop characteristic but now places several closed loop gains on it and shows the amplifier bandwidth (f_{BW}) for each.

Closed loop gain line A (60 dB) can be compensated on curve 2 because it is a large closed loop gain. This produces a bandwidth of 400 kHz.

Closed loop gain lines B (40 dB) and C (20 dB) are compensated on curve 3, which produces a bandwidth of 100 kHz for line B and 1 MHz for line C, where the excess loop gain for each becomes unity.

The only reason that line A has a larger bandwidth than line B is that it is compensated with a smaller capacitor, thus producing a lighter degree of compensation.

5.2.10 Constant Gain – Bandwidth Product

Anywhere along an open-loop-gain line with a slope of -20 dB/decade, the gain–bandwidth product (f_{GB}) is a constant. This is true because of the $1/f$ term in Eq. (5-1) for region 3. As the bandwidth increases, the gain decreases by a like ratio, and vice versa.

To find the gain–bandwidth product along a slope of -20 dB/decade, pick a point (any point) and find the gain (in ratio only—not dB) and the frequency at this point. Multiply these two together and the result is the gain–bandwidth product. Since any point can be used, it is best to pick a point of known gain (say gain = 1, 0 dB) and find the corresponding frequency; or find a known frequency (say 1 kHz)

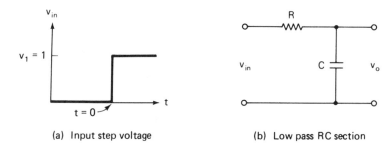

(a) Input step voltage

(b) Low pass RC section

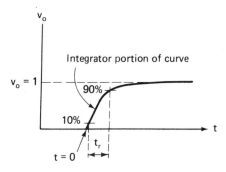

(c) Output voltage rise with time

Fig. 5-11 Rise time of integrator.

and find the corresponding gain.

This technique is *valid only* in regions of -20 dB/decade, and gain–bandwidth values on other slopes are incorrect. Also, it does not matter how long or short the region of -20 dB/decade is; the technique is valid anywhere on this slope.

5.3 Op Amp Slew Rate

The slew rate of an operational amplifier is rate-limited on the output; it can only proceed from one output voltage to another at a given rate but not faster. The slew rate of an op amp is divided into two distinct regions:

1. The op amp is constant-current-limited and the output voltage slews from one voltage to another in a straight line.
2. The op amp is gain–bandwidth-limited and slews from one output voltage point to another in an exponential curve, as shown in Fig. 5.11.

Slew rate is a term applied to an operational amplifier to specify how fast in volts per microsecond the output can slew from one output voltage to another when the input experiences an instantaneous voltage-step change (in essentially zero time).

5.3.1 Output Voltage of Low-Pass *RC* Section with Step Input

To understand slew rate, one must first understand rise time. Rise time is the length of time it takes a voltage to go from one defined point to another. Two separate definitions of rise time will be used in this chapter:

1. The initial rise of the *RC* charge curve projected to the input voltage level (see Fig. 5-11c) is the *initial* rise time.
2. The time it takes the *RC* charge curve to go from the 10% to the 90% points is defined as the *average* rise time; this is the definition usually found in texts on the subject.

The equation for the output-voltage *RC* charge curve of a low-pass *RC* filter section when a step voltage is applied is

$$v_o = v_1(1 - \varepsilon^{-t/RC}) \tag{5-10}$$

and is illustrated in Fig. 5-11c. The initial portion of the *RC* charge curve is nearly straight and approximates the constant current (straight line) charge of an "integrator." The slope of the initial part of the rise is v_1/RC volts/second and the *initial* rise time when v_1 is 1 V and v_o proceeds from 0 V to 1 V is

$$\text{initial } t_r = RC \quad \text{seconds} \tag{5-11a}$$

The average slope between the 10% and 90% output-voltage points is found by solving Eq. (5-10) at two points:

$$\text{when } v_o = 0.1v_1: \quad t = 0.1RC$$
$$\text{when } v_o = 0.9v_1: \quad t = 2.3RC$$

The *average* rise time is the difference between the two, or

$$\text{average } t_r = 2.2 \ RC \tag{5-11b}$$

For a low-pass *RC* section with a 1 V step on the input, the output will begin slewing at

$$\text{initial SR} = \frac{v_1}{RC} \quad \text{volts/second} \tag{5-12a}$$

for the straight-line "integrator" portion of the curve. The average slew rate is found using Eq. (5-11b) and the output voltage change over this time, which is 0.8 V for a 1 V step, or

$$\text{average SR} = \frac{0.8 \text{ V}}{t_r} = \frac{0.8 \text{ V}}{2.2 \ RC} = (0.36)\left(\frac{v_1}{RC}\right) \quad \text{volts/second} \tag{5-12b}$$

If a low-pass *RC* section has a time constant (*RC*) of 1 μs (pole frequency is 159 kHz), the initial slew rate is then 1 V/μs and the average slew rate is 0.36 V/μs.

5.3.2 Rise Time and Slew Rate in Terms of Gain – Bandwidth

Equation (5-11a), which defines the *initial* rise time, can be rewritten as

$$t_r = \frac{1}{\omega_{GB}} = \frac{1}{2\pi f_{GB}} = \frac{0.159}{f_{GB}} \quad \text{seconds} \tag{5-13a}$$

and for the *average* rise time, Eq. (5-11b) can be written as

$$t_r = \frac{2.2}{\omega_{GB}} = \frac{2.2}{2\pi f_{GB}} = \frac{0.35}{f_{GB}} \quad \text{seconds} \qquad (5\text{-}13b)$$

where f_{GB} is the gain–bandwidth product of the low-pass RC section or op amp.

The op amp would normally slew according to this equation, but at low closed loop gains (less than 10), the op amp slews in a constant current (straight-line) fashion. This constant-current slew (SR_{cc}) is the maximum current that the collector of a transistor can deliver to the dominate pole (first rolloff) capacitor; it is the rate published in the specification sheet. This slew rate is always slower than the maximum possible slew rate given by

$$SR_{max} = RC = 2\pi f_{GB} \times 10^{-6} \, V/\mu s \qquad (5\text{-}14a)$$

which is the theoretical maximum slew rate. This rate is not obtainable in an op amp, as the constant current slew rate prevents it from occurring, but it is useful for determining the slew rate at *any* "apparent" closed loop gain given by

$$SR = \frac{SR_{max}}{\sqrt{(SR_{max}/SR_{cc})^2 + (A_{VCL})^2}} V/\mu s \qquad (5\text{-}14b)$$

In the region where the apparent noninverting closed loop gain (A_{VCL}) is greater than 10, the slew rate becomes an exponential rise, for a step input, as shown in Fig. 5-11a and the slew rate can be predicted by Eq. (5-14b). Figure 5-12 illustrates a

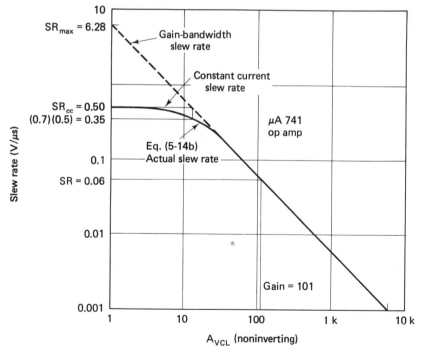

Fig. 5-12 Curve for calculating slew rate.

sample curve for the μA 741 op amp, constructed using Eqs. (5-14a) and (5-14b). Example 5-7 illustrates the construction of a curve such as Fig. 5-12 and its use to predict slew rates at various apparent closed loop gains.

EXAMPLE 5-7

A μA 741 op amp is being used as an inverting summer circuit (Fig. 5-13a) with $R_F = 10$ kΩ and 10 inputs, each using a 1 kΩ resistor (R_1, R_2, etc.). Determine the slew rate of this circuit.

Solution

The published typical slew rate is 0.5 V/μs and the gain–bandwidth product is 1 MHz. $SR_{max} = 2\pi(10^6)(10^{-6}) = 6.28$ V/μs, which is the beginning point of the slew rate line at $A_{VCL} = 1$. The published constant current slew rate is $SR_{cc} = 0.5$ V/μs, and this is the beginning of a horizontal line that intersects the slew-rate line. The apparent input resistance to the summer circuit is $R_I = 1$ k$\Omega/10 = 100$ Ω, and the apparent noninverting closed loop gain is

$$A_{VCL} = \frac{10 \text{ k}\Omega + 100}{100} = 101$$

Finding 101 on the A_{VCL} axis and projecting up to the slew-rate line and then across to the slew-rate axis, we find that the actual slew rate is

$$SR = 0.06 \text{ V}/\mu s$$

using Eq. (5-14b), we find the slew rate,

$$SR = \frac{6.28(10^{-6})(10^6)}{\sqrt{(6.28/0.5)^2 + (101)^2}} = \frac{6.28}{101.7} = 0.0617 \text{ V}/\mu s$$

which affirms the graphical value. The summer circuit is severely degraded in performance. This value is the *initial* rate of the actual slew rate. The average rate is $(0.36)(0.06) = 0.022$ V/μs. The output slew waveform is exponential (not straight) in shape.

5.3.3 Effects of Slew-Rate Limiting

Operational amplifiers typically come with slew rates from 1 V/μs for a poor one to 20 V/μs for a good one. Usually, a family of op amps will have a given and specified slew rate. Op amps with slew rates of up to 1000 V/μs can be found, but these are usually special-purpose types and are expensive.

Three situations exist when an operational amplifier is slew-rate-limited and has a square-wave input signal as shown in Fig. 5-13a.

1. The slew rate is sufficiently fast that the op amp output attains feedback control before the input waveform changes direction. Figure 5-13b illustrates this situation. The equation for this situation is

$$t_s = \frac{V_2 - V_1}{\text{slew rate}} \ \mu s \tag{5-15}$$

An example of the use of Eq. (5-15) is given in the following example.

Op Amp Slew Rate   **129**

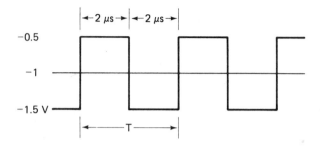

(a) Input voltage waveform

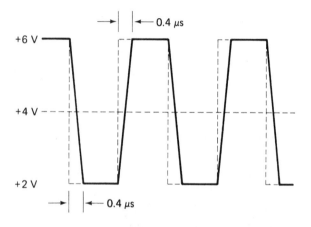

(b) Output voltage reaches feedback control

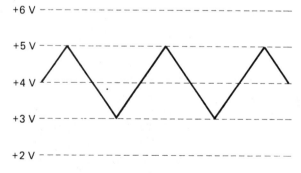

(c) Output voltage never reaches feedback control

Fig. 5-13 Voltage-follower response time.

EXAMPLE 5-8

An operational amplifier with a slew rate of 10 V/μs is connected as an inverting amplifier with a gain of −4. It has a 250 kHz square wave on the input with a 0.5 V peak and a −1 V dc average voltage.

    a. Draw the output waveform.

    b. Find the time (t_s) it takes to reach feedback control.

Solution

    a. The input waveform appears as shown in Fig. 5-13a. The output waveform is drawn like that shown in Fig. 5-13b, with the ideal output waveform dashed in for time reference. The output waveform has a 4 V peak-to-peak amplitude; therefore, $V_2 - V_1$ is 4 V.

    b. Using Eq. (5-15) to find t_s,

$$t_s = \frac{4\ \text{V p} - \text{p}}{10\ \text{V}/\mu\text{s}} = 0.4\ \mu\text{s}$$

    which is less than the 2 μs of the half-period; thus the op amp achieves feedback control before the input square wave changes polarity.

2. The input frequency is high enough that the op amp output never reaches feedback control; it just continues to slew at its maximum slew rate, first in one direction, then in the other. Figure 5-13c shows this situation.

The period of the input waveform and the slew rate are known; thus the peak-to-peak voltage may be calculated from

$$V_B - V_A = (\text{slew rate})\left(\frac{T}{2}\right) \qquad \text{V p} - \text{p} \tag{5-16}$$

Example 5-9 illustrates the use of Eq. (5-16).

EXAMPLE 5-9

An op amp with a slew rate of 1 V/μs is connected as an inverting amplifier with a gain of −4. It has a square wave on the input which has a dc average of −1 V, a peak voltage of 0.5 V, and a frequency of 250 kHz, as illustrated in Fig. 5-13a. Draw the output waveform.

Solution

An inverting amplifier is dc-coupled, so that the −1 V dc average of the input waveform will appear on the output as +4 V dc. The period of the 250 kHz input waveform is $T = 4$ μs and the half-period is $T/2 = 2$ μs. The output waveform can only slew for 1 μs before the input waveform changes direction. The amplifier only has 1 μs to slew, so it cannot reach the full-peak voltage before reversing direction; thus

$$V_B - V_A = (1\ \text{V}/\mu\text{s})(2\ \mu\text{s}) = 2\ \text{V p} - \text{p}$$

It thus reaches half of the desired 4 V p − p in each direction. The output waveform is illustrated in Fig. 5-13c, where $V_B = +5$ V and $V_A = +3$ V.

3. A sine-wave voltage is impressed on the input, which has a maximum slew rate greater than that of the op amp. The maximum slew rate of a sine wave occurs

at the zero crossing (Appendix A9) and is given by

$$\max \sin(\text{SR}) = 2\pi(10^{-6})(V_p)(f_{\text{Hz}}) \text{ V}/\mu\text{s} \qquad (5\text{-}17)$$

where V_p is the peak voltage of the *output* sine-wave voltage. If Eq. (5-17) yields a slew rate greater than that of the op amp, the output will be slew-rate-limited. The resulting waveform is illustrated in Appendix A6.

EXAMPLE 5-10

An operational amplifier is contemplated for use in a high-fidelity preamplifier circuit. The op amp is guaranteed to have a slew rate of 0.5 V/μs. The preamplifier is to have a closed loop gain of 100 and the input signal is $v_{\text{in}} = 0.05 \sin(1.25 \times 10^5 t)$. Will the preamplifier be able to pass the signal without slew-rate limiting?

Solution

The output signal is

$$(0.05)100 = 5 \text{ V peak}$$

at

$$\frac{125{,}000}{6.28} \text{ Hz} = 19.9 \text{ kHz}$$

$$\max \sin(\text{SR}) = 2\pi \times 10^{-6} \times 5 \times 19.9 \times 10^3 = 0.63 \text{ V}/\mu\text{s}$$

The preamplifier will possibly slew-rate-limit, as the required slew rate is greater than the maximum specified slew rate.

5.3.4 Response Time

Operational amplifier manufacturers specify the speed of their devices under feedback control by connecting them in a voltage-follower mode, applying a square wave to the input and measuring the resulting slew rate on the output. This they call response time. Figure 5-14 shows a typical output waveform from a voltage follower with a square-wave input.

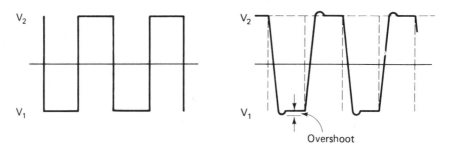

(a) Input square wave   (b) Output waveform

Fig. 5-14 Slew-rate-limited output.

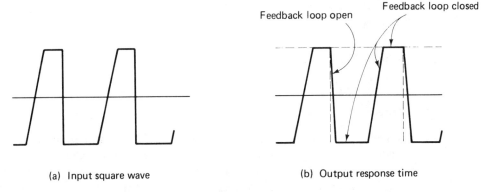

(a) Input square wave   (b) Output response time

Fig. 5-15 Feedback loop: open versus closed.

5.3.5 Feedback Loop: Open versus Closed

It is important for a circuit user to know when the op amp circuit is under feedback control and is following the closed-loop-gain equations (loop closed) and when it is not under feedback control (loop open). These two terms should not be confused with open loop gain and closed loop gain.

Figure 5-15 illustrates an op amp circuit with an unusual input waveform. The op amp can follow the input waveform for part of the cycle, but becomes slew-rate-limited for another part of the cycle. Thus three definitions will be given:

1. When the op amp output is following the input waveform and is related to it by the closed loop-gain factor, the *feedback loop is closed*.

2. When the op amp output cannot be related to the input waveform through the closed-loop-gain equation, the *feedback loop is open*.

3. When the summing junction (inverting op amp input) has a potential equal to that of the noninverting input ($+$), the *feedback loop is closed*. When the summing junction voltage is not equal to the noninverting input voltage, the *feedback loop is open*.

The situation in definition 3 is an excellent measurement method of determining the operating condition of an op amp circuit. An oscilloscope is attached to the summing junction and the waveform observed. As long as the feedback loop is closed, the voltage will be exactly equal to that in the noninverting input. *If the feedback loop opens for any reason, such as a faulty op amp or slew-rate limiting, the summing junction voltage will differ sharply from the noninverting input voltage; it does not make any difference what the signal voltage waveform is on the noninverting input.*

5.4 Noise

Operational amplifiers are composed of active devices (transistors, FETs) and resistors. These constituent devices are inherently noisy, each in its own way, and contribute to a total noise characteristic of the op amp.

Op amps experience three basic types of noise:

1. Shot (Schottky) noise.
2. Thermal (Johnson) noise.
3. Flicker ($1/f$) noise.

5.4.1 Shot Noise

Any active semiconductor device appears to be conducting a continuous dc current, but is actually carrying a current that is the result of a series of recombinations between electrons and holes (positive charges). Each individual recombination creates a slight sudden change in the current which contributes to the noise riding the average dc current. When a multitude of these recombinations occur, the result is a Schottky-type noise. This noise is represented by a change in average current given by

$$I_n = \sqrt{2qI_{dc}(\text{BW})} \tag{5-18}$$

where
$q = 1.6 \times 10^{-19}$ coulomb (charge on one electron)
I_{dc} = average current in semiconductor
BW = frequency bandwidth over which the noise is considered

When I_n flows through a resistor R, the resulting voltage is $V_n = I_n R$.

5.4.2 Thermal Noise

Johnson noise is created by the random motion of charge carriers within a conductor. This generates noise that is related to the temperature of the conductor and is given by

$$V_n = \sqrt{4kTR(\text{BW})} \tag{5-19}$$

where
$k =$ 1.38×10^{-23} joule/° K (Boltzmann's constant)
$T =$ degrees Kelvin (degrees Celsius + 273°)
$R =$ resistance (of wirewound or metal film resistors) (*Note:* This equation is not for carbon composition resistors; their noise is very large and unpredictable)
$\text{BW} =$ hertz (frequency bandwidth)

5.4.3 Flicker Noise

This $1/f$ noise is observed as a change in the dc current over a period of time. The variations can be as low as one cycle every few seconds. It is hard to observe and is usually mistaken for power-supply or input-signal variations. The reader is referred to RCA Application Note ICAN-6732, "Measurement of Burst ("Popcorn") Noise in Linear Integrated Circuits."

5.4.4 Combination of Noise on the Op Amp Input

Since noise is an rms type of function, it cannot be added in the first power of the voltage or current; it must be added as the second power or mean square (not root-mean-square) voltage.

I_n appears on the op amp input like a bias current and thus appears on the op amp output as

$$V_{noi} = I_n R_F \qquad (5\text{-}20)$$

V_n appears on the op amp input as an ac "offset" voltage and thus appears on the op amp output as

$$V_{nov} = V_n \left(\frac{R_F}{R_I} + 1 \right) \qquad (5\text{-}21)$$

The total output voltage V_{no} then appears as

$$V_{no} = \sqrt{V_{noi}^2 + V_{nov}^2} \qquad (5\text{-}22)$$

The flicker noise is independent of this equation and appears as a totally separate phenomenon.

An example will help to clarify the use of Eqs. (5-18) to (5-22).

EXAMPLE 5-11

A noninverting amplifier with a closed loop gain of 6 uses a 50 kΩ metal film resistor for R_F; R_I is a 10 kΩ metal film resistor. The op amp has a dc bias current of 150 nA. The positive input is terminated in a 10 kΩ metal film resistor connected to common. The op amp is frequency-compensated for a bandwidth of 200 kHz and is operated at room temperature (20° C). What is the output noise for zero signal input?

Solution

$$I_n = \sqrt{2(1.6 \times 10^{-19})(150 \times 10^{-9})(200 \times 10^3)}$$

$$= 98 \times 10^{-12} \, \text{A rms}$$

$$V_n = \sqrt{4(1.38 \times 10^{-23})(293)(10 \times 10^3)(200 \times 10^3)}$$

$$= 5.68 \, \mu\text{V}$$

$$V_{noi} = 98 \times 10^{-12}(50 \times 10^3) = 4.9 \, \mu\text{V}$$

$$V_{nov} = 5.68(6) = 34 \, \mu\text{V}$$

$$V_{no} = \sqrt{(4.9)^2 + (34)^2} \, \mu\text{V}^2 = 34.3 \, \mu\text{V}$$

The noise voltage on the amplifier output can be represented as voltage generators on the op amp input by dividing the output noise voltage by the noninverting gain. Through this means, these voltage generators can be placed on the op amp inputs and the relative magnitude of the various disturbance voltage generators representing offsets, CMRR, PSRR, and noise can be compared for performance decisions.

PROBLEMS

1. A low-pass RC section has $R = 20$ kΩ and $C = 0.047 \, \mu$F.

   a. Find the pole frequency.

   b. Find the response and the phase at the pole frequency.

   c. Find the response and the phase at $1/100$ of the pole frequency.

   d. Find the response and the phase at 100 times the pole frequency.

   e. Find the response and the phase at 1000 times the pole frequency. How much lower is this response than that of part d? What is the rate of decline as the frequency increases?

2. An RC integrator has a time constant $(T = RC)$ of 1.2 ms. Determine the magnitude and phase of V_o/V_i at frequencies of 2 Hz, 200 Hz, 2 kHz, and 20 kHz using Eq. (5-1).

3. Repeat Problem 2 for the case where an inverting amplifier with a gain of -1 is driving the integrator. The input (V_i) is now at the inverting amplifier input. Determine the slope in dB/decade above 2 kHz.

4. An operational amplifier has the following internal characteristics: differential amplifier pole at 250 Hz with a gain of 300; level shifter pole at 30 kHz with a gain of 250; output-stage pole at 400 kHz with a gain of 0.98.

   a. Write the expression of open loop gain in the form of Eqs. (5-2) and (5-3).

   b. Find the open loop gain and phase shift at 150 kHz.

   c. Find the frequency where the phase is within $\pm 0.5°$ of a $0°$ phase shift.

5. An operational amplifier has three internal stages with the following gain and phase relationships:

Stage	Gain	Pole Frequency
DA	200	220 Hz
LS	175	25 kHz
OS	0.97	390 kHz

   a. Determine the approximate frequency where the open loop gain becomes unity.

   b. Find the approximate frequency where the phase becomes zero degrees.

6. Where will the amplifier of Problem 4 oscillate if the op amp is connected as an inverting amplifier with a gain of -1?

7. An operational amplifier is strapped for a closed loop gain of 95. At 15 kHz, the open loop gain has decreased to 50 dB, and the phase shift has become $360°$ $(0°)$. What is likely to happen to this amplifier circuit?

8. An op amp with a variable closed loop gain has an open loop gain of 40 dB and an open loop phase of $0°$ at 20 kHz. What closed loop gain must be present at 20 kHz if the amplifier is to remain stable?

9. Construct a Bode plot of the open loop gain curve defined in Problem 4.

   a. Locate the frequency where a dominate-pole capacitor should be placed to provide a 90° phase shift down to 0 dB open loop gain.

   b. How does this compare to the compensated curve of the LM 208 with a 30 pF capacitor?

10. Why does the phase shift, as specified for the LM 208, not agree with the phase shift in problem 4?

11. A properly compensated op amp has its dominate pole at 30 Hz, and the open loop gain is 90 dB at dc.

    a. What would the open loop gain be at 1 kHz?

    b. What is the actual value and phase of the excess loop gain at 10 kHz if the amplifier now has an inverting gain of −10?

    c. What is the actual value of the closed loop gain and phase shift at 10 kHz?

12. A noninverting amplifier is strapped for a gain of 25. The correctly compensated open loop gain is $j135$ at 35 kHz.

    a. Find the actual closed loop gain and phase at 35 kHz. Calculate the percent error.

    b. Now calculate the percent error using Eq. (5-9). What is the reason for the two not agreeing?

13. Find the excess loop gain for Problem 12 at 3.5 kHz.

14. The amplifier in Problem 12 is required to have no more than 3° of closed-loop-gain phase shift at 10.25 kHz.

    a. Does the amplifier meet its requirements?

    b. What is the error in gain at this frequency?

15. Determine the gain–bandwidth product for the op amp used in Problem 12.

16. Calculate the gain–bandwidth product for the op amp used in Problem 4 at 15 kHz.

17. What is the gain–bandwidth product for the μA 741 operational amplifier?

18. Calculate the gain–bandwidth product for the LM 208 operational amplifier with the following capacitor values:

    a. $C_1 = 3$ pF

    b. $C_1 = 30$ pF

19. An LM 208 is compensated using a 3 pF capacitor. The inverting amplifier is strapped for a gain of 40 dB.

    a. Determine the gain–bandwidth product of the amplifier.

    b. What is the bandwidth of the amplifier?

20. The open loop gain of an op amp at dc is 31,623. It has poles at $f_1 = 31.62$ kHz, $f_2 = 3.162$ MHz, and $f_3 = 31.62$ MHz.

    a. What is the value of the reference gain?

    b. At what frequency does the 0° loop phase occur?

c. What closed loop gain will produce a 20 dB gain margin?

d. What is the phase margin for part (c)?

21. Find the initial and average rise times for a low-pass RC section with $R = 15 \text{ k}\Omega$ and $C = 0.047 \, \mu\text{F}$.

22. Find the initial and average slew rates for the low-pass RC section in Problem 20 with a step-input voltage of 2 V.

23. Find the gain–bandwidth product for the low-pass RC section in Problem 21 using Eq. (5-13).

24. Find the maximum slew rate for a LM 208 with a 30 pF compensation capacitor. Locate the constant current slew rate in the specification sheet. Hint: Look at voltage follower pulse response. Construct a curve similar to Fig. 5-12.

25. Predict the average slew rate for a LM 208 inverting summer with six 2 $\text{k}\Omega$ input resistors and a 20 $\text{k}\Omega$ feedback resistor. Refer to your curve from Problem 24.

26. Compare your answer from Problem 25 with the slew rate calculated using Eq. (5-14b).

27. An op amp with a slew rate of 0.75 V/μs is connected to a unity-gain inverting amplifier. The amplifier has a 20 kHz square wave on its input at an amplitude of 15 V p – p. Draw the output waveform to scale.

28. The same op amp of Problem 27 has its square-wave frequency increased to 50 kHz. Draw the output waveform to scale.

29. The inverting summer of Problem 25 has all of its inputs grounded except for one. The input has a 10 kHz square-wave input with an amplitude of 0.55 mV p – p. Draw the output waveform to scale.

30. An op amp with a slew rate of 0.8 V/μs is used as one stage in a hi-fi preamplifier. This stage has a closed loop gain of 12 and is amplifying an input signal of 250 mV p – p. Will the preamplifier stage slew-rate-limit at 15 kHz?

31. The op amp of Problem 30 is also used in one of the final driving stages. The output voltage must be 10 V peak.

a. Will this stage operate without limiting at 15 kHz?

b. If not, at what frequency does the op amp start to limit?

32. Show where the feedback loop is open and closed for Problem 27.

33. Show where the feedback loop is open and closed for part (a) of Problem 31.

34. A μA 74½A op amp is connected as a noninverting amplifier. Metal film resistors are used in the circuit; $R_1 = 2 \text{ k}\Omega$ and $R_f = 50 \text{ k}\Omega$. The positive input is terminated with a 2 $\text{k}\Omega$ resistor to common. How much Schottky and Johnson noise appears on the output at 40°C with the input grounded? What is the total noise?

35. Repeat Problem 34 for the LM 208 op amp with a 3 pF compensation capacitor.

36. A power-supply regulator uses the amplifier from Problem 34. The input ripple on the supply is 2 V p – p. Draw the schematic of the amplifier with the following input voltage generators: voltage offset, current offset, V_{PSRR}, and total noise. Compare these, and evaluate which is most prevalent.

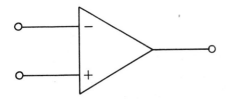

6 Linear Circuit Applications

6.1 Introduction

Operational amplifiers are used in almost every piece of electronics equipment manufactured today. Even in the digital computer, portions such as the power supplies contain feedback-type controllers which are either op amps or operate in a similar manner.

Some engineers have said that the linear field is dead now that digital devices have come of age. But that is far from true. Linear devices will be around for decades and will play an important and possibly even greater role than they play now. The laws of nature are linear, not digital.

This chapter presents a collection of circuits, using operational amplifiers, which are basic to the development of more complex circuits to be covered in later chapters. These circuits are somewhat independent, so that the natural flow of development, enjoyed in the last five chapters, will be lost in this chapter. At this point you know the fundamentals; let's apply them to some circuits.

6.2 Differential-Mode Amplifiers

Differential-mode instrumentation amplifiers are used where a large external common-mode noise is expected to exist which must be rejected from the input signal. They are also called "line receivers."

Two examples of their use are as follows:

1. As the null sensing circuit for a bridge measuring instrument.
2. As the amplifier to a transducer that is connected via a long twisted pair of wires laid in a noisy environment.

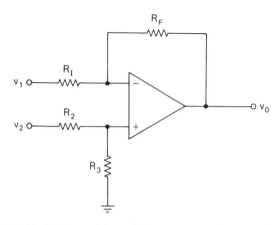

If $R_1 = R_2 = R_3 = R_F$, then

$v_0 = v_2 - v_1$

(see Section 3.4.2)

Fig. 6-1 Differential-mode instrumentation amplifier.

This family of circuits is similar in operation to the scaling subtractor discussed in Section 2.4.2. It uses one or more op amps connected in a circuit which takes the difference between the two input voltages and yields an output that is proportional to this difference. Most of the common-mode input voltage is rejected.

6.2.1 Difference Amplifier

Figure 6-1 is a repeat of Fig. 2-8b, where the conditions for a difference amplifier will be interpreted. Equations (2-18) to (2-20) still apply. When the conditions for Eq. (2-20) are imposed, namely $R_1 = R_2 = R_3 = R_F$, we have a difference amplifier with the same gain for both inputs, but the input impedance is low: R_1 for v_1, $R_2 + R_3$ for v_2, and $R_1 + R_2$ for $v_2 - v_1$ (see Appendix A7).

The amplifier is dependent upon the equality of the resistor values for the circuit common-mode rejection, not the op amp CMRR; but CMRR is the limiting factor in a circuit with perfectly matched resistors. Therefore, the resistors must be precision and probably equal in value to within 0.1% or better.

6.2.2 Differential-Mode Instrumentation Amplifier

The problem of low input impedance can be eliminated by placing a voltage follower ahead of each input, as illustrated in Fig. 6-2. The circuit in Fig. 6-2 requires three operational amplifiers to construct the high input impedance differential-mode amplifier. Equations (2-18) to (2-20) still apply, as the only change is the impedance transformation on the two inputs. The input impedance of the two voltage followers is extremely high.

The same high common-mode advantage can be obtained with two op amps by using the circuit in Fig. 6-3. The equation for this circuit as a high-impedance

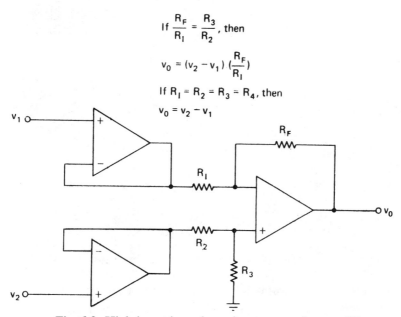

$$\text{If } \frac{R_F}{R_I} = \frac{R_3}{R_2} \text{, then}$$

$$v_0 = (v_2 - v_1)\left(\frac{R_F}{R_I}\right)$$

$$\text{If } R_1 = R_2 = R_3 = R_4 \text{, then}$$

$$v_0 = v_2 - v_1$$

Fig. 6-2 High-input-impedance instrumentation amplifier.

difference amplifier is

$$v_o = \left(\frac{R_F}{R_I} + 1\right)(v_2 - v_1) \tag{6-1}$$

for the case where $R'_I = R_F$ and $R'_F = R_I$; the derivation for Eq. (6-1) is in Appendix A8.

Again, this circuit requires precision-matched resistors for good common-mode rejection.

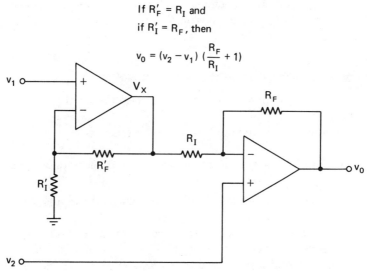

$$\text{If } R'_F = R_I \text{ and}$$
$$\text{if } R'_I = R_F \text{, then}$$

$$v_0 = (v_2 - v_1)\left(\frac{R_F}{R_I} + 1\right)$$

Fig. 6-3 Two-op-amp instrumentation amplifier.

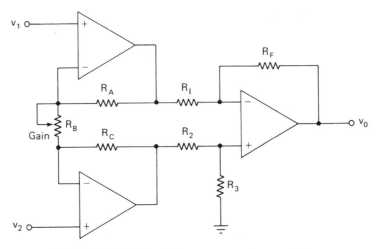

Fig. 6-4 Variable-gain instrumentation amplifier.

6.2.3 Variable-Gain Differential-Mode Instrumentation Amplifiers

On occasion it is desirable to be able to vary the gain on a differential-mode amplifier by adjusting the value of one resistor; Fig. 6-4 shows such a circuit. The equation for output voltage is (Appendix A9)

$$v_o = \left(\frac{2R_A}{R_B} + 1 \right)\left(\frac{R_F}{R_I} \right)(v_2 - v_1) \qquad (6\text{-}2)$$

when $R_2 = R_I$, $R_3 = R_F$, and $R_A = R_C$; the derivation appears in Appendix A9. The gain is adjusted by varying R_B; as R_B increases in value, the gain decreases. When R_B is infinite, the circuit reverts to that in Fig. 6-2. When R_B is zero, the gain approaches A_{VOL} for the op amps. The National Semiconductor LH0038 contains all but R_B, internally, and has an excellent CMRR.

6.3 Op Amp Circuits Using Two-Port Networks

An inverting amplifier using two-port networks for the input and feedback paths is illustrated in Fig. 6-5.

It is instructive to note two factors in this circuit:

1. The output voltage of network A and the input voltage of network B are "virtually" zero.

2. The output current of network A equals the input current of network B.

One would first think that the circuit gain can be found by taking the ratio of the two network transfer functions (v_o/v_{in}); this is not true because the transfer-function equation of a network presumes a source voltage on the input and an output voltage. Each network used in the op amp lacks one of these requirements.

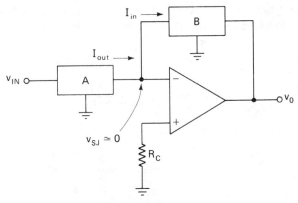

Fig. 6-5 Op amp circuit using two-port networks.

The *transfer admittance* of each network must be found [1] and are

$$y_{21} = -\frac{i_{\text{SJ}}}{v_{\text{in}}} \quad \text{when } v_{\text{SJ}} = 0 \text{ V} \quad \text{for network A}$$

$$y_{12} = -\frac{i_{\text{SJ}}}{v_o} \quad \text{when } v_{\text{SJ}} = 0 \text{ V} \quad \text{for network B}$$

[*Note:* The minus signs exist because of opposing currents (see Fig. 6-6).] Then take the reciprocal of each to yield

$$\frac{v_o}{v_{\text{in}}} = -\frac{(1/y_{12})_{\text{B}}}{(1/y_{21})_{\text{A}}} \tag{6-3}$$

[*Note:* The minus sign exists because it is an inverting amplifier (see Fig. 6-5).] This leads us to Rule 6 for operational amplifier circuits.

Rule 6: *If either the input or feedback circuit is a two-port network, the negative reciprocal of the transfer admittance for that network is used in place of the pure resistance element.*

For passive (R, L, and C only) circuits, $y_{12} = y_{21}$; since they are equal, the most easily computed method may be used for the transfer admittance in Eq. (6-3).

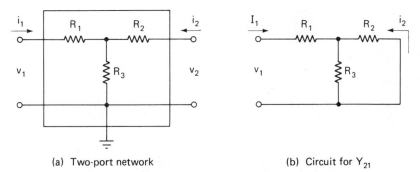

(a) Two-port network   (b) Circuit for Y_{21}

Fig. 6-6 Three-resistor T two-port network.

A good example of a two-port network commonly used in op amp circuits is the T network formed from three resistors. This network is shown in Fig. 6-6a.

Since $y_{12} = y_{21}$, let us calculate $y_{21} \cdot y_{21} = i_2/v_1$ when $v_2 = 0$ (output-shorted), in Fig. 6-6b, yields the transfer admittance (see Appendix A10)

$$y_{12} = -\frac{R_3}{R_1R_2 + R_1R_3 + R_2R_3} = y_{21}$$

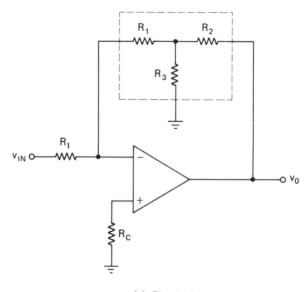

(a) Fixed gain

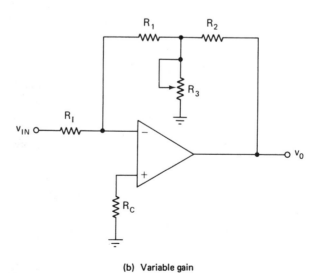

(b) Variable gain

Fig. 6-7 Inverting amplifier with resistor T in feedback path.

Thus the negative reciprocal of the transfer admittance is

$$-\frac{1}{y_{12}} = +\frac{R_1 R_2 + R_1 R_3 + R_2 R_3}{R_3} \qquad (6\text{-}4)$$

When this T network is used in the feedback path of an inverting amplifier such as that shown in Fig. 6-7a, the closed loop gain becomes (Appendix A10)

$$\frac{v_o}{v_{\text{in}}} = -\frac{R_1 R_2 + R_1 R_3 + R_2 R_3}{(R_3)(R_I)} \qquad (6\text{-}5)$$

as $-1/y_{21} = +R_I$ for the input resistor.

Two additional questions arise concerning this circuit:

1. What equivalent resistance should be used for R_F when calculating the output offset voltage (V_{OI}) due to bias or offset current?
2. What equivalent R_F should be used when calculating the value of the current-compensation resistor (R_C)?

The answer to both these questions is $-1/y_{12}$, from **Rule 6**. Thus Eq. (3-5) for voltage out due to bias current would be written

$$V_{OI} = \left(-\frac{1}{y_{21}}\right)(I_B-) = \left(\frac{R_1 R_2 + R_1 R_3 + R_2 R_3}{R_3}\right)(I_B-) \qquad (6\text{-}6)$$

and Eq. (3-7) for the value of R_C would be written

$$R_C = \frac{R_I}{R_I(-y_{21}) + 1} = \frac{R_I}{\dfrac{R_I R_3}{R_1 R_2 + R_1 R_3 + R_2 R_3} + 1} \qquad (6\text{-}7)$$

6.4   High-Gain Amplifier

A circuit that is often used to obtain very high gains without using large feedback resistors is shown in Fig. 6-7a. An example will help to clarify use of equations involving y_{12}.

EXAMPLE 6-1

An inverting amplifier is to have a gain of 100. The input resistor (R_I) must be 10 kΩ to prevent loading the previous stage.

a. Develop a T feedback network using two 10 kΩ resistors for R_1 and R_2 which will yield a gain of 100.
b. Find the value of R_C that must be used on this amplifier.
c. What is the equivalent value of R_F for use in calculating the effects of offset current?

Solution

a. $\dfrac{-1}{y_{12}} = \dfrac{(10\text{ k}\Omega)^2 + R_3(10\text{ k}\Omega + 10\text{ k}\Omega)}{R_3}$. Thus

$$100 = \dfrac{10^8 + R_3(20\text{ k}\Omega)}{R_3(10\text{ k}\Omega)} \quad \text{from Eq. (6-5)}$$

and $R_3 = 102\ \Omega$; this makes $-1/y_{12} = 1\text{ M}\Omega$.

b. $R_C = \dfrac{(10\text{ k}\Omega)(1\text{ M}\Omega)}{10\text{ k}\Omega + 1\text{ M}\Omega} = 9.9\text{ k}\Omega$

c. The equivalent value of R_F for current offset calculations is $1\text{ M}\Omega$.

6.4.1 Variable-Gain Amplifier

If R_3 in Fig. 6-7a is changed to a variable resistor, the gain of the circuit can be varied by adjusting the value of R_3 as illustrated in Fig. 6-7b. The adjustment of R_3 produces:

1. A circuit gain of $(R_1 + R_2)/R_I$ when R_3 is infinite.
2. A circuit gain approaching A_{VOL} as R_3 approaches zero ohms. (*Note:* To prevent the second condition from occurring, R_3 is usually composed to a fixed and a variable resistor in series so that R_3 always has a nonzero value.)

6.4.2 Variable-Gain Differential-Mode Instrumentation Amplifier

A second type of variable-gain differential-mode instrumentation amplifier is composed of two voltage followers and a differential-mode amplifier with a T network in the feedback path as shown in Fig. 6-8. The first type was given in Fig. 6-4

6.5 Constant Current Amplifier

Constant current amplifiers are a family of circuits that exhibit the characteristics of the Norton equivalent circuit: namely, the output impedance is nearly infinite. This means that as the load impedance varies, the load current remains constant over a wide output-voltage range. Figure 6-9 illustrates a constant current amplifier for which the output impedance (Appendix A11) is given by

$$R_O = r_o + (A + 1)R_{\text{CS}} \tag{6-8}$$

where r_o is the op amp output impedance and R_{CS} is the "current sense" resistor, to be defined in Section 6.5.1. The load current is given by

$$i_L = \dfrac{v_{IN}}{R_{\text{CS}}}\left[\dfrac{1}{1 + \dfrac{R_L + r_o}{(A + 1)R_{\text{CS}}}} \right] \tag{6-9}$$

from which it can be seen that the load current diminishes as the open loop gain (A) falls off in magnitude.

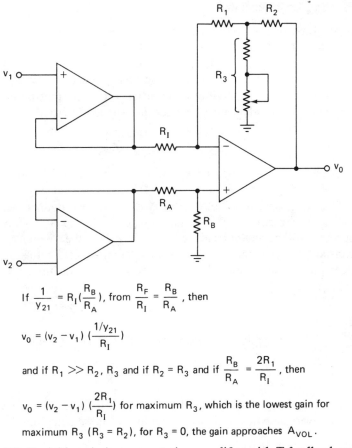

If $\dfrac{1}{y_{21}} = R_I(\dfrac{R_B}{R_A})$, from $\dfrac{R_F}{R_I} = \dfrac{R_B}{R_A}$, then

$$v_0 = (v_2 - v_1)\,(\dfrac{1/y_{21}}{R_I})$$

and if $R_1 \gg R_2$, R_3 and if $R_2 = R_3$ and if $\dfrac{R_B}{R_A} = \dfrac{2R_1}{R_I}$, then

$$v_0 = (v_2 - v_1)\,(\dfrac{2R_1}{R_I})$$ for maximum R_3, which is the lowest gain for

maximum R_3 $(R_3 = R_2)$, for $R_3 = 0$, the gain approaches A_{VOL}.

Fig. 6-8 Variable-gain instrumentation amplifier with T feedback path.

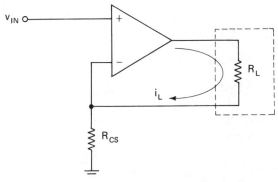

Fig. 6-9 Constant current amplifier with floating load resistor.

6.5.1 Floating Load Resistor

A variation of the noninverting amplifier (Section 2.4) is formed when R_F of this circuit is used as a "floating load" resistor (R_L) and R_I is used as a current sense resistor (R_{CS}). The circuit, as used in this fashion, is illustrated in Fig. 6-9. The voltage drop across the current sense (R_{CS}) resistor must equal the input voltage (v_{IN}) by **Rule 1**; it therefore follows that for a constant input voltage, the current through R_L and R_{CS} must be constant. The load current is given by

$$i_L = \frac{v_{IN}}{R_{CS}} \tag{6-10}$$

and this current is *constant* regardless of the value of R_L within the operating range of the op amp.

When the op amp output voltage saturates, the amplifier can no longer maintain the current constant. The voltage range of the op amp output, over which the current will be maintained constant, is called the "voltage compliance" of the constant current amplifier. The voltage compliance for the circuit in Fig. 6-9 sets the value of the largest permissible load resistor, which is given by

$$R_{L_{max}} = \frac{V_{sat} - v_{IN}}{i_L} \tag{6-11}$$

while the transconductance ratio (i_L/v_{IN}) is given by (Appendix A12)

$$\frac{i_L}{v_{IN}} = \frac{1}{R_{CS}} \tag{6-12}$$

which represents the amount of load current for a given input voltage.

An example will help illustrate the use of these equations.

EXAMPLE 6-2

A constant current amplifier uses a μA 741 op amp with a supply voltage of ± 15 V. The load resistor will vary from 200 to 750 Ω. The amplifier must deliver 15 mA dc to the load when the input voltage is 2 V dc. The op amp output voltage will saturate at ± 13 V.

    a. Determine the value of R_{CS}.

    b. Determine if the voltage compliance is sufficient.

Solution

Using Eq. (6-12), we obtain

    a. $\dfrac{1}{R_{CS}} = \dfrac{15\ mA}{2\ V} = 7.5\ m_S,\ R_{CS} = 133\ \Omega.$

    b. By **Rule 1**, the voltage at the junction of R_L and R_{CS} is 2 V. The maximum voltage (V_{sat}) from the op amp is 13 V.

$$R_{L_{max}} = \frac{13\ V - 2\ V}{15\ mA} = 733\ \Omega$$

The voltage compliance is insufficient to handle a 750 Ω load resistance.

6.5.2 Ground-Referenced Load Resistor

In certain circumstances, the load resistor must not float but must be ground-referenced. When this is true, the circuit of Fig. 6-10a is used. Although this circuit contains two op amps, it possesses some unique qualities. First, the input voltage may be at any common-mode voltage above ground and the current will be unaffected. The only voltage that affects the output current is the differential input voltage. Second, the load current is not limited as in other forms of grounded load resistor constant current amplifiers. When all resistors, except R_{CS} and R_L, are equal, the load current is again given by (Appendix A12)

$$i_L = \frac{v_{IN}}{R_{CS}} \qquad (6\text{-}13)$$

where all the R values are equal; R_{CS} sets the transconductance ratio and R_L may have any value from zero to $R_{L_{max}}$. The voltage compliance establishes the value of $R_{L_{max}}$ and is given by

$$R_{L_{max}} = \frac{V_{sat}}{i_L} - R_{CS} \qquad (6\text{-}14)$$

where R_{CS} has been determined from Eq. (6-10). The op amp output voltage for any load and sense resistor is given by

$$V_O = i_L(R_{CS} + R_L) \qquad (6\text{-}15)$$

while the maximum output voltage for $R_L = R_{L_{max}}$ is

$$V_{O_{max}} = i_L(R_{CS} + R_{L_{max}}) \qquad (6\text{-}16)$$

An example will help illustrate the use of these equations.

EXAMPLE 6-3

A grounded load constant current amplifier uses an output op amp that saturates at ± 14 V. The output op amp also includes an emitter-follower current driver that will deliver 100 mA to a load when the differential input voltage is $+3$ V.

    a. Determine the value of R_{CS}.
    b. Determine the value of the largest load resistor.
    c. Determine the range of current output voltages as the load resistor varies from 0 Ω to $R_{L_{max}}$.

Solution

    a. $R_{CS} = \dfrac{3\text{ V}}{100\text{ mA}} = 30.0\ \Omega$

    b. $R_{L_{max}} = \dfrac{14\text{ V}}{100\text{ mA}} - 30\ \Omega = 110\ \Omega$

    c. At $R_L = 0\ \Omega$, $V_0 = (30\ \Omega)(100\text{ mA}) = +3$ V
       At $R_L = 110\ \Omega$, $V_0 = (110 + 30\ \Omega)(100\text{ mA})$
            $= 14$ V as expected from part (b)

Constant Current Amplifier

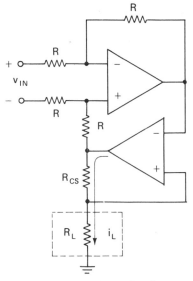

(a) Grounded load resistor*

(b) Instrument using patented circuit in part (a).

Fig. 6-10 Grounded load resistor constant current amplifier.

*Patent No. 4,091,333; Valhalla Scientific Inc., San Diego, Calif., May 23, 1978; Guy Carlyle Thrap, inventor.

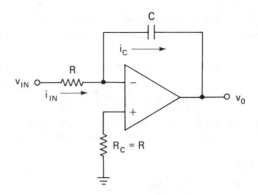

Fig. 6-11 Active integrator circuit.

6.6 Integrator

In Chapter 5, several low-pass RC filter sections were presented. These passive circuits integrate only at frequencies above 10 times the pole frequency. A circuit that integrates from nearly dc to the op amp's unity open-loop-gain frequency is illustrated in Fig. 6-11. The pole frequency for this circuit is at [1, p. 294]

$$f_p = \frac{1}{2\pi A_{\text{VOL}} RC}$$

Since the summing junction is zero volts, $i_{\text{IN}} = v_{\text{IN}}/R$, and since the inverting input impedance is very high, $i_C = i_{\text{in}}$. The output voltage of an integrator is given by

$$v_O = -\frac{1}{C} \int_0^t i_C(t)\, dt + v_C(0) \quad \text{but } i_C(t) = \frac{v_{\text{IN}}}{R}$$

Thus

$$v_O = -\frac{1}{RC} \int_0^t v_{\text{IN}}\, dt + v_C(0) \tag{6-17}$$

where $v_C(0)$ is the voltage across C at the *beginning* of the integration period; at $t = 0$, $v_O = v_C(0)$.

6.6.1 DC Input Voltages

Equation (6-17), integrated for a constant v_{in} with the initial value of v_C (see Fig. 6-11) equal to zero (0), is

$$\Delta v_O = -\frac{1}{RC}(v_{\text{IN}})\, \Delta t \tag{6-18}$$

or

$$change \text{ in } v_O = -\frac{1}{RC}(v_{\text{IN}})\, (elapsed \text{ time}) \tag{6-19}$$

A useful rearrangement of Eq. (6-19) produces the change in output voltage per unit time; it is

$$\frac{\Delta v_O}{\Delta t} = -\frac{1}{RC}(v_{\text{IN}})\ \text{V/s} \tag{6-20}$$

A good way to analyze the operation of this integrator for a dc input voltage involves solving Eq. (6-19) for a given input and then comparing this to an intuitive solution.

The following example should serve to accomplish this.

EXAMPLE 6-4

An integrator has an RC time constant of 0.1 s ($R = 100$ kΩ, $C = 1$ μF). It has a dc voltage on the input of $+0.2$ V. What is the output voltage with respect to time?

Solution

Mathematical solution:

$$v_O = -\frac{1}{0.1} \int (0.2 \text{ V}) \, dt$$

since the 0.2 V is a constant. Equation (6-18) yields $\Delta v_O = -10(0.2) \, \Delta t = -2 \, \Delta t$; Δt is in seconds, so the output voltage decreases by 2 V each second, producing a straight-line ramp. The ramp begins at zero volts at $t = 0$ and continues to drop forever.

Intuitive solution: Since $v_{IN} = +0.2$ V dc,

$$i_{IN} = \frac{v_{IN}}{R} = \frac{0.2 \text{ V}}{100 \text{ k}\Omega} = 2 \text{ }\mu\text{A}$$

The current through the capacitor has no way of changing, so it must be the same as i_{in}; thus $i_C = 2$ μA and it is constant. The equation for voltage across the capacitor is given by Eq. (4-1), as $\Delta V_C = I_C / C \Delta t$. Equation (4-1) states that the *change* in capacitor voltage (ΔV_C) is the ratio of capacitor current to capacitance, multiplied by the *elapsed* time (Δt).

$$\text{change in } V_C = \frac{I_C}{C} \text{ (elapsed time)}$$

$I_C = 2$ μA and $C = 1$ μF; thus the output-voltage change each elapsed second of time is

$$\text{change in } v_C = -\frac{2 \text{ }\mu\text{A}}{1 \text{ }\mu\text{F}} (1 \text{ s}) = -2 \text{ V}$$

The minus sign comes as a result of the current flowing into the capacitor from the summing junction. The output voltage is dropping in a straight-line ramp by -2 V each second. In this case it stops when the op amp output saturates at the negative power-supply voltage.

The two methods yield the same result.

Equation (4-1), for a constant capacitor current, is repeated as

$$\text{change in } V_O = \frac{i_C}{C} \text{ (elapsed time)} \tag{6-21}$$

and $V_O = v_C$, since the summing junction voltage is zero (Fig. 6-11).

The two important words here are *change* and *elapsed*. This means that if the input voltage is zero, the capacitor current will be zero and the output voltage will *remain* where it last was and *stay* there until the input voltage regains a voltage value [see v_C in Eq. (6-17)].

A second example will help to clarify the use of Eq. (6-17).

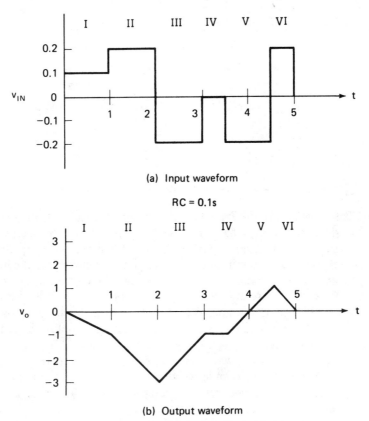

(a) Input waveform

RC = 0.1s

(b) Output waveform

Fig. 6-12 Input and output waveforms for integrator.

EXAMPLE 6-5

An integrator with an *RC* time constant of 0.1 s has the input waveform in Fig. 6-12a. The output waveform is given in Fig. 6-12b. Verify that the output waveform is correct.

Solution

The input waveform is divided into regions of constant voltage; in this case there are 6. Equation (6-19) is solved for each region to obtain the output voltage.

Region		Change in v_c
I	$v_O = -10(0.1)(1) =$	-1 V
II	$v_O = -10(0.2)(1) =$	-2 V
III	$v_O = -10(-0.2)(1) =$	$+2$ V
IV	$v_O = -10(0)(0.5) =$	0 V
V	$v_O = -10(-0.2)(1) =$	$+2$ V
VI	$v_O = -10(0.2)(0.5) =$	-1 V

Integrator

Now let us see if that is what happened.

Region	Ending Time	Output Changed by:	$v_C(0)$	Output Became
	$t = 0$		0	0 V (beginning)
I	1 s	−1	0	−1
II	2 s	−2	−1	−3
III	3 s	+2	−3	−1
IV	3.5 s	0	−1	−1
V	4.5 s	+2	−1	+1
VI	5 s	−1	+1	0

The waveform in the example above began and ended at the same location: zero. This does not always happen. The integrator is a "conservative system" (i.e., averages the positive and negative halves of the input waveform area). There are an equal number of area units above the zero line as below; thus the integrator ends the cycle at the same voltage at which it began. It happened to begin at zero, thus it ends at zero. If there are more area units above the zero line, the integrator ends below its starting point, and vice versa. To prove this in your own mind, stop the input waveform in Fig. 6-12a at various times and visualize the output remaining at the present voltage.

Integrators are used wherever a ramp voltage with nearly perfect linearity is desired. An excellent example for the use of an integrator is the horizontal sweep voltage in an oscilloscope. This is a ramp having perfect voltage linearity with time and a switchable time constant for changing sweep rates.

The ramp is generated by inserting a constant dc voltage at the integrator input and allowing the output voltage to slew until the desired end point has been reached; the integrator capacitor is then discharged by an external device, and the ramp begins again.

6.6.2 AC Input Voltages

The integrator takes the time integral of ac voltages as well as dc. It will integrate both, simultaneously, if they are present on the input.

The ac input voltage will be represented by

$$v_{IN} = V_p \sin(\omega t) \tag{6-22}$$

and the output voltage is

$$v_O = -\frac{1}{RC} \int V_p \sin(\omega t)\, dt$$

which, when integrated, becomes (Appendix A13)

$$v_O = +\frac{V_p}{RC\omega} \sin\left(\omega t + \frac{\pi}{2}\right) \tag{6-23}$$

An example will help to clarify the use of equation (6-23).

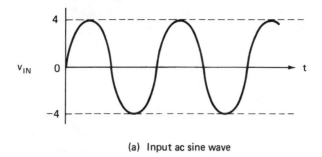

(a) Input ac sine wave

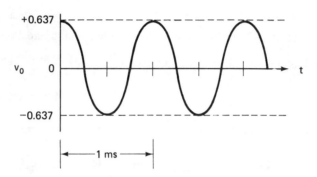

(b) Steady state output ac cosine wave

Fig. 6-13 Input and output sine wave for integrator.

EXAMPLE 6-6

An integrator with $R = 1$ kΩ and $C = 0.1$ μF has an ac sine-wave signal of 1 kHz on the input with a peak voltage of 4 V. Draw the output amplitude and phase in time synchronism with the input waveform.

Solution

The answer is given in Fig. 6-13 and a detailed solution is performed in Appendix A13.

6.7 Differentiator

The differentiator is a circuit that takes the time derivative of the input voltage.

6.7.1 High-Pass *RC* Filter Section

To place this circuit in the same perspective as the integrator, a quick analysis of the high-pass *RC* section will be presented. The high-pass *RC* section, together with its Bode plot, are shown in Fig. 6-14.

The transfer-function equation for this high-pass *RC* section is (see Appendix A14)

$$\frac{v_o}{v_{\text{in}}} = j\frac{f}{(1/2\pi RC) + jf} \tag{6-24}$$

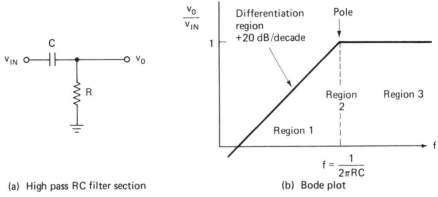

(a) High pass RC filter section   (b) Bode plot

Fig. 6-14 High-pass RC filter section and Bode plot.

which has three separate areas of operation:

1. $f \ll \dfrac{1}{2\pi RC}, \dfrac{v_o}{v_{in}} = \omega RC \underline{/+90°}$ differentiation region

2. $f = \dfrac{1}{2\pi RC}, \dfrac{v_o}{v_{in}} = 0.707 \underline{/+45°}$ amplitude at "pole" frequency

3. $f \gg \dfrac{1}{2\pi RC}, \dfrac{v_O}{v_{IN}} = 1 \underline{/0°}$ constant amplitude region

Notice in region 1 (the differentiation region) that the amplitude constantly increases with frequency.

Recall from Section 5.4 on noise that the bandwidth for this differentiation region is essentially unlimited. Since the output amplitude is greater at higher frequencies, the output noise could be greater than the output signal being differentiated.

6.7.2 Op Amp Differentiator

Figure 6-15 illustrates an op amp differentiator. Since the summing junction is at zero volts,

$$i_{IN} = C_D \frac{d(v_{IN})}{dt} \quad \text{and} \quad i_R = i_{IN} = -\frac{v_O}{R_D}$$

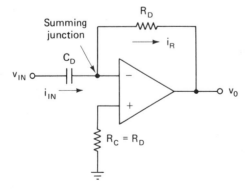

Fig. 6-15 Active differentiator circuit.

156

Thus

$$v_O = -i_{\mathrm{IN}}R_D$$

It therefore follows that

$$v_O = -R_DC_D\frac{d(v_{\mathrm{IN}})}{dt} \tag{6-25}$$

The "pole" for this circuit is essentially at the frequency [1, p. 295] $f_p = A_{\mathrm{VOL}}/2\pi R_DC_D$, presuming that the op amp has a wide bandwidth. This means that this circuit will differentiate down in frequency to nearly dc. It also means that the gain is always larger for higher frequencies over the whole differentiation region. This circuit is unstable, and oscillates at high frequencies thus is essentially useless to a circuit designer. But these circuits are used; and let us see how.

The bandwidth must be limited while maintaining the differentiator characteristics over the desired frequency range.

6.7.3 Practical Differentiator

To overcome the bandwidth and resulting noise problems, an integrator is placed on the differentiator with a short time constant. This limits the bandwidth and stabilizes the differentiator operation so that a usable circuit is realized. Figure 6-16 illustrates such a circuit together with its Bode plot.

The differentiator time constant (R_DC_D) is greater than the integrator time constant (R_IC_I) by at least 10^4 and probably 10^5 times. But two other time constants exist which are not apparent at first; they are R_DC_I and R_IC_D. Their value is the geometric mean between the differentiator and integrator time constants given by

$$R_DC_I = R_IC_D = \frac{R_DC_D}{\sqrt{R_DC_D/R_IC_I}} \tag{6-26}$$

These two time constants are made equal to each other for reasons that become apparent only when the circuit equation is derived (Appendix A15); the result of this derivation is [1]

$$\frac{v_O}{v_{\mathrm{IN}}} = -(j\omega R_DC_D)\left[\frac{1}{(1 + j\omega R_DC_I)^2}\right] \tag{6-27}$$

An example will help to clarify the development of a practical differentiator:

EXAMPLE 6-7

A differentiator with a differentiation time constant of 10 s is to have a stabilizing integrator with a time constant of 10^{-3} s. In Fig. 6-16, $C_D = 4$ μF and $R_D = 2.5$ MΩ. Calculate the values of R_I and C_I for the integrator portion.

Solution

$R_DC_D = 10$ s; $R_IC_I = 10^{-3}$ s. From Eq. (6-26) $R_DC_I = R_IC_D = 10$ s$/100 = 0.1$ s and an equal contribution can be obtained from R_I and C_I. Thus $R_I = 2.5$ M$\Omega/100 = 25$ kΩ, and

$C_I = 4\ \mu\text{F}/100 = 0.04\ \mu\text{F}$. Then $(25\ \text{k}\Omega)(4\ \mu\text{F}) = (2.5\ \text{M}\Omega)(0.04\ \mu\text{F}) = 0.1$ s. Then

$$f_D = 0.016\ \text{Hz}$$
$$f_{ID} = 1.6\ \text{Hz}$$
$$f_I = 1600\ \text{Hz}$$
$$f_{GB} = 10^6\ \text{Hz} \qquad \text{for a } \mu\text{A 741}$$

Thus $A_{\text{VEX}} = 625$ or 56 dB.

Several clarifications of Eq. (6-27) are in order:

1. The j in the magnitude portion of Eq. (6-27) is the $\underline{/+90°}$ of Eq. (6-24) in the differentiation region (area 1).

2. The magnitude portion of the equation defines the "differentiator equation."

3. The second or "modifier" term defines the circuit operation over the whole frequency range. When ω is small, this term goes to unity, leaving only the differentiator equation, as expected. When ω is large, the circuit becomes an integrator and the amplitude decreases with frequency, as desired.

4. In the integration region, v_O/v_{IN} becomes $1/\omega R_I C_I \underline{/+90°}$ as is expected from area 3 of Eq. (5-1). The sign reversal between the two equations is a result of the 180° phase reversal in the active integrator. [*Note:* $(R_D C_I)^2 = (R_D C_I)(R_I C_D)$.]

A differentiator is designed in the following manner:

1. Select an $R_D C_D$ using Eq. (6-25) as a basis for the value of RC; $R_D C_D$ in this case determines the differentiator "sensitivity" or gain to an input signal.

2. To determine where this selected $R_D C_D$ appears on the op amp open-loop gain characteristic curve, the magnitude portion only of Eq. (6-27) ignoring the bracketed term, is set equal to unity and solved for ω after inserting the numerical value of $R_D C_D$. The ω obtained in this manner will yield the frequency $f = \omega/2\pi$ where the $+20$ dB/decade slope of the differentiator frequency response crosses the unity (0 dB)-gain line on the open-loop-gain characteristic curve.

3. A compensating integrator $(R_I C_I)$ is selected which yields an excess loop gain of at least 10 (20 dB) and preferably 100 (40 dB) and at the same time has a time constant of 10^{-4} or 10^{-5} the value of the differentiating time constant. The frequency response of the compensating integrator crosses the unity-gain line at $f_I = 1/2\pi R_I C_I$, as illustrated in Fig. 6-16b. The open loop gain of the op amp will inherently provide a compensating integrator, but this is not desirable, as the feedback loop is open along the open-loop-gain frequency-response line and the op amp has no ability to control.

4. The highest frequency that can be effectively differentiated is at 0.1 of the intersection frequency between the differentiator frequency response $(+20$ dB/decade) and the compensating integrator frequency response $(-20$ dB/decade); this intersection occurs at $f_{ID} = 1/2\pi R_I C_D$, as illustrated in

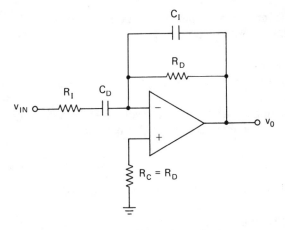

(a) Practical differentiator circuit

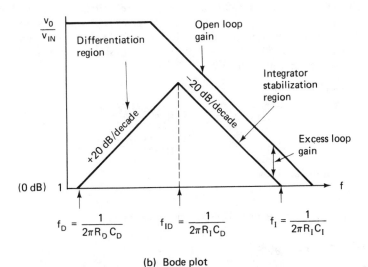

(b) Bode plot

Fig. 6-16 Practical differentiator and Bode plot.

Fig. 6-16b. The values of R_I and C_I are determined using Eq. (6-26) and the constraints of step 3.

6.7.4 Differentiator Operation

The output voltage of a differentiator is the time derivative of the input voltage multiplied by the time constant. Thus the laws of differentiation apply. They are:

1. If the input voltage is a constant (of any magnitude), the output is zero volts.
2. If the input is a ramp voltage, the output is a constant voltage of opposite polarity.

3. If the input is a second-order (parabolic)-type function, the output is a ramp of the opposite polarity.

The differentiator is used most extensively for case 2, a ramp voltage input. It is used to measure the rate of change of input voltage and yields an output voltage proportional to this rate.

The equation for a positive ramp voltage is defined as $v = (v_{IN})t$, where v_{IN} is in volts/second. For this case, Eq. (6-27) becomes

$$v_O = -R_D C_D \frac{d(v_{IN}t)}{dt} = -R_D C_D v_{IN} \qquad (6\text{-}28)$$

An example will help to clarify the use of Eq. (6-28).

EXAMPLE 6-8

A train-speed measurement instrument operates by placing a constant current through the rails and measuring the resulting voltage across the rails at a road crossing. When a train approaches, the wheels short the rails and the voltage begins to diminish as the train approaches the crossing. The rate of decline in voltage is proportional to the train speed. A differentiator with a time constant of 10 s is used to measure the train speed. When the train is approaching at 50 mi/h, the voltage is diminishing at 250 mV/s. What is the output voltage from the differentiator?

Solution

$$v_O = -R_D C_D \frac{d[(V/s)t]}{dt}$$

$$= -10 \frac{d[(-250 \text{ mV/s})t]}{dt}$$

Equation (6-28) yields

$$v_O = -10(-250 \text{ mV}) = +2.5 \text{ V}$$

Thus $+2.5$ V represents a 50 mi/h oncoming train.

6.7.5 AC Input Voltages

The differentiator takes the time derivative of only ac or changing dc voltages. It will differentiate both simultaneously if they are present on the input.

The ac input voltage will be represented by

$$v_{IN} = V_p \sin(\omega t) \qquad (6\text{-}29)$$

and the output voltage is

$$v_O = -R_D C_D \frac{d[V_p \sin(\omega t)]}{dt}$$

which, when differentiated, becomes

$$v_O = \omega R_D C_D V_p \sin\left(\omega t - \frac{\pi}{2}\right) \qquad (6\text{-}30)$$

The derivation of Eq. (6-30) may be found in Appendix A16.

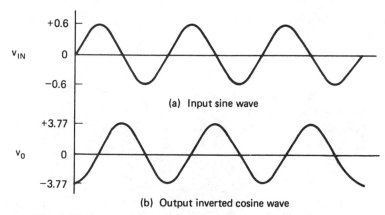

Fig. 6-17 Input and output ac waveforms for differentiator.

EXAMPLE 6-9

A differentiator with $R_D = 10$ kΩ, $C_D = 0.1$ μF has an ac sine-wave signal of 1 kHz, 0.6 V peak applied. It may be presumed that the compensating integrator is in place.

Solution

Only the answer is given in Fig. 6-17. The detailed solution is performed in Appendix A16.

6.8 Solving Differential Equations

Any physical system incorporating energy storage elements must be represented by a differential equation. Energy storage elements are capacitors, inductors, mass, springs, heat, and so on. Each energy storage element added to a system requires the increase of the differential equation order by 1; that is, a circuit with a single capacitor can be represented by dv/dt; one with a capacitor and inductor must be represented by d^2v/dt^2.

Although the study of differential equations is beyond the scope of this text, it is relevant to present the electrical simulation of differential equations through the use of operational amplifiers. These are used in the industry in a variety of ways, some of which the reader will no doubt encounter. Analog computers are electronic networks that solve differential equations; these employ operational amplifiers as integrators and inverting summers.

6.8.1 Differential Equation of Mechanical Systems

A mechanical system is shown in Fig. 6-18a. Its differential equation is

$$F(t) = m\frac{d^2x}{dt^2} + f\frac{dx}{dt} + Kx$$

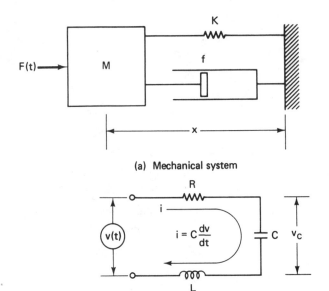

(a) Mechanical system

(b) Electrical system

Fig. 6-18 Mechanical versus electrical system analogies.

which is usually rewritten in the form

$$\frac{d^2x}{dt^2} = -\frac{K}{M}x - \frac{f}{M}\frac{dx}{dt} + \frac{1}{M}F \qquad (6\text{-}31a)$$

for state-variable* evaluation, where

$$\dot{y}_2 = -\frac{K}{M}y_1 - \frac{f}{M}y_2 + \frac{1}{M}F \qquad (6\text{-}31b)$$

where

$$y_2 = \frac{dy_1}{dt} = \frac{dx}{dt} \quad \dot{y}_2 = \frac{dy^2}{dt}$$

6.8.2 Differential Equation of Electrical Systems

An electrical system is shown in Fig. 6-18b. Its differential equation is

$$v(t) = LC\frac{d^2v}{dt^2} + RC\frac{dv}{dt} + V$$

which is usually written in the form

$$\frac{d^2v}{dt^2} = -\frac{1}{LC}v - \frac{R}{L}\frac{dv}{dt} + \frac{1}{LC}e(t) \qquad (6\text{-}32a)$$

*A "state variable" is the term used to define the effect of an energy-storing element in a physical system. There is one state variable for each energy storage element represented by capacitor voltages and inductor currents.

162

Table 6-1

Constant	Mechanical	Electrical	Definition
ω_n	$\sqrt{\dfrac{K}{M}}$	$\sqrt{\dfrac{1}{LC}}$	Resonant frequency
ζ	$\dfrac{f}{2\sqrt{KM}}$	$\dfrac{R}{2}\sqrt{\dfrac{C}{L}}$	Damping ratio

for state-variable evaluation, where

$$\dot{y}_2 = -\frac{1}{LC}y_1 - \frac{R}{L}y_2 + \frac{1}{LC}e(t) \qquad (6\text{-}32b)$$

is in state-variable form.

6.8.3 Differential Equation of Damped, Oscillatory Systems

The general equation for a damped, oscillatory system is

$$F(t) = \frac{d^2y}{dt^2} + 2\zeta\omega_n\frac{dy}{dt} + \omega_n^2 y$$

which is written in the form

$$\frac{d^2y}{dt^2} = -\omega_n^2 y - 2\zeta\omega_n\frac{dy}{dt} + F \qquad (6\text{-}33a)$$

and in state-variable form as

$$\dot{z}_2 = -\omega_n^2 z_1 - 2\zeta\omega_n z_2 + F \qquad (6\text{-}33b)$$

A comparison of Eqs. (6-31) and (6-32) with (6-33) not only illustrates the similarity but also allows direct comparison of coefficients. Table 6-1 gives the mechanical and electrical coefficients in terms of the general equation coefficients. The effect of increased damping is illustrated in Fig. 6-19b.

6.8.4 Analog-Computer Simulation of Differential Equation

Any differential equation can be simulated by a network of integrators, inverters, and inverting summers. The simulation of Eq. (6-33) by an analog-computer network is illustrated in Fig. 6-19a. The reason for the arrangement of equations (6-31) to (6-33) can now be justified:

1. The second derivative (d^2y/dt^2) term is at the *input* to the *first integrator*.
2. The *output* of the first integrator yields the first derivative (dy/dt), which must then, through some multiplying factor, be fed back to the *input* of the *first integrator*, since it is one of the equating terms.

Solving Differential Equations　　　　**163**

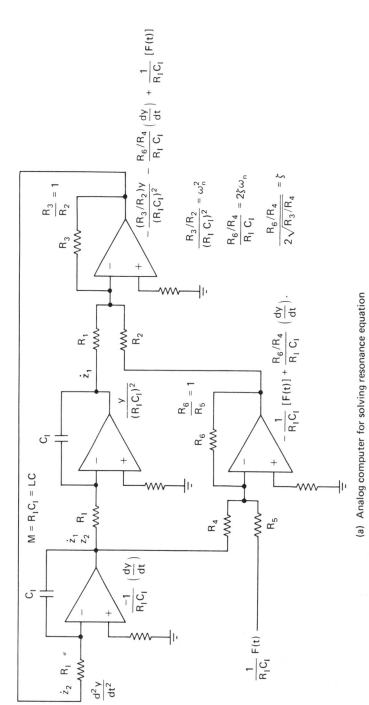

(a) Analog computer for solving resonance equation

(b) Waveforms, showing decay rate, for large and small damping factors

Fig. 6-19 Analog computer for solving resonance differential equations.

164

3. The zero-order (y) term is available at the output of the second integrator. It, too, is scaled by a multiplying factor and fed back to the *input* of the *first integrator*.

4. The forcing function $f(t)$ is summed with the zero-order term and fed back to the *input* of the *first integrator*. The first integrator *input* is now the *sum* of the three fed-back functions.

5. The RC time constants of all integrators in the system are equal.

6. The *outputs* of the two integrators are the state variables of the system and are labeled as follows:

z_2 output of first integrator

z_1 output of second integrator

The equation for the analog-computer circuit is developed and, by comparison with one of the three equations, the constants are determined. The analog-computer output voltages are exact representations of the mechanical movement or electrical circuit parameters. The development of such technology as that of springs and shock absorbers for the automobile industry employs analog computers.

Many industrial instruments and industrial controllers use analog-computer circuits to solve differential equations. One example of an industrial use is the Grade Crossing Predictor* for the railroad industry, which is illustrated in Example 6-10.

EXAMPLE 6-10

A Grade Crossing Predictor is a device that solves the differential equation

$$x - \left(\frac{dx}{dt} \right) t = 0$$

t is the time a constant-speed train takes to reach the crossing, regardless of speed. x is distance of the train from the crossing (see Fig. 6-20a). It is usually 22 s (i.e., the crossing will begin to "ring" 22 s before the train reaches it, regardless of train speed). Thus the differential equation becomes $x = 22(dx/dt)$. Presume that a train is approaching the crossing at 40.91 mi/h (60 ft/s); at how many feet away will it "ring" the crossing?

Solution

$$22 \text{ s } (60 \text{ ft/s}) = 1320 \text{ ft from crossing}$$

$$x = 1320 \text{ ft}$$

The circuit to perform this calculation is shown in Fig. 6-20b.

This circuit, plus many of the circuits in Chapters 2 and 5, is contained on the circuit board pictured in Fig. 6-21.

*U.S. Patent 3,246,143, Safetran Systems Corporation, Louisville, Kentucky.

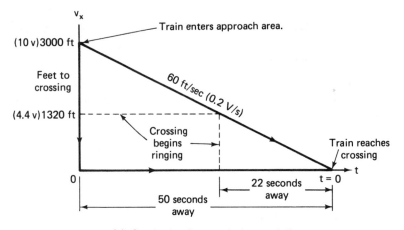

(a) Graph, showing speed, time, and distance

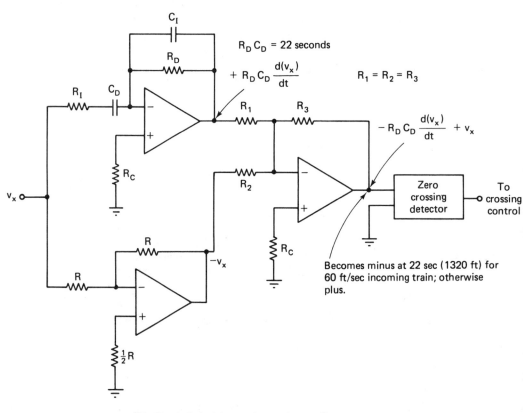

(b) Circuit for solving grade crossing predictor differential equation.

Fig. 6-20 Grade Crossing Predictor.

Fig. 6-21 Circuit board employing op amp circuits. (Courtesy Safetran Systems Corporation, Louisville, Ky.)

REFERENCE

1. **Philbrick Researches, Inc.,** *Application Manual for Computing Amplifiers* (George A. Philbrick Researches, Inc., 1966).

PROBLEMS

1. A differential-mode amplifier (Fig. 6-1) uses 20 kΩ resistors. The tolerances are all 1%. R_2, R_3, and R_f are all 0.5% low (19.9 kΩ), and R_I is 0.5% high (20.1 kΩ).

   a. Find the common-mode voltage when both V_1 and V_2 have the same 2 V p-p sine wave as input voltages. [*Hint:* use Eq. (2-18).]

   b. Calculate the common-mode rejection ratio (CMRR) of the circuit.

2. Repeat Problem 1 for a tolerance error of 0.05% low and high, respectively. Convert the CMRR to CMR(dB), and compare this value to the standard value of 85 dB available in an operational amplifier.

3. The op amp of Problem 2 has an output signal of 1 V p-p, and an output common-mode noise 20 dB lower than the output signal voltage.

   a. Find the input common-mode noise level.
   b. Find the output signal-to-noise ratio.
   c. Find the input signal-to-noise ratio.

4. A differential-mode amplifier (Fig. 6-1) is required to have an input impedance of 15 kΩ to both inputs, and a differencing gain of 4. Determine the values of all resistors in the circuit, and select the nearest 1% value from the table in Appendix B1.

5. The high-input-impedance instrumentation amplifier shown in Fig. 6-2 is to have R_F equal to 50 kΩ and unity gain. Determine the value of all resistors, and select the nearest 1% value from Appendix B1.

6. Determine the amount of common-mode signal that passes through to the output using Eq. (2-18) and the actual resistor values chosen in Problem 4. The common-mode input voltage is $5\sin(2\pi)t$. Is this a good differential amplifier if the differential input is 200 mV rms? How could the amplifier be improved?

7. The two-op-amp instrumentation amplifier shown in Fig. 6-3 is to have a gain of 8 and an input resistance of 4 kΩ. Select the other resistors from Appendix B1 to provide the lowest common-mode gain possible.

8. The variable-gain instrumentation amplifier with a T network (Fig. 6-8) has $R_I = 2$ kΩ, $R_1 = 150$ kΩ, and $R_2 = 250$ Ω. Determine the other resistor values and the circuit gain.

9. A T network is placed in the feedback loop to form a high-gain inverting amplifier. If R_I, R_1, and R_2 all equal 25 kΩ.

   a. Find the value of R_3 for a gain of -45.
   b. Find the value of R_c.

10. A high-gain inverting amplifier has a bias current of 100 nA. Find the amount of change in output voltage due to the bias currents if $R_I = 2.2$ kΩ, $R_1 = R_2 = 12$ kΩ, $R_3 = 275$ Ω, and $R_c = 0$ Ω.

11. A variable-gain differential-mode instrumentation amplifier uses a T network in its feedback loop. If $R_I = 18$ kΩ, $R_1 = 175$ kΩ, and $R_A = 2.2$ kΩ:

   a. Find the values for most of the resistors in the circuit for a gain of 35.
   b. Find the range of values of R_3 if the gain is to vary from 10 to 25.

12. A floating-load constant current amplifier uses a μA 741 op amp with ± 5 V supplies. The load resistance can vary between 50 and 2.5 kΩ. If the load current must be 2 mA, with an input voltage of $+0.5$ V:

   a. Find the value of the current-sense resistor.
   b. How high can the load resistance become before the voltage compliance is exceeded?

13. What is the largest input voltage for the current-sense resistance found in part (a) of Problem 12 if the load resistance does not exceed 1.5 kΩ?

14. A 15 W permanent magnet loudspeaker has a nominal impedance of 10 Ω at 2.5 kHz. Each time the frequency doubles, the impedance is raised by 1 Ω. The speaker is being driven as a floating load in a constant current amplifier containing a power driver as its output stage. The power supplies are at ±30 V, and the input voltage is 3.2 V rms for 15 W out.

   a. Calculate the value and the power rating of the current-sense resistor.

   b. With the speaker operating at full power, at what frequency will the voltage compliance be exceeded? Remember to convert to peak current.

15. A grounded-load constant current amplifier uses an op amp that saturates at ±10 V. The amplifier must deliver 2 mA to the load resistor with an input voltage of −6 V.

   a. Find the value of R_{cs}.

   b. Determine the value of the largest load resistor.

   c. Determine the range of the op amp output currents that can be expected.

   d. If the op amp can deliver 15 mA, will it need a current driver?

16. An integrator has $R = 2$ MΩ and $C = 1$ μF. The input voltage is a two-polarity dc voltage that switches from −250 mV to +250 mV in the following sequence: +250 mV for 10 s; −250 mV for 20 s; +250 mV for 30 s; −250 mV for 20 s; +250 mV for 30 s; −250 mV for 10 s. The polarities then reverse, and the pattern repeats itself. Draw the output waveform to scale beginning at 0 V.

17. An integrator with $R = 10$ kΩ and $C = 0.1$ μF has the input waveform shown in Fig. P6-17. Plot the output waveform to scale.

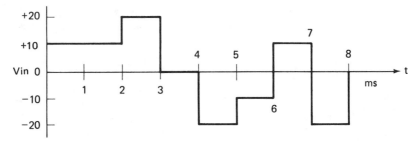

Fig. P6-17

18. An integrator with $R = 20$ kΩ and $C = 0.022$ μF has a 150 Hz sine wave with a 5 V peak amplitude on its input. Find the amplitude and phase of the output voltage.

19. An integrator with a time constant of 0.001 s uses an op amp with an open loop gain of 75,000.

   a. Find the pole frequency.

   b. What is the approximate lowest frequency of integration?

20. A high-pass RC section has $R = 100$ kΩ and $C = 0.22$ μF.

   a. Find the gain and phase at 2 Hz.

   b. Find the gain and phase at 20 Hz.

   c. Find the rate of gain rise vs. frequency using the results of parts a. and b.

   d. Find the gain and phase at 200 Hz.

   e. Find the phase and gain at 20 kHz.

21. A differentiator has $R = 200$ kΩ and $C = 2$ μF. The input voltage is a ramp with a slope of $+10.5$ V/s. What are the magnitude and polarity of the output voltage?

22. If the differentiator of Problem 20 uses an op amp with an open loop gain of 10,000, find the closed loop pole frequency.

23. A differentiator has $R = 150$ kΩ and $C = 20$ μF. The differentiator employs a stabilizing integrator with a time constant of 20 ms to prevent ringing on the leading and trailing edges of any amplitude steps caused by sudden input slope changes.

   a. Calculate the values of R_I and C_I.

   b. Calculate the frequencies of f_D, f_{ID}, and f_I.

   c. Calculate the excess loop gain if an op amp with 1.5 MHz gain–bandwidth is used.

24. In part (c) of Problem 23 it was determined that the integration was too heavy, and was excessively rounding the steps produced by input-slope changes. In an attempt to reduce this rounding, the excess loop gain will be reduced to 80 dB.

   a. Determine the new value of f_I.

   b. Calculate the new values R_I and C_I.

   c. Find the new value for f_{ID}.

25. A differentiator must produce an output voltage of -2V with an input ramp voltage of 250 mV/s. If $R_D = 2$ MΩ, 1%, calculate the value of C_D, and select a value from Appendix B2.

26. A differentiator with $R = 20$ kΩ and $C = 0.47$ μF has a 50 Hz sine wave with a peak amplitude of 10 V applied to the input. Determine the output amplitude and phase.

27. A Grade Crossing Predictor is installed on a 1000 ft approach to a street crossing. The input voltage to the predictor without a train in the approach is 5 V. The crossing bell must ring 50 s before a train reaches the street at 35 mph.

   a. Calculate the differentiator time constant.

   b. If the train is moving 35 mph toward the crossing, what is the output voltage from the differentiator?

   c. What is $-v_x$ at 500 ft?

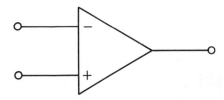

7 Nonlinear Applications

7.1 Introduction

Any time the output waveform and the input waveform are not related by a linear equation, the circuit is said to be operating in a nonlinear fashion. Several circuit connections create nonlinear operation: a nonlinear element such as a diode may be used which causes the circuit to operate in a series of limited linear regions; the feedback loop may be open during all or part of the input waveform cycle. This chapter deals with many of these applications where nonlinear operation is occurring.

7.2 Precision Rectifier

A precision rectifier is a circuit that will rectify ac signals appearing at the input. Diodes will normally only rectify at voltages above their threshold level, but when placed in a feedback loop, the threshold effect of the diode is almost entirely eliminated because of the large open loop gain of the op amp. Thus precision rectifiers will rectify signals at millivolt voltage levels with no difficulty. The rectification produced in a precision rectifier is sharp and clean with no mushy effects near the threshold point; the output voltage is exactly zero during the reverse portion of the input waveform.

7.2.1 Half-Wave Precision Rectifier

The half-wave precision rectifier is formed, beginning with an inverting amplifier, by adding a diode in series with the output end of R_F, as illustrated in Fig. 7-1a. The

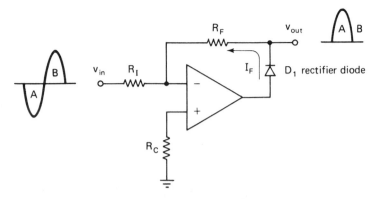

(a) Basic half wave rectifier circuit

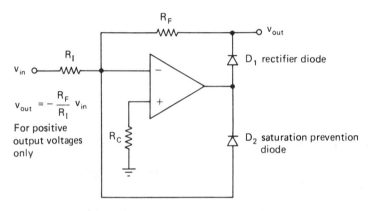

$$v_{out} = -\frac{R_F}{R_I} v_{in}$$

For positive output voltages only

(b) Half wave precision rectifier circuit

Fig. 7-1 Half-wave precision rectifier.

circuit output is at the junction of R_F and the diode. The polarity of rectification is determined by the diode direction. As can be seen in Fig. 7-1a, the direction of current through R_F is determined by the diode direction, and this current determines the output voltage polarity. A second diode, shown in Fig. 7-1b, between the op amp output and the summing junction prevents the op amp from saturating during the unwanted half-cycle (or portion thereof if not sinusoidal). Figure 7-1b illustrates the complete circuit. The voltage gain between v_{in} and v_{out} is $-(R_F/R_I)$, as though the circuit were an inverting amplifier. The circuit gain can be changed by altering the value of either resistor, that is, R_I or R_F in the half-wave precision rectifier stage.

7.2.2 Full-Wave Precision Rectifier

A full-wave precision rectifier (absolute value circuit) is formed by combining two or more op amps in a circuit which, essentially, sums the outputs of two half-wave precision rectifiers of opposite polarity.

172

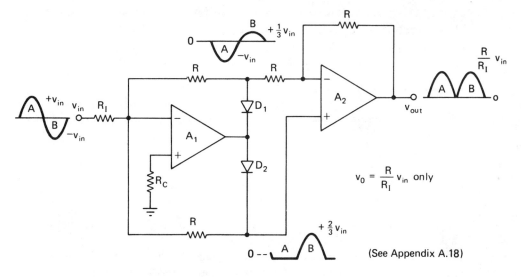

$$v_0 = \frac{R}{R_I} v_{in} \text{ only}$$

(See Appendix A.18)

(a) Dual polarity rectifier and differential - mode amplifier

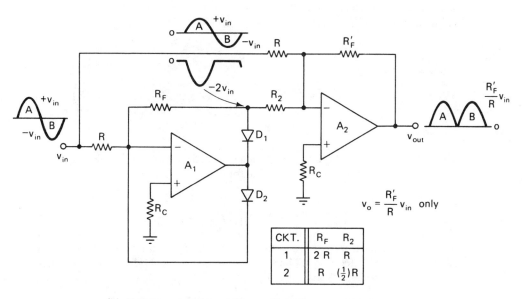

$$v_o = \frac{R'_F}{R} v_{in} \text{ only}$$

CKT.	R_F	R_2
1	2 R	R
2	R	$(\frac{1}{2})$R

(b) Half wave precision rectifier and inverting summer

Fig. 7-2 Full-wave precision rectifier.

Two circuits will be presented here, each employing only two op amps to perform the necessary full-wave rectification. Figure 7-2a illustrates the first of the two circuits. The first stage is a dual-polarity half-wave precision rectifier. The second stage is a scaling subtractor (see Section 2.4.2) with the input divider ratio equal to unity. The waveforms on the circuit show the operation. The gain of this circuit may be changed by adjusting the value of R_I; all other resistors must remain equal in value. The circuit yeilds unity gain when all resistors are equal. The waveforms on the circuit in Fig. 7-2a illustrate the operation, and a derivation for voltage relationships is given in Appendix A17.

An example will illustrate selecting appropriate resistor values.

EXAMPLE 7-1

Select the resistor values for a full-wave precision rectifier of the type shown in Fig. 7-2a. $R = 21.0$ kΩ and the circuit should have a gain $= 2$.

Solution

$$R_I = \frac{21.0 \text{ k}\Omega}{2} = 10.5 \text{ k}\Omega, 1\%$$

The second of the two circuits is illustrated in Fig. 7-2b. It is composed of a half-wave precision rectifier followed by an inverting summer (see Section 2.7.2). There are two sets of resistor values for implementing this circuit, outlined in the accompanying table. The waveforms on the circuit illustrate the operation. The gain of this circuit may be changed from unity by modifying the value of the feedback resistor on the inverting summer (see Fig. 7-2b).

An example will illustrate the selection of resistor values for a specific voltage gain.

EXAMPLE 7-2

A 100 mV peak sine wave is to be full-wave-rectified and raised in signal level to 1.00 V peak on the output. Select resistor values for this circuit from the 1% table in Appendix B1.

Solution

Circuit 1:   A value for R_F must be selected which has a standard 1% value just twice the size of R. Select the 1.05 kΩ value for R.

$$R_F = 2R = 2.10 \text{ k}\Omega$$

$$R_2 = R = 1.05 \text{ k}\Omega$$

The feedback resistor in the inverting summer must then be $10R$ or 10.5 kΩ.

Circuit 2:   A value for R must be selected that has a standard 1% value just half its size. Select the 3.48 kΩ value for R.

$$R_F = R = 3.48 \text{ k}\Omega$$

$$R_2 = \tfrac{1}{2}R = 1.74 \text{ k}\Omega$$

The feedback resistor in the inverting summer must then be $10R$ or 34.8 kΩ.

The output waveform for a sinusoidal input, for both circuits, is a full-wave rectified waveform; the waveform can be either positive or negative, depending on the diode direction selected. On occasion, alternate half-cycles will differ in amplitude (i.e., one peak will be high, the next peak low, the next high, and so on). There are three situations that can cause this to occur:

1. The gain resistors are not matched in value (or ratio); this phenomenon causes the alternate peaks to be uneven at *all* voltage levels.
2. The input signal voltage has a dc offset. This phenomenon is usually observed at smaller amplitudes, where the offset is a larger proportion of the peak signal amplitude. If this is suspected, the input signal should be checked for a dc offset on an oscilloscope.
3. The input offset voltage on the op amp of the half-wave precision rectifier is not properly nulled out. The input offset voltage of the combining (second) op amp affects the zero voltage level of the combined waveform without affecting the alternate peak relationship to any great degree.

7.3 Curve Shapers

A curve shaper is a circuit that exhibits a varying gain as the input voltage increases. There are two types of curve shapers:

1. Gain changes in discrete steps as the input voltage increases.
2. Gain changes continuously as the input voltage increases.

These circuits are of the two-quadrant type (i.e., they will only operate properly with input voltages of *one polarity*). Four-quadrant devices will operate with input voltages of either polarity.

7.3.1 Discrete Step Curve Shapers

Discrete step curve shapers are circuits that use resistors and zener diodes in series and fall into two categories: those in which the gain decreases as the input voltage increases and those in which the gain increases as the input voltage is increased. The first case has the series combinations in the feedback loop; in the second case they are outside the feedback loop.

A curve shaper with series zener diodes and resistors in the feedback loop and its gain curve are illustrated in Fig. 7-3a and b, respectively. The circuit has three distinct regions of operation, each following **Rule 2** of an inverting amplifier but with separate gains in each region. The usual development method is to draw several straight-line segments approximating the actual curve. The slope of each segment is the gain over that region; the connecting points of the segments yield the zener diode voltages. Two or three slopes will usually define a curve quite accurately.

In the lowest output-voltage region of Fig. 7-3b, the gain is greatest and is given by **Rule 2**. When the output voltage reaches the avalanche point of D_2, R_2 is placed in parallel with R_F to yield a total feedback resistance R_{F_2}, and the circuit gain reduces accordingly. The value of R_{F_2} is found from Fig. 7-3b by determining the circuit gain from the straight-line slope in Region 2. The value of R_2 is found using

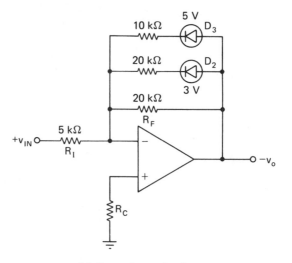

(a) Curve shaper circuit

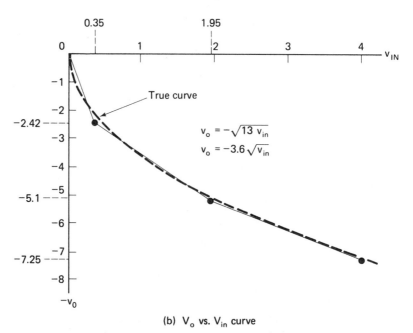

$$v_o = -\sqrt{13\, v_{in}}$$

$$v_o = -3.6\sqrt{v_{in}}$$

(b) V_o vs. V_{in} curve

Fig. 7-3 Curve shaper for concave curve.

the reverse parallel resistor formula,

$$R_2 = \frac{(R_F)(R_{F_2})}{R_F - R_{F_2}} \qquad (7\text{-}1)$$

As the output voltage increases to the avalanche value of D_3, R_3 enters the circuit and becomes in parallel with the R_F and R_2 (the parallel combination is R_{F_2}). The circuit gain in this region is determined using the shallowest slope in Fig.

7-3b. The value of the total equivalent feedback resistance (R_{F_3}), the parallel combination of R_F, R_2, and R_3, is used in a like manner to the preceding case to determine the value of R_3 from

$$R_3 = \frac{(R_{F_2})(R_{F_3})}{R_{F_2} - R_{F_3}} \tag{7-2}$$

An example will help to clarify the development of this type of circuit.

EXAMPLE 7-3

Construct a discrete curve-shaper circuit that will simulate the equation $v_{in} = \frac{1}{13}(v_o)^2$. The circuit and the curve are illustrated in Fig. 7-3.

Solution

Three straight lines are drawn which closely approximate the true curve. The lines intersect at $v_o = -2.42$ V and at $v_o = -5.1$ V, and the last line terminates at $v_o = -7.25$ V. The three slopes (v_o/v_{in}) are

$$1. \quad -\frac{2.42 - 0}{0.35 - 0} = -6.92$$

$$2. \quad -\frac{5.1 - 2.42}{1.95 - 0.35} = -1.67$$

$$3. \quad -\frac{7.25 - 5.1}{4.0 - 1.95} = -1.05$$

Let R_I be 1 kΩ, which immediately sets the desired value of R_F at 6.92 kΩ; select the 698 kΩ from the 1% table in Appendix B1. R_2 and R_3 are then found using the reverse parallel resistor formula as follows:

$$R_2 = \frac{(6.98 \text{ k}\Omega)(1.67 \text{ k}\Omega)}{6.98 \text{ k}\Omega - 1.67 \text{ k}\Omega} = 2.19 \text{ k}\Omega$$

Select 2.21 kΩ from Appendix B1. The actual parallel impedance is then

$$\frac{(6.98 \text{ k}\Omega)(2.21 \text{ k}\Omega)}{6.98 \text{ k}\Omega + 2.21 \text{ k}\Omega} = 1.68 \text{ k}\Omega$$

$$R_3 = \frac{(1.68 \text{ k}\Omega)(1.05 \text{ k}\Omega)}{1.68 \text{ k}\Omega - 1.05 \text{ k}\Omega} = 2.80 \text{ k}\Omega$$

which is a standard 1% tolerance-value resistor. The zener diode voltages are equal to the lowest break-point voltage for each slope selected; namely the D_2 zener for R_2 is 2.45 V and D_3 for R_3 is 5.1 V.

The overall circuit gain can be adjusted by selecting an appropriate value of R_I with respect to R_F, R_2, and R_3.

The second type of curve shaper is illustrated in Fig. 7-4a. The basic circuit configuration is that of a noninverting amplifier with several zener diode and series resistor networks paralleling the R_I shunt resistor. As the output voltage increases, zener diodes D_2 and D_3 avalanche and insert respective resistances in parallel with R_I, making the total equivalent shunt resistance smaller; the circuit gain thus increases in steps according to *Rule 3* as the output voltage increases. Again, the actual curve is approximated by several straight-line segments; three segments will

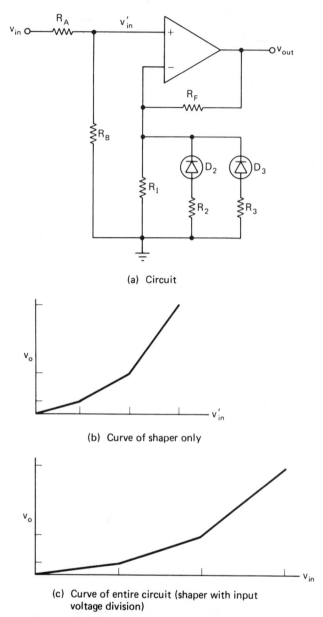

(a) Circuit

(b) Curve of shaper only

(c) Curve of entire circuit (shaper with input
voltage division)

Fig. 7-4 Convex curve shaper.

usually represent the curve quite accurately. This circuit, being a noninverting amplifier, always has a closed loop gain greater than unity. The circuit must usually be preceded by a voltage divider (R_A and R_B) to reduce the overall gain to a usable value that does not saturate the op amp in the higher-output-voltage regions.

In the lowest-output-voltage region of Fig. 7-4b, the gain is least and is governed by **Rule 3**. When the output voltage reaches the avalanche point of D_2, R_2 is placed in parallel with R_I, to yield a total shunt resistance of R_{I_2} and the circuit gain increases accordingly. The value of R_{I_2} is found from Fig. 7-4b by determining the circuit gain from the straight-line slope in region 2 and using the formula

$$R_{I_2} = \frac{R_F}{\text{region 2 gain} - 1} \tag{7-3}$$

R_2 is then determined by using the reverse parallel resistance formula,

$$R_2 = \frac{(R_I)(R_{I_2})}{R_I - R_{I_2}} \tag{7-4}$$

As the output voltage increases to the point where D_3 avalanches, R_3 is placed in parallel with the equivalent shunt resistance and it decreases again; the gain increases accordingly. The equivalent shunt resistance in region 3 is found from

$$R_{I_3} = \frac{R_F}{\text{region 3 gain} - 1} \tag{7-5}$$

R_3 is then determined by

$$R_3 = \frac{(R_{I_2})(R_{I_3})}{R_{I_2} - R_{I_3}} \tag{7-6}$$

The values of the input voltage divider can be determined by selecting an arbitrary value for R_A and calculating the value of R_B using the equation

$$R_B = R_A \left(\frac{\text{division ratio}}{1 - \text{division ratio}} \right) \tag{7-7}$$

where the division ratio is a number less than unity. An example will help to clarify the development of this type of circuit.

EXAMPLE 7-4

A circuit must be constructed to simulate a curve shape having the equation $v_o = 6 - 6 \cos[(\pi/16)v_{in}]$. In (Fig. 7-5), $(\pi/16)v_{in}$ is found from the equation $(2\pi/T)t$, where $T = 4v_{in_{max}}$ and $t = v_{in}$, which extends from 0 to $+8$ V.

Solution

The gain in the lowest output voltage region is $(0.45 - 0)/(2.1 - 0) = 0.21$, but we know that the gain of the shaper, from v_{in} to v_o, must be greater than unity. Let $R_i = R_F = 10\text{ k}\Omega$ to create a gain of 2 in the lowest-gain region. Since the circuit gain from v_{in} to v_o must be 0.21, the voltage divider must have a division ratio of $0.21/2 = 0.105$. Let R_A be $10\text{ k}\Omega$ and calculate R_B using Eq. (7-7):

$$R_B = 10\text{ k}\Omega \left(\frac{0.105}{1 - 0.105} \right) = 1.173\text{ k}\Omega$$

Select a 1.18 kΩ resistor from a 1% table in Appendix B1: the shaper gain is now $0.21/0.105 = 2$, which is the desired gain. The circuit gain in region 2 must be $(2.1 - 0.45)/(4.45 - 2.1) = 0.70$; the shaper gain must be $0.70/0.105 = 6.7$. From Eq. (7-3),

$$R_{I_2} = \frac{10\text{ k}\Omega}{6.7 - 1} = 1.75\text{ k}\Omega \qquad R_2 = \frac{(10\text{ k}\Omega)(1.75\text{ k}\Omega)}{10\text{ k}\Omega - 1.75\text{ k}\Omega} = 2.12\text{ k}\Omega$$

From Appendix B1, select a 2.15 kΩ resistor for R_2. The circuit gain in the highest-gain

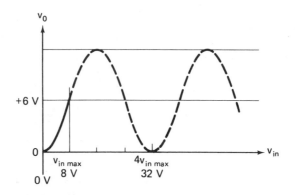

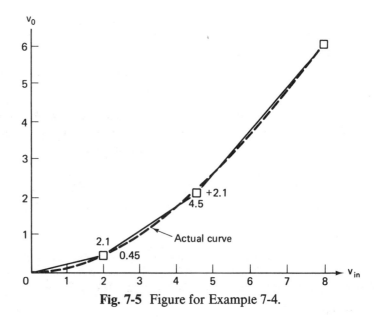

Fig. 7-5 Figure for Example 7-4.

region is $(6 - 2.1)/(8 - 4.45) = 1.1$; the shaper gain is then $1.1/0.105 = 10.5$. From Eq. (7-5), determine the equivalent resistance in the highest-gain region,

$$R_{I_3} = \frac{10 \text{ k}\Omega}{10.5 - 1} = 1.05 \text{ k}\Omega$$

R_3 is then found from Eq. (7-6) to be

$$R_3 = \frac{(1.75 \text{ k}\Omega)(1.05 \text{ k}\Omega)}{1.75 \text{ k} - 1.05 \text{ k}\Omega} = 2.62 \text{ k}\Omega$$

Select a 2.61 kΩ resistor from Appendix B1 for R_3. The zener diode break voltages must be $D_2 = 0.45$ V and $D_3 = 2.1$ V. Forward-biased silicon or hot carrier diodes may be used for D_2, and a 2.1 V zener diode (TRW) may be used for D_3 (Appendix B4).

180

7.3.2   Continuous Change Curve Shapers

This type of circuit usually employs a nonlinear device in the feedback loop without an associated resistor; the nonlinear device is usually a diode or transistor.

A logarithmic amplifier is a continuous change curve shaper. It has a semiconductor diode in the feedback which causes the output voltage to be proportional to the natural logarithm of the input current; the circuit is illustrated in Fig. 7-6a. The diode is operating in its forward-biased region; thus the output voltage is quite small and usually possesses an offset. The log amplifier must usually be followed by another amplifier, with both gain and offset adjustable. The gain of this amplifier establishes the "base" of the logarithm and large gains represent small bases. The offset acts as a "multiplier" on the input voltage (number for which the logarithm is being taken). Through the use of these two controls, the output voltage can be altered to fit the user's need. When a transistor is used in the feedback loop, it acts as a diode and the result is the same as for a diode.

Most semiconductor diodes are only good for logarithmic conversion for a few orders of magnitude. Diodes are manufactured with logarithmic use as the intended purpose. A 1N914 diode will operate over five to six orders of magnitude with good accuracy; if a selected diode is used, it will operate over eight orders of magnitude. Log diodes, as with any diode, are temperature-dependent; therefore, the diode and probably both the diode and amplifier must be contained in a temperature-controlled oven to eliminate any drift problems.

EXAMPLE 7-5

The logarithmic amplifier illustrated in Fig. 7-6 is being used to drive the vertical axis of an X-Y plotter to plot the dB signal amplitude of a dc voltage. The plotter should correspond as follows:

Input Voltage	dB Indicated	X-Y Plotter Input Voltage
10 V	60	6.0 V
1 V	40	4.0 V
0.1 V	20	2.0 V
10 mV	0	0 V

From the curve in Fig. 7-6e it can be seen that the output diode voltage is 0.97 V for a 10 mA input current, and that the output voltage is 0.5 V for a 10 μA input current. The output voltage range (970 mV $-$ 500 mV) must be expanded to the X-Y plotter input voltage range (6 V $-$ 0 V). Thus the amplifier must have a closed loop gain of 6 V/470 mV = 12.76. With this gain, the lowest log amplifier output voltage will be (12.76)(0.5 V) = 6.38 V; this must be offset to zero volts in the output amplifier. Both op amp circuits are of the inverting type; thus the output voltage will be positive. The input resistor to the log circuit must be R_I = 10 V/10 mA = 1 kΩ. If μA 741 op amps are used, the input bias current to the inverting input can be as large as 0.5 μA; this is 5% of the input summing current and creates a 5% error in the diode current at that level.

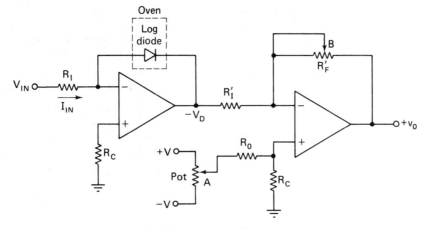

(a) Log circuit with base and coefficient correction

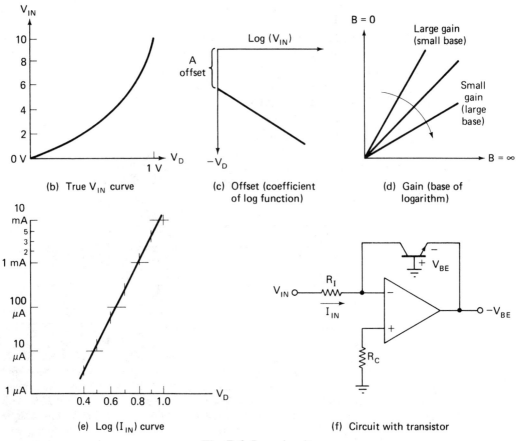

(b) True V_{IN} curve

(c) Offset (coefficient of log function)

(d) Gain (base of logarithm)

(e) Log (I_{IN}) curve

(f) Circuit with transistor

Fig. 7-6 Log circuit.

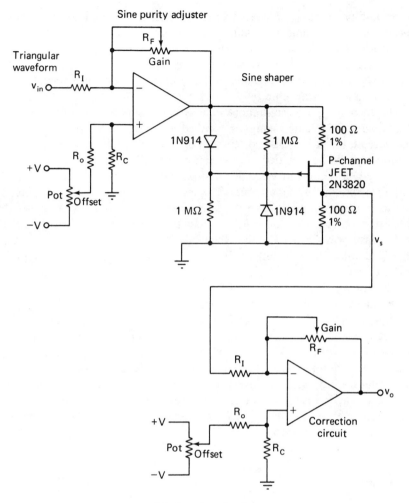

(a) Complete circuit

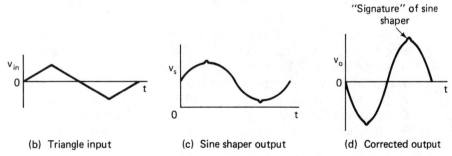

(b) Triangle input

(c) Sine shaper output

(d) Corrected output

Fig. 7-7 Sine shaper.

An antilog amplifier is created when the position of the log device and the input resistor are interchanged. A multiplier circuit can be created by connecting two logarithmic amplifiers into a summing circuit which is followed by an antilogarithmic amplifier. The output voltage is then $\log^{-1}(\log v_1 + \log v_2) = (v_1)(v_2)$ after proper scaling with gains and offsets. The Motorola 1594L is a good example of this circuit in a monolithic chip.

The second type of continuous change curve shaper is the field-effect-transistor (FET) sine-wave shaper, commonly called a "FET sine shaper." This is *not* an op amp circuit, but it is used so often with op amp-type function generators that it is a natural topic to be included in this section. The FET sine shaper is a four-quadrant device (i.e., it will operate equally well on input voltages of either polarity). Figure 7-7a illustrates a FET sine shaper. The input amplifier requires two potentiometers, one for voltage amplitude and the other for voltage offset. By adjusting these pots, the distortion can be reduced to less than 2% total harmonic distortion (THD). The input waveform for a FET sine shaper is the triangular waveform. At the slope-change point of the triangular waveform, the sine shaper fails to respond correctly and produces a small projection at the center peak of the sine waveform on both top and bottom. This projection is the "signature" of a FET sine shaper and can be observed on many commercial function generators.

Typical input voltages for the FET sine shaper circuit only are a 6 V peak-to-peak triangular waveform with a $+0.4$ V dc offset; the output voltage is typically 200 mV peak to peak with a -10 mV dc offset when it is adjusted for the best sine wave. The method found most successful and least time-consuming is to adjust the shaper on a dual-trace oscilloscope overlaid with a sine wave of known good quality. The amplifier that follows the sine shaper raises the voltage amplitude to the desired level and removes the dc offset. The triangular waveform is usually obtained by integrating a square wave.

7.4 Clippers

A clipper, or amplitude limiter as it is sometimes called, is an inverting amplifier circuit with a back-to-back zener diode and signal diode in the feedback loop. The circuit for a clipper is illustrated in Fig. 7-8a. The gain of this circuit is -1, for ease of understanding. The input signal, a 10 V-peak sine wave, would be a 10 V-peak sine wave on the output except for the zener diode. The semiconductor diode forward-biases (0.7 V) and the zener diode avalanches ($+4.3$ V) at an output voltage of $+5$ V. The output is "clipped" or "limited" to this voltage until the input signal causes the output voltage to fall below $+5$ V; the circuit then acts as an inverting amplifier during the balance of this cycle. When the output voltage becomes negative, the zener diode would normally become forward-biased and limit the output at -0.7 V, but the semiconductor diode reverse-biases in this region and the diodes do not affect the inverting amplifier operation.

A clipper that limits in both directions can be constructed using two opposing zener diodes, as illustrated in Fig. 7-8b. The clip level of this circuit is the forward-bias voltage of one zener diode plus the avalanche voltage of the second

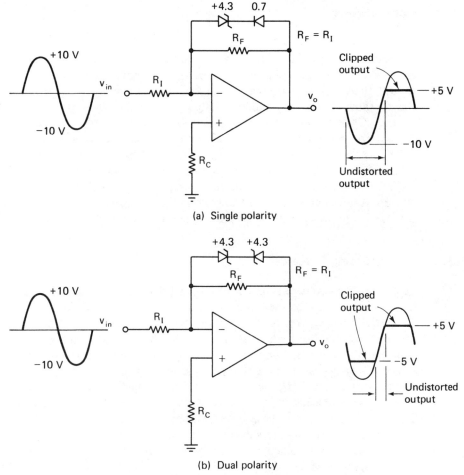

(a) Single polarity

(b) Dual polarity

Fig. 7-8 Voltage clipper.

zener diode. This circuit only acts as an inverting amplifier at output voltages between the two clip levels.

An example will help to illustrate the use of limiter circuits.

EXAMPLE 7-6

Develop a circuit with an inverting gain of -20 and an input impedance of 1 kΩ which will limit an input sine-wave signal of 0.5 V peak at $+0.375$ V referred to the input.

Solution

$$(0.5 \text{ V})(-20) = -10 \text{ V peak on output}$$
$$(+0.375 \text{ V})(-20) = -7.5 \text{ V on output}$$

The circuit is limited at -7.5 V but passes all output voltages which are more positive than -7.5 V as an inverting amplifier. The semiconducter diode has a 0.7 V forward drop across it, so 7.5 V − 0.7 V = 6.8 V. Choose the 6.8 V zener diode from Appendix B4. Both the zener diode and the semiconducter diode are reversed in direction from that shown in Fig. 7-8a.

7.5 Comparators

A comparator, as a generalization, is an operational amplifier without any negative feedback. Op amps may be used as comparators, but a true comparator differs from an operational amplifier in several respects. A comparator has a slew rate as much as 100 times faster than that in an operational amplifier. It is not frequency-compensated and thus would probably be unstable if negative feedback were applied. CMRR and PSRR are not always specified in comparators. In the 710 comparator, the input common-mode voltage is limited to ± 5 V and the output voltage is limited to -0.5 V and $+3$ V. The 311 comparator has better specifications, larger common-mode input voltage range, larger output voltage range, 100 times as fast as an op amp, but the output is "open collector" and an external pull-up resistor must be used for proper operation. The output voltage of a comparator has only two states: positive saturation and negative saturation.

7.5.1 Zero-Crossing Detector

One form of a comparator is a zero-crossing detector (op amp or comparator) with the inverting lead grounded and the input signal applied to the noninverting lead. When the input voltage is slightly more positive than the zero "reference voltage" on the inverting lead, the output slews to positive saturation (see Fig. 7-9a). Conversely, when the input signal voltage is slightly more negative than the reference voltage (zero volts), the output slews to negative saturation. The crossover point is at the zero reference voltage; thus it is called a zero-crossing detector.

A zero-crossing detector that inverts the input signal is illustrated in Fig. 7-9b. This circuit is a handy phase inverter when a signal polarity inversion is desired. Zero-crossing detectors are subject to "chatter" at the crossing point. This usually occurs when a noise voltage is present on the signal. A chatter situation is illustrated in the waveform of Fig. 7-9b.

Hysteresis is an effect where the present state input is affected by the past state output; it is created by the application of positive feedback. A zero-crossing detector with hysteresis is illustrated in Fig. 7-10a. Because of the voltage divider between output and the reference voltage (zero volts) with the division point connected to the noninverting input lead, the comparator does not switch states when the signal input voltage is *at* zero volts but instead switches at a voltage slightly following the reference voltage. This prevents "chatter" around the crossing point for a noisy input voltage. The hysteresis voltage is represented by a small change on each side of the reference voltage (see Fig. 7-10a) and is given by the equation

$$v_{\text{ZH}} = \left(\frac{R_I}{R_F} \right) V_S \tag{7-8}$$

where $R_F \gg R_I$ and V_S is the comparator saturation voltage.

An example will help to illustrate the use of Eq. (7-8).

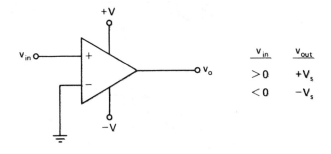

v_{in}	v_{out}
> 0	$+V_s$
< 0	$-V_s$

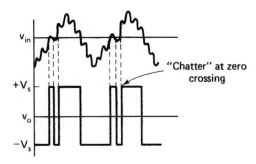

"Chatter" at zero crossing

(a) Zero crossing detector

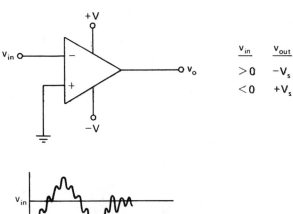

v_{in}	v_{out}
> 0	$-V_s$
< 0	$+V_s$

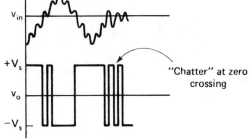

"Chatter" at zero crossing

(b) Inverted zero crossing detector (inverter)

Fig. 7-9 Zero-crossing detectors.

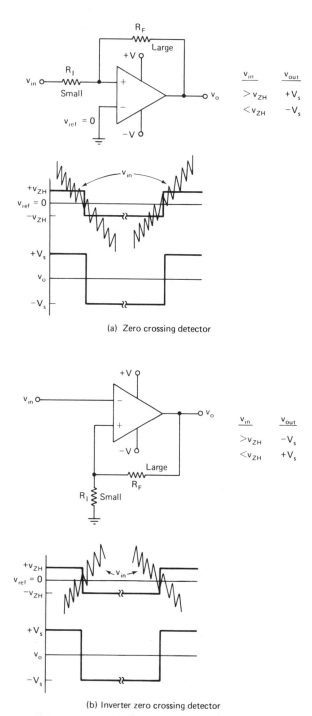

(a) Zero crossing detector

(b) Inverter zero crossing detector

Fig. 7-10 Zero-crossing detectors with hysteresis.

EXAMPLE 7-7

A voltage comparator that saturates at ± 15 V is connected as a zero-crossing detector. The input is a 5 V-peak sine wave with a 200 mV peak-to-peak noise voltage riding on it. If $R_I = 1$ kΩ, determine the value of R_F to prevent chatter at the zero crossings.

Solution

V_{ZH} is the hysteresis voltage on *each* side of the O V reference voltage. Thus $2(v_{ZH}) = 200$ mV and $v_{ZH} = 100$ mV.

$$100 \text{ mV}\left(\frac{1 \text{ k}\Omega}{R_F}\right)(15 \text{ V}) \qquad R_F = \frac{15 \text{ V}}{100 \text{ mV}}(1 \text{ k}\Omega) = 150 \text{ k}\Omega$$

An inverting zero-crossing detector with hysteresis (Fig. 7-10b) operates in the same manner. The equation for the hysteresis voltage is now given by

$$v_{ZH} = \left(\frac{R_I}{R_F + R_I}\right)V_S \tag{7-9}$$

which, if $R_F \gg R_I$, is nearly equal to Eq. (7-8).

EXAMPLE 7-8

Calculate the hysteresis voltage for an inverting zero-crossing detector with ± 15 V saturation voltages, $R_I = 1$ kΩ, and $R_F = 150$ kΩ.

Solution

Using Eq. (7-9),

$$v_{ZH} = \left(\frac{1 \text{ k}\Omega}{150 \text{ k}\Omega + 1 \text{ k}\Omega}\right)(15 \text{ V}) = 99.3 \text{ mV}$$

which is only 0.6% from the value of v_{ZH} in Example 7-7.

7.5.2 Voltage Comparator

A voltage comparator is a circuit where one input of the comparator is attached to a *reference voltage*. The circuit for a voltage comparator is illustrated in Fig. 7-11. When the input voltage becomes more positive than the reference voltage, the output slews to positive saturation; when the output becomes more negative than the reference voltage, the output slews to negative saturation. The inverted voltage comparator illustrated in Fig. 7-11b works in a similar manner. When the input voltage is more positive than the reference voltage, the output slews to negative saturation; when the input voltage is more negative than the reference voltage, the output voltage slews to positive saturation.

Hysteresis can be placed on the voltage comparator to prevent "chatter" at the crossing point. The circuit for the in-phase voltage comparator is given in Fig. 7-11c. The *switch* voltage or center point between the two hysteresis voltages (above and below) is given by

$$v_{SW} = \left(\frac{R_F + R_I}{R_F}\right)V_R \tag{7-10}$$

while the hysteresis voltage (v_{CH}) band on each side of the switch voltage (v_{SW}) is

Comparators   **189**

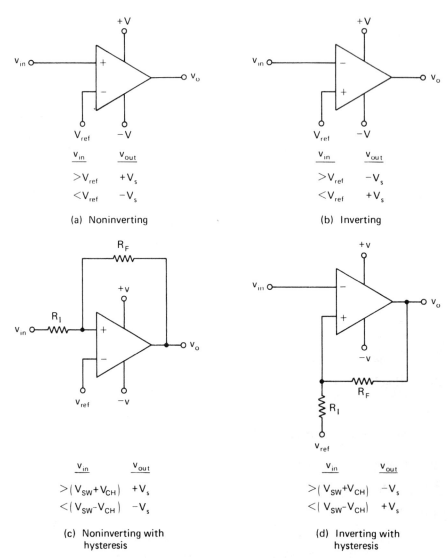

Fig. 7-11 Voltage comparator.

defined by

$$v_{CH} = \pm \left(\frac{R_I}{R_F} \right) V_S \qquad (7\text{-}11)$$

An example will help to illustrate the use of these equations.

EXAMPLE 7-9

A voltage comparator uses an op amp with ± 15 V supplies. The comparator is in phase and should switch at $+5$ V $\pm$ 50 mV; $R_I = 1$ kΩ. Determine the value of R_F and the reference voltage.

Solution

Using Eq. (7-11) first, we have

$$5 \text{ mV} = \left(\frac{1 \text{ k}\Omega}{R_F}\right)(15 \text{ V}) \qquad R_F = 300 \text{ k}\Omega$$

Then from Eq. (7-10),

$$V_R = \left(\frac{300 \text{ k}\Omega}{301 \text{ k}\Omega}\right)(5 \text{ V}) = 4.98 \text{ V}$$

The circuit for the inverting voltage comparator is given in Fig. 7-11d. The switch voltage or center point between the two hysteresis voltages (above and below) is given by

$$v_{\text{SW}} = \left(\frac{R_F}{R_F + R_I}\right)V_R \qquad (7\text{-}12)$$

and the hysteresis band on each side of the switch voltage is defined by

$$v_{\text{CH}} = \pm\left(\frac{R_I}{R_F + R_I}\right)V_S \qquad (7\text{-}13)$$

EXAMPLE 7-10

An inverting voltage comparator uses an op amp with saturation voltages of ± 12 V, $R_I = 2$ kΩ, and $R_F = 390$ kΩ. The reference voltage is -3 V. Determine the two voltages at which the comparator will switch.

Solution

Using Eq. (7-12), we obtain

$$v_{\text{SW}} = \left(\frac{390 \text{ k}\Omega}{392 \text{ k}\Omega}\right)(-3 \text{ V}) = -2.985 \text{ V}$$

and from Eq. (7-13),

$$v_{\text{CH}} = \pm\left(\frac{2 \text{ k}\Omega}{392 \text{ k}\Omega}\right)(12 \text{ V}) = \pm 61 \text{ mV}$$

Thus the two voltages are -2.985 V $+ 0.061$ V $= \underline{-2.924 \text{ V}}$ and -2.985 V $- 0.061$ V $= -3.046$ V.

7.6 Schmitt Trigger

The Schmitt trigger circuit, implemented using op amps, is the same circuit as the inverting zero-crossing detector with hysteresis or the inverting voltage comparator with hysteresis. The major difference between the Schmitt trigger and the two comparator circuits is in the ratio of resistor values used. The two resistors (R_I and R_F) have more nearly the same value in a Schmitt trigger, where in the comparators the ratio between the two is quite large. The circuit for an inverting Schmitt trigger is illustrated in Fig. 7-12a.

The inverting Schmitt trigger has two "threshold" voltages: one that is more positive and one that is more negative. The more positive one is labeled V_{tH} and the more negative one is labeled V_{tL}; they may both be negative, both positive, or of

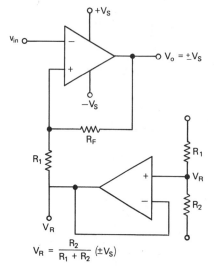

$$V_R = \frac{R_2}{R_1 + R_2}(\pm V_S)$$

(a) Inverting circuit diagram

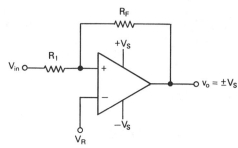

(d) Noninverting circuit diagram

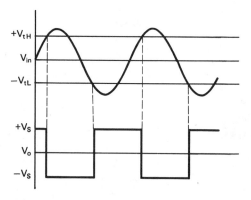

(b) Output vs. input for $V_R = 0$

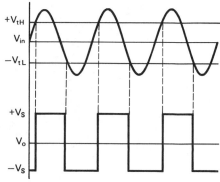

(e) Output vs. input for $V_R = 0$

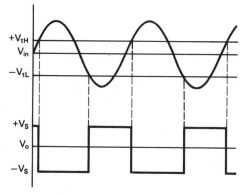

(c) Output vs. input for $V_R \neq 0$ V

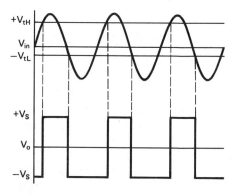

(f) Output vs. input for $V_R \neq 0$

Fig. 7-12 Output versus input voltages for Schmitt trigger.

opposite signs. The relationship for V_t is given by

$$V_t = \left(\frac{R_F}{R_F + R_I} \right) V_R \pm \left(\frac{R_I}{R_F + R_I} \right) V_S \qquad (7\text{-}14)$$

where V_{tH} is given by the *sum* and V_{tL} is given by the *difference*. In the most usual circumstance, V_S is the saturation voltage of the op amp or comparator and is a function of the power supplies, which are usually a specific voltage. The *spread* between the two threshold voltages is given by

$$|V_{tH} - V_{tL}| = 2 \left(\frac{R_I}{R_F + R_I} \right) |V_S| \qquad (7\text{-}15)$$

where V_{tH} and V_{tL} are usually of opposite signs. The reference voltage used in the circuit establishes the average *position* of the two threshold voltages (see Fig. 7-12b and c); the reference voltage is given by

$$V_R = \left(\frac{V_{tH} + V_{tL}}{2} \right) \left(\frac{R_F + R_I}{R_F} \right) \qquad (7\text{-}16)$$

Through the use of Eqs. (7-15) and (7-16), an inverting Schmitt trigger can usually be set to trigger at specific voltages. An example will illustrate the use of these equations.

EXAMPLE 7-11

Determine the reference voltage and resistor sizes to cause an inverting Schmitt trigger to switch at $+3$ V and -5 V. The op amp saturates at ± 15 V and $R_I = 10 \text{ k}\Omega$.

Solution

The spread is $+3 - (-5) = 8$ V; using Eq. (7-15) yields

$$8 \text{ V} = 2 \left(\frac{10 \text{ k}\Omega}{R_F + 10 \text{ k}\Omega} \right) (15 \text{ V}) \qquad R_F = 27.5 \text{ k}\Omega$$

Then from Eq. (7-16),

$$V_R = \left[\frac{+3 + (-5)}{2} \right] (1.36) = -1.36 \text{ V}$$

A noninverting Schmitt trigger is shown in Fig. 7-12b and has the following equations, corresponding to (7-14), (7-15), and (7-16):

$$V_t = \left(\frac{R_F + R_I}{R_F} \right) V_R \pm \left(\frac{R_I}{R_F} \right) V_S \qquad (7\text{-}17)$$

for the threshold voltage,

$$|V_{tH} - V_{tL}| = 2 \left(\frac{R_I}{R_F} \right) |V_S| \qquad (7\text{-}18)$$

for the spread, and

$$V_R = \left(\frac{V_{tH} + V_{tL}}{2} \right) \left(\frac{R_F}{R_F + R_I} \right) \qquad (7\text{-}19)$$

for the reference voltage. The equations work just like those for the inverting Schmitt trigger.

An example will help to illustrate the process.

EXAMPLE 7-12

Determine the resistor ratios and reference voltage to cause a noninverting Schmitt trigger to switch at $+6$ and $+2$ V. The circuit uses an op amp with saturation voltages of ± 12 V.

Solution

The spread is $+6 - (+2) = 4$ V; using Eq. (7-18) yields

$$4 \text{ V} = 2\left(\frac{R_I}{R_F}\right)(12) \quad \frac{R_I}{R_F} = \frac{1}{6} \quad R_F = 6R_I$$

Then from Eq. (7-19),

$$V_R = \left(\frac{+6 + 2}{2}\right)\left(\frac{6}{7}\right) = +3.43 \text{ V}$$

When the reference voltage (V_R) is zero in Eq. (7-14), the two threshold voltages are equally spaced above and below the zero input voltage. In this case, the inverting threshold voltage is given by

$$V_{tH} = -V_{tL} = \left(\frac{R_I}{R_F + R_I}\right)V_S \qquad (7\text{-}20)$$

Likewise, Eq. (7-17) is used for the noninverting Schmitt trigger. An example will serve to illustrate the use of Eq. (7-20).

EXAMPLE 7-13

An integrator circuit is being driven by a square-wave input voltage. The output is a triangular waveform of 20 V peak to peak centered around zero volts. The integrator is followed by a Schmitt trigger with a zero reference voltage. If the Schmitt trigger uses an op amp that saturates at ± 15 V, calculate the resistor ratio that will produce threshold voltages of ± 6 V.

Solution

Using Eq. (7-20), we obtain

$$6 \text{ V} = \left(\frac{R_I}{R_F + R_I}\right)(15 \text{ V})$$

$$\frac{R_I}{R_F + R_I} = 0.4 = \frac{1}{1/0.4} = \frac{1}{2.5} = \frac{1}{1.5 + 1}$$

$$R_F = 1.5R_I$$

7.7 Peak Holding Circuit

A circuit that will retain the largest voltage, of one polarity, appearing on the input is called a peak holding circuit. The schematic for a peak holding circuit is shown in Fig. 7-13a. If an electronic switch is used on the input signal, the circuit is then a peak sample-and-hold circuit, as illustrated in Fig. 7-13b. The two circuits operate

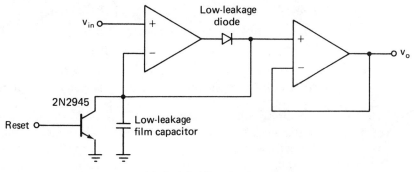

(a) Peak holding circuit

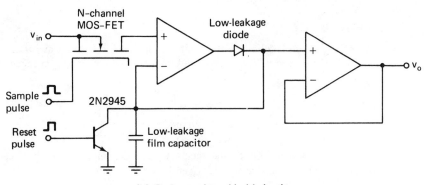

(b) Peak sample and hold circuit

Fig. 7-13 Analog voltage memories.

in an identical fashion when the switch on the sample-and-hold circuit is closed. The circuit is basically a voltage follower with a diode in the feedback loop in place of R_F, and a capacitor in place of R_I. The resulting circuit operates in the following manner:

1. As a positive voltage signal enters the input (noninverting op amp input), the input voltage is greater than the capacitor voltage and the op amp output tries to saturate, forwarding-biasing the diode and charging the capacitor. The capacitor charges to the value of the input voltage.

2. When the capacitor has charged to a value slightly greater than the input voltage, or the input voltage become less than the current capacitor voltage, the op amp output slews to negative saturation, reverse-biasing the diode and leaving the capacitor with a charge equivalent to the highest input voltage encountered.

3. The capacitor then slowly discharges through five paths:

   a. The op amp inverting input.

   b. The reverse-biased diode.

c. The reset transistor.

d. The internal leakage of the capacitor.

e. The noninverting input of the output voltage follower.

4. A positive pulse applied to the reset transistor will dump the charge in the capacitor and the peak holding or sampling is ready to begin again.

This type of circuit exhibits several problems in circuit operation. The charging rate of the capacitor is limited by the amount of current which the op amp, in saturation, can deliver through the diode. This charging rate can be expressed by Eq. (4-1). For a large capacitor, this can be a significant time and limits the ability of the circuit to follow fast-rising input voltages.

An example will help to illustrate this limitation.

EXAMPLE 7-14

A peak holding circuit is following the positive half of a 1 kHz sine wave with a peak voltage of 5 V. The op amp is able to deliver 20 mA to a load; the memory capacitor is 1 μF.

Solution

From Eq. (4-3), rearranged, we obtain

$$\frac{\Delta V}{\Delta t} = \frac{20 \text{ mA}}{1 \text{ } \mu\text{F}} = 0.02 \text{ V}/\mu\text{s}$$

is the maximum rate of rise of the memory capacitor in the peak holding circuit. The maximum slew rate of the sine wave is given by Eq. (5-16).

$$\text{max sin(SR)} = 2\pi(10^{-6})(5 \text{ V})(10^3 \text{ Hz}) = 0.031 \text{ V}/\mu\text{s}$$

Thus the peak holding circuit is unable to follow the maximum rate of the sine wave. It will, however, catch the peak as both the sine wave and the peak holding circuit reach $+5$ V in 250 μs; thus if only the peak voltage, and not an intermediate one is desired, the peak holding memory charges at the required rate.

If the capacitor is made smaller to compensate for this effect, the five leakage currents cause the capacitor to have an excessive decay rate. An example will show this effect.

EXAMPLE 7-15

The peak holding memory of Example 7-14 uses an op amp with bias currents of 50 nA, a diode with a reverse leakage current of 30 nA, and a reset transistor with a leakage current of 20 nA. The self-leakage of the film capacitor is 1 pA. Determine the voltage on the memory capacitor 10 s after is has reached 5 V, but before it is reset.

Solution

The total leakage (discharge) current from the capacitor is

$$50 \text{ nA} + 50 \text{ nA} + 30 \text{ nA} + 20 \text{ nA} + 0.001 \text{ nA} = 150 \text{ nA}$$

Using Eq. (4-1), we obtain

$$\Delta V = \frac{150 \text{ nA}}{1 \text{ } \mu\text{F}}(10 \text{ s}) = 1.5 \text{ V drop in voltage}$$

The capacitor voltage is then 5 V $-$ 1.5 V $=$ 3.5 V. The need for a FET input op amp and low-leakage diode and reset transistor becomes immediately apparent.

The third problem manifests itself just after reset. The capacitor should remain at zero volts following reset; it would be were it not for a phenomenon called "dielectric absorption." This occurs because a film capacitor, after being charged to a large voltage and then discharged, retains a small portion of the charge in the dielectric, which appears as a few millivolts of voltage on the capacitor just after reset. The amount of dielectric absorption depends on the film being used; polycarbonate film has the least amount of dielectric absorption. Dielectric absorption is usually specified as a percentage of the voltage on the capacitor before reset. A double reset tends to reduce this problem.

PROBLEMS

1. A 35 mV peak ac signal voltage with a 0 V dc average is to be half-wave rectified and amplified to a peak of 570 mV. The output peaks must be positive. The input resistor is 15 kΩ.

   a. Determine the size of R_f and R_c. Select values from Appendix B1.

   b. Draw the circuit diagram.

2. The input signal to a half-wave precision rectifier is
$$V_{in} = 0.3 + 0.02 \sin(100\pi t)$$
   The gain through the rectifier is -5, and the output peaks must be positive. Draw the waveforms of the input and output voltages to scale.

3. A full-wave precision rectifier of the type shown in Fig. 7-2a is used to rectify a triangular waveform with a 10 V peak amplitude. The output waveform should have -20 V peaks.

   a. If $R = 37.4$ kΩ, 1%, select values for R_I and R_c.

   b. Draw the circuit diagram.

   c. Show the output waveform to scale.

4. A full-wave precision rectifier of the type shown in Fig. 7-2b is used to rectify a 2 V peak sine wave. The output should be 2 V peak; $R = 7.5$ kΩ.

   a. Select resistor values for circuit 1.

   b. Select resistor values for circuit 2.

5. A full-wave precision rectifier uses op amps with a slew rate of 0.8 V/μs. The input waveform has a 15 kHz square wave with a 10 V peak amplitude. If the rectifier has unity gain with positive output peaks:

   a. Draw the output waveform to scale.

   b. Show on your drawing where the feedback loop is open and closed.

6. A full-wave precision rectifier of the type shown in Fig. 7-2b (circuit 1) has unity gain. A sine wave with dc offset having the equation $V_{in} = 0.75 + 2.5 \sin(10\pi t)$ appears on the input. The output peaks are positive. Draw all intermediate waveforms and the output waveform to scale.

7. A precision rectifier of the type shown in Fig. 7-2b, circuit 1, has $R = 12$ kΩ and $R_f = 24$ kΩ. R_2 should be 12 kΩ, but it has been chosen in error to be 11 kΩ. The

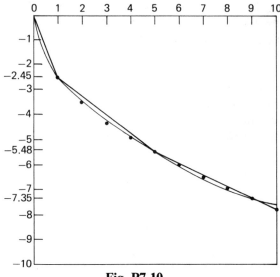

Fig. P7-10

output peaks are positive if $V_{in} = 5 \sin(20\pi t)$ V. Draw all intermediate and the output waveforms to scale.

8. The output offset of A_1 in Fig. 7-2b is $+25$ mV, and A_2 has no offset. Circuit 2 is used where $R = 15$ kΩ. If $V_{in} = 0.25 \sin(200\pi t)$ V, draw all significant waveforms to scale.

9. Repeat Problem 8 for the circuit in Fig. 7-2a where $R = 7.5$ kΩ.

10. The curve shown in Fig. P7-10 is a plot of $V_o = -\sqrt{(6V_{in})}$. The circuit is shown in Fig. 7-3a. Calculate the values of all circuit components if $R_I = 5.49$ kΩ, 1%.

11. A curve shaper must simulate the equation $V_o = 1/5(V_{in})^2$.

   a. Build a table of V_o vs. V_{in} in steps of 1 V from 0 to 10 V.

   b. Plot the curve on a grid.

   c. Simulate the curve with two breakpoints.

   d. Select a gain for the lowest region and calculate a voltage-divider ratio.

   e. Calculate the circuit values, and select resistors from Appendix B1.

12. Develop the circuit for a log converter that will convert the input voltage range of 150 mV to 15 V to an output voltage range of $+2$ to $+5$ V. Calculate resistor values for all circuits, and select values from Appendix B1 (use Fig. 7-6e); $R_F = 30.1$ k, $V_s = \pm 12$ V.

13. The output voltage from a FET sine shaper is a 25 mV peak with an offset of $+2$ mV. Select values for a correction circuit that will provide a 2 V peak signal with a ± 0.5 V offset capability. The supply voltages are ± 5 V; $R_F = 40.2$ k 1%.

14. Select component values for a single-polarity voltage clipper that has a gain of -2 and clips a 5.2 V peak input sine-wave voltage equivalent to a 2.05 V input voltage. $R_F = 2$ k.

15. A dual-polarity voltage clipper using two 8.2 V zener diodes must clip a sine wave between the intervals of 30 to 150° and 210 to 330°. If the forward drop across each

Nonlinear Applications

zener diode is 0.7 V:

   a. Find the peak value of the unclipped sine wave.
   b. Find R_F if $R_I = 2\ k\Omega$ and the input sine wave is 500 mV peak.
   c. Indicate by degrees of the sine-wave cycle where the output is undistorted.

16. A noninverting zero-crossing detector uses a μA 741 op amp with an open loop gain of 92 dB. If the power supply voltages are ± 15 V and the input offset is 0 V:

   a. Find the input voltage range that causes the output voltage to just slew between the power-supply rails.
   b. If a 1 mV sine wave is on the input, draw the output waveform in time synchronism with the input waveform.

17. An inverting zero-crossing detector uses ± 10 V power supplies. $R_I = 2.2\ k\Omega$, and the input voltage is a 4.5 V peak triangular waveform with 250 mV p-p riding on it. Find the value of R_F to prevent chatter on the detector output.

18. A noninverting zero-crossing detector has $R_I = 5\ k\Omega$ and $R_F = 500\ k\Omega$. If the supplies voltages are ± 10 V, find the amount of peak noise before output chatter appears on the input signal.

19. A voltage comparator uses an LM 208 op amp as an active element. If the noninverting input is connected to -5.4 V, draw the output waveform in time synchronism with a 10 V peak input sine wave. The voltage supplies are ± 12 V.

20. An in-phase voltage comparator uses ± 10 V supplies and $R_I = 2.2\ k\Omega$. The comparator must be able to tolerate a 52 mV peak noise without chattering, and switch at an average value of -5 V. Find the values of R_F and V_R.

21. An inverting voltage comparator has ± 12 V supply voltages; $R_I = 3.3\ k\Omega$ and $R_F = 300$ $k\Omega$. Find the two voltages at which the comparator switches for a reference voltage of -2.2 V.

22. A Schmitt trigger must switch at $+4$ V and -5 V. The op amp output saturates at ± 15 V. Find R_F and V_R for an $R_i = 2\ k\Omega$.

23. A Schmitt trigger must switch at $+7$ V and -1 V. If the supply voltages are ± 12 V, find the ratio of R_F to R_I and the reference voltage.

24. A Schmitt trigger must switch at $+8$ V and $+2$ V. The supply voltages are at ± 20 V. Find V_R and R_F for an $R_I = 4.7\ k\Omega$. Select values from Appendix B1.

25. A Schmitt trigger must switch at -6 V and -1.5 V. The power supplies are at ± 10 V. Find the ratio of R_F to R_I, and V_R. Select values for R_F and R_I from Appendix B1.

26. An integrator circuit is being driven by a square-wave input voltage. The output triangular waveform has a ± 3 V peak. A Schmitt trigger that must switch at ± 1 V follows the integrator. If both circuits operate on ± 5 V power supplies.:

   a. Calculate the ratio of R_F to R_I.
   b. For $R_I = 15\ k\Omega$, select R_F from Appendix B1.

27. A peak memory circuit with a $0.47\ \mu$F memory capacitor uses an op amp that can deliver 45 mA. Determine the time for the memory to charge the capacitor to $+15$ V if a $+15$ V step voltage suddenly appears in the input.

28. The peak-holding circuit of Problem 27 has a 4 V peak sine wave on the input. Determine the highest attainable frequency.

29. A peak-sample-and-hold circuit uses the op amp and memory capacitor of Problem 27. The input is gated such that the positive-half cycle of a 4 V peak sine wave is permitted to enter. Determine the highest frequency of the sine wave if the memory capacitor must charge at the maximum slew rate of the sine wave.

30. A peak-holding memory uses a FET input op amp for both the peak holding and voltage follower circuits. The memory capacitor is 2.2 μF and it has been charged to 12 V. The component current leakages are as follows:

$$\text{Op amps:} \qquad I_B = 30 \text{ pA}$$
$$\text{Diode:} \qquad I_R = 10 \text{ pA}$$
$$\text{Reset transistor:} \quad I_{co} = 5 \text{ pA}$$

Find the length of time it takes the memory to discharge to 11.990 V.

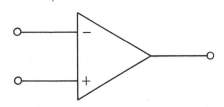

8 Digital Applications

8.1 Introduction

As was mentioned earlier, few systems are purely analog or digital. There is a marriage of the two technologies, each being used where it performs most effectively. This chapter is an example of the combined use of analog and digital devices.

8.2 Interfacing Analog and Digital Devices

In every section in this chapter, an analog device (op amp or comparator) is required to drive a digital logic gate, or a digital logic gate must drive an analog device or resistor network. Both analog and digital devices have input and output limitations in the maximum voltages and currents that the device can tolerate or deliver. These limitations vary, to some degree, with the manufacturer of the device. Even though several devices carry equivalent part numbers, their internal structure can be different and have their own peculiar limitations. At the time of the writing of this work, TTL devices are well established and COS/MOS devices are gaining an ever-increasing demand in the industry. The COS/MOS devices do not switch as fast as TTL devices; thus high-speed systems such as minicomputers are not employing COS/MOS to any large degree. But in systems where high speed is not the prime consideration, COS/MOS devices are very desirable because of their broader power-supply voltage range and their low-power consumption. The term "gate" used in connection with COS/MOS may lead one to believe that registers and other devices operate differently than gates until one remembers that these more complex devices are composed of gates in their internal structure.

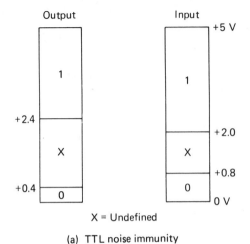

X = Undefined

(a) TTL noise immunity

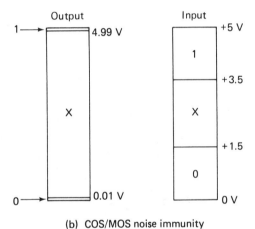

(b) COS/MOS noise immunity

Fig. 8-1 Gate input- and output-voltage levels.

Figure 8-1a illustrates the input and output voltage levels for the TTL gate. In Fig. 8-1a, the maximum LO output voltage is 0.4 V and the maximum LO input voltage is 0.8 V. The minimum HI output voltage is 2.4 V, and the minimum HI input voltage is 2 V.

Figure 8-1b illustrates the input and output voltage levels for the COS/MOS gate. The minimum HI output voltage is 4.99 V (5 V supply voltage) and the minimum HI input voltage is 3.5 V. Similarly, the maximum LO output voltage is 0.01 V, the maximum LO input voltage is 1.5 V (5 V supply voltage).

The real problem in interfacing analog devices to digital devices manifests itself in that the inputs to digital devices are almost always driven to the point where the input voltage limitations of the gates are exceeded. This would normally destroy or permanently alter the performance of the digital device. Thus methods to prevent this from occurring must be established.

A second problem comes in interfacing a digital device to an analog device or resistor network with its output voltage and current limitations. These limitations will be discussed separately for TTL and COS/MOS devices.

8.2.1 Interfacing of Analog and TTL Devices

The input to a TTL gate comprises both voltage and current limitations. Figure 8-2a illustrates the input circuit of a bipolar TTL gate; gate types are divided into bipolar and the integrated Schottky–barrier diode-clamped transistor.* The two gate types

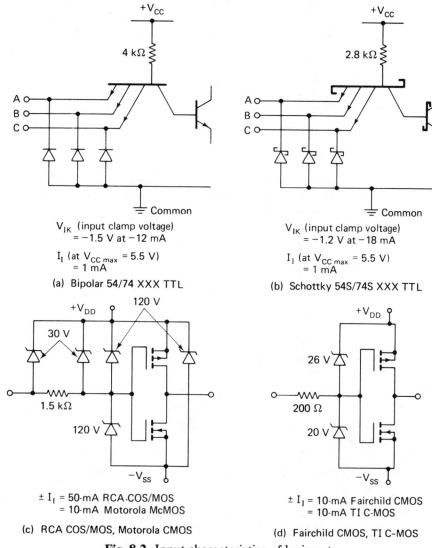

Fig. 8-2 Input characteristics of logic gates.

*U.S. Patent 3,363,975, Texas Instruments, Inc., Dallas, Texas.

have slightly different input characteristics. Figure 8-2b illustrates the "Schottky" type of input circuit. The input specification in Fig. 8-2a states that the negative clamp voltage is -1.5 V at -12 mA for the 54/74 family and it is -1.2 V at -18 mA for the 54S/74S family of TTL devices. This means that the input current must be limited to 12 mA when the input leads are driven negative.

I_I is the maximum input current when the input leads are driven more positive than the positive supply voltage. 1 mA is the limiting current, as it is the most current that can be tolerated for either input voltage polarity when the input is being overdriven.

If an op amp or comparator is driving a TTL gate, the input current to the TTL device must be limited to 1 mA through the use of a series-limiting resistor connected to the gate input lead. The minimum size of the limiting resistor is given by

$$R_{min} = \frac{+V_S - 5.5 \text{ V}}{1 \text{ mA*}} \tag{8-1}$$

where $+V_S$ is the positive saturation voltage of the op amp or comparator.

An example will help to illustrate the use of Eq. (8-1).

EXAMPLE 8-1

A comparator with supply voltages of ± 12 V is driving one input lead of the AND gate on an analog to digital converter. Determine the minimum value of the series-limiting resistor between the comparator and the TTL gate.

Solution

$$R_{min} = \frac{+12 \text{ V} - 5.5 \text{ V}}{1 \text{ mA}} \cong 6.8 \text{ k}\Omega$$

Use 10 kΩ, as it causes slightly less input current to flow; 6.8 kΩ would be very adequate. The current in the negative direction is then

$$-I = \frac{-12 \text{ V} - (-1.5 \text{ V})}{6.8 \text{ k}\Omega^\dagger} = -1.5 \text{ mA}$$

which is well under the -12 mA maximum.

The minimum HI output voltage of a totem pole TTL is 2.4 V, while the maximum LO output voltage is 0.4 V. This is intolerable if the register or gate is driving a digital-to-analog converter (DAC) resistor network. The solution to this dilemma is found in Section 8.2.2.

8.2.2 Interfacing of Analog Devices and COS / MOS Gates

The COS/MOS gate has both better input and output characteristics than TTL gates. Each input lead is protected with a diode-resistor network shown in Fig. 8-2c and d. The diode saturation voltages are 0.9 V and various manufacturer's gates can

---

*For 54L/74L the maximum current is 100 μA.
†For a 54L/74L family device, use 68 kΩ.

sink different input currents; the RCA COS/MOS gate will sink 50 mA and the Fairchild and Motorola CMOS gates will sink 10 mA. To form a generalization on COS/MOS gates, it would be safe to limit the input current to 10 mA for any COS/MOS gate used. Thus the equation for the minimum value of series-limiting resistor between an op amp or comparator and a COS/MOS gate is

$$R_{min} = \frac{V_{SS} - (-V_S) - 0.9 \text{ V}}{10 \text{ mA}} \tag{8-2}$$

where V_{SS} is the negative supply voltage on the COS/MOS device (normally zero volts) and $-V_S$ is the negative saturation voltage of the op amp or comparator. The negative voltage is chosen to determine the resistor value, as it produces the largest-value series resistor.

An example will help to illustrate the use of Eq. (8-2).

EXAMPLE 8-2

A COS/MOS AND gate with ± 5 V supplies is being driven by a comparator with ± 15 V supplies. Determine the minimum value of series-limiting resistor to use in the circuit.

Solution

$$R_{min} = \frac{-5 \text{ V} - (-15 \text{ V}) - 0.9 \text{ V}}{10 \text{ mA}} = 910 \ \Omega$$

Use a 1 kΩ resistor; 10 kΩ would be better as there is no need to draw 10 mA from the comparator.

The output of a COS/MOS gate is always within 10 mV of one of the two supply voltages; this is true regardless of the value of the supply voltages. Thus a COS/MOS gate is usually chosen to drive the resistors of a ladder network. The COS/MOS gate is also the "switch" that is used on the output of a TTL device to eliminate its undesirable characteristics.

8.3   Digital-to-Analog Converter

The digital-to-analog (D/A) converter (DAC) is discussed as the first circuit in this topic, as it forms the basis for many other circuits. A DAC is formed by driving an op amp circuit with a digital counter register. The output voltage of the analog circuit is proportional of the value of the binary number in the "ripple counter." Each segment (flip-flop) of the binary counter is either SET (HI = 1) or RESET (LO = 0). These segment outputs are connected to a variable-gain op amp circuit with "weighted" (different-value) gain resistors. As the binary number in the register increases, the input voltage to each segment resistor goes HI (1) or LO (0) and the circuit output voltage changes accordingly.

8.3.1   Inverting Summer DAC

The most easily understood DAC is formed by connecting the input resistors of an inverting summer circuit to the counter segment outputs. The smallest-value input resistor (largest gain) is connected to the most significant bit (MSB), and the largest-value input resistor (smallest gain) is connected to the least significant bit (LSB) in the register. The size of the input resistors begin with the smallest value and double in value for each next-larger resistor; this doubling in value is the "weighting" previously described. The feedback resistor of the inverting summer is chosen to yield the desired output voltage when the register is at "full count" (all 1's). The circuit for this DAC is illustrated in Fig. 8-3a. This inverting summer DAC

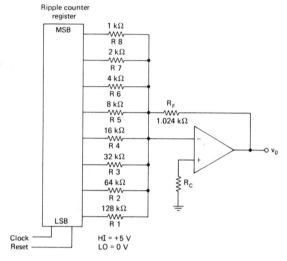

(a) Inverting summer DAC circuit

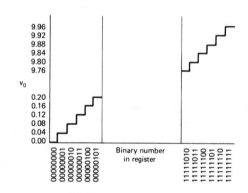

(b) DAC output voltage vs. register contents

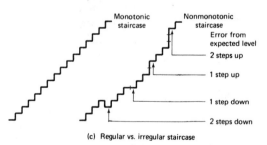

(c) Regular vs. irregular staircase

Fig. 8-3 Inverting summer DAC.

has several problems which cause it to be the least desirable of the available circuits:

1. If a TTL* register is used, each segment must be followed by some form of electronic switch which exhibits an extremely low ON impedance and a virtually infinite OFF impedance. This characteristic is usually obtained by using some form of MOS-FET switch. COS/MOS registers can be used directly, without any switch, as the HI output voltage is within 10 mV of the supply voltage and the LO output voltage is within 10 mV of zero volts.

2. The input resistors must be of high precision or the staircase voltage (Fig. 8-3b) will not increase "monotonically." A monotonically changing staircase voltage rises or falls without interruption, and all steps have an equal change in voltage. If the smallest input resistor (MSB) has a tolerance error equal to the percentage change of one step, the current it produces will cause an output voltage that will either add to or subtract from the expected LSB voltage-step change. This will cause the staircase voltage to be irregular as illustrated in Fig. 8-3c. It is then not monotonic but produces an error voltage on the output. This is intolerable in a DAC, as several digital numbers produce the same output voltage. To yield some idea of the problem, consider the following:

   a. A 6-bit DAC has 64 voltage levels. Each voltage level is $\frac{1}{64}$, or 1.5% of the total range; the most-significant-bit resistor must have a tolerance error better than $\pm 0.5\%$ to be barely acceptable.

   b. A 7-bit DAC has 0.78% steps and must use $\pm 0.2\%$ resistors.

   c. An 8-bit DAC has 0.4% steps and must use $\pm 0.1\%$ resistors.

   When building a 10 or 12-bit DAC, careful judgment must be used in the selection of resistors. Several manufacturers sell monolithic resistor networks which are laser-trimmed to within $\pm 0.01\%$ of the desired values. Fig. 8-4 illustrates several networks for this purpose.

3. If the feedback resistor (R_F) is large in comparison to R_8 (MSB), the ratio of the parallel combination of resistors on the input to the feedback resistor of the inverting summer produces the phenomenon alluded to in Section 2.7.2, where the excess loop gain is reduced. The parallel combination of input summing resistors is never less than half the value of R_8. The decreased excess loop gain may cause the op amp of the DAC to exhibit a reduced slew rate, described in Section 5.3.2. Digital-to-analog converters must usually operate at very high speeds, and this is inconsistent with a reduced slew rate; thus the value of R_F should be maintained in the same order of magnitude as the value of R_8 (MSB resistor).

The *hypothetical*[†] output voltage of a DAC at 2^N counts is V_N, where N is the number of segments in the register. The least-significant-bit voltage is given by

$$V_{\text{LSB}} = \frac{V_N}{2^N} \qquad (8\text{-}3)$$

*Transistor–transistor logic.
[†] The largest output voltage from a DAC occurs at $2^N - 1$ counts.

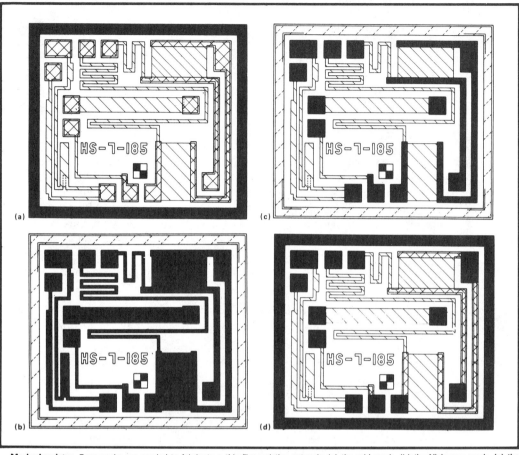

Masked resistor. Four masks are needed to fabricate a thin-film resistive network: (a) the grid mask, (b) the Nichrome mask, (c) the aluminum mask, and (d) a glass passivation mask. All are the size of the resistor and are stepped and repeated across the wafer.

(a)

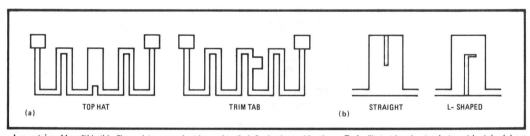

Laser trim. Monolithic thin-film resistors may be trimmed to their final values with a laser. To facilitate trimming top hats or trim tabs (a) are added to the resistor pattern. The actual laser trim may be a straight or L-shaped cut (b), but usually the latter has much greater effect.

(b)

Fig. 8-4 $R/2R$ resistor network for DAC: (a) mask set for resistors; (b) laser trimming of networks. (Courtesy McGraw-Hill Publishing Company from article in *Electronics Magazine* by Hybrid Systems Corporation, Bedford, Mass., August 3, 1978.)

SPECIFICATIONS

Ladder Resistance (R) 5K, 10K, 25K, 50K

Temperature Coefficient .... − 75 to − 125 ppm/°C

T.C. of Conversion Accuracy .... ± ¼ LSB

Setting Time .... TO 20 ns

Noise .... < − 25dB

Temperature Range .... − 55°C to + 125°C

Accuracy .... ± ½ Least Significant Bit

Switch Compensation up to − 5%

Fig. 8-4 (cont.) (c) (Courtesy TRW/IRC Network Operation, Burlington, Iowa.)

where V_{LSB} is the amount of output voltage change for each step. All digital-to-analog converters have one characteristic in common; the output voltage at full count is one step less than V_N; Thus the output voltage at full count is represented by V_{FC} and is given by the equation

$$V_{FC} = V_{LSB}(2^N - 1) \tag{8-4}$$

or

$$V_{FC} = V_N \left(1 - \frac{1}{2^N}\right) \tag{8-5}$$

An example will help to place these various quantities in proper perspective.

EXAMPLE 8-3

A 4-bit DAC has a hypothetical output voltage of 8.0 V at $2^4 = _{16 \text{ counts}}$ (step 17).

   a. Determine V_{LSB} and V_{FC}.

   b. Show the output voltage at each step.

Digital-to-Analog Converter

Solution

a. $V_{LSB} = \dfrac{8.0\ V}{16} = 0.5\ V$ per step

b. $V_{FC} = 0.5\ V(16 - 1) = 7.5\ V$

Step	Count	Output Voltage	Step	Count	Output Voltage
1	0000	0.0 V	9	1000	4.0 V
2	0001	0.5 V	10	1001	4.5 V
3	0010	1.0 V	11	1010	5.0 V
4	0011	1.5 V	12	1011	5.5 V
5	0100	2.0 V	13	1100	6.0 V
6	0101	2.5 V	14	1101	6.5 V
7	0110	3.0 V	15	1110	7.0 V
8	0111	3.5 V	16	1111	7.5 V

The development of a DAC follows a logical process, employing the equations and techniques just presented. The output voltage, produced by the effect of the most-significant-bit (MSB) input resistor (R_8), is given by

$$V_{MSB} = \frac{V_N}{2} \qquad (8\text{-}6)$$

where the MSB input voltage is applied only to R_8; all other input voltages are zero. Now, with the help of these equations, we are prepared to develop a DAC.

An example will help to illustrate the process.

EXAMPLE 8-4

An 8-bit DAC should have an output voltage (V_N) of -10.000 V at 256 counts. $R_F = 1\ k\Omega$ and the HI input voltage to each input resistor is $+5.000$ V.

a. Determine the output voltage at full count.

b. Determine the ideal and actual 1% value of each resistor in the circuit.

c. Determine the actual output voltage at count 10110011.

Solution

Refer to Fig. 8-3a

$$V_{MSB} = \frac{-10.00}{2} = -5.00\ V$$

$$-5.00\ V = +5\ V\left(-\frac{1\ k\Omega}{R_8}\right) \qquad R_8 = 1\ k\Omega$$

$$V_{LSB} = \frac{-10.00\ V}{256} = -39.063\ mV$$

a. $V_{FC} = 39.063$ mV $(256 - 1) = 9.961$ V

b. DAC resistor values:

Resistor	Ideal	Actual
R_8	1.00 kΩ	1.00 kΩ
R_7	2.00 kΩ	2.00 kΩ
R_6	4.00 kΩ	4.02 kΩ
R_5	8.00 kΩ	8.06 kΩ
R_4	16.0 kΩ	15.8 kΩ
R_3	32.0 kΩ	32.4 kΩ
R_2	64.0 kΩ	63.4 kΩ
R_1	128 kΩ	127 kΩ

c. 1 0 1 1 0 0 1 1
   MSB   LSB

$$+5V\left(-\frac{1\text{ k}\Omega}{1\text{ k}\Omega}\right) = -5.000 \text{ V}$$

$$+5\text{ V}\left(-\frac{1\text{ k}\Omega}{4.02\text{ k}\Omega}\right) = -1.244 \text{ V}$$

$$+5\text{ V}\left(-\frac{1\text{ k}\Omega}{8.06\text{ k}\Omega}\right) = -0.620 \text{ V}$$

$$+5\text{ V}\left(\frac{1\text{ k}\Omega}{63.4\text{ k}\Omega}\right) = -0.079 \text{ V}$$

$$+5\text{ V}\left(-\frac{1\text{ k}\Omega}{127\text{ k}\Omega}\right) = -0.039$$

$$V_O = -(5.000 + 1.244 + 0.620 + 0.079 + 0.039)$$

$$= -6.982 \text{ V actual}$$

$$V_O(\text{ideal}) = -6.992 \text{ V for } 0.14\% \text{ error}$$

EXAMPLE 8-5

a. How much error in output voltage will occur if the MSB resistor (1 kΩ) is 1% high in value for part (c) of Example 8-4?

b. Determine the output voltage at 01111111 and at 10000000.

Solution

a. $+5$ V$(-1$ kΩ$/1.01$ kΩ$) = -4.95$ V or 50 mV below the value in part (c) of Example 8-4. The output voltage will be 50 mV low or $-6.982 - (-0.05$ V$) = -6.932$ V. If each step is 30 mV, this will appear as count 10110010 in voltage and is thus one step lower than the count indicates.

b. The output voltage at 01111111 is $-(2.5 + 1.244 + 0.620 + 0.316 + 0.154 + 0.079 + 0.039) = -4.952$ V. The output voltage at 10000000 is -4.950 V (from Example 8-4). Thus both the 01111111 and 10000000 steps have essentially the same output voltage because the staircase is not monotonic.

8.3.2 *R/2R* Ladder DAC

The $R/2R$ ladder DAC is formed by connecting the bit outputs (or switches) of the register to the inputs of a resistive ladder network, as illustrated in Fig. 8-5. Thus if an inverting amplifier is used, the $2R$ termination resistor is ungrounded and used as R_I for the inverting amplifier as shown in Fig. 8-5b. In either case, the ladder appears to be terminated in $2R$ and the output voltage is a function of only the ladder resistors. This ladder circuit is probably most widely used in DACs for the following reasons:

1. Only two values of resistors are used; R and $2R$.
2. The actual value used for R is relatively unimportant as long as extremely large values, where stray capacitances enter the picture, are not employed; only the *ratio* of resistor values is critical.
3. $R/2R$ ladder networks are available in monolithic chips, as indicated in Fig. 8-4. These are laser-trimmed to be within 0.01% of the desired ratios.
4. The staircase voltage is more likely to be monotonic as the effect of the most-significant-bit resistor is not many times greater than that for the least-significant-bit resistor. The error in the staircase will be just over one-half step for 13-bit DAC with 0.01% resistors, and a 12-bit DAC has the first nearly monotonic staircase (one-third step error).
5. If the closed loop gain of the output amplifier is less than 10, the slew rate is usually the maximum possible and is the rate indicated in the "voltage-follower pulse response" for the op amp.

Figure 8-6a illustrates a 4-bit $R/2R$ ladder network. By beginning at the bottom (LSB) end of the ladder network, it can be seen that the last $2R$ resistor and the $2R$ resistor to V_1 are in parallel and form an equivalent value of R; this R is in series with the R between the V_1 and V_2 sections and forms an equivalent value of $2R$ to common. Thus each section of the ladder has the same load impedance.

The impedance to the next higher section is also $2R$. Thus each register output voltage (V_n) is loaded by a voltage divider composed of $2R$ and R to common. The voltage appearing at the division point (V_L) is *always* $\frac{1}{3}V_n$ and is half that value at the adjacent section division point in either direction. We are only concerned with the direction toward the ladder output. As each section is passed, the effective voltage due to a particular V_n voltage diminishes by half its value at the previous section. This effect can be observed by following this phenomenon in Fig. 8-6. Thus the $R/2R$ ladder output voltage is dependent upon the distance, in sections, away from the V_n voltage and is given by

$$v_L = \left(\frac{1}{2}\right)^{(N-n)}\left(\frac{1}{3}\right)V_n \qquad (8\text{-}7)$$

(For the noninverting amplifier, see Fig. 8-5a.) where N is the number of bits in the DAC and n is the segment number.

An example will help to illustrate the use of Eq. (8-7).

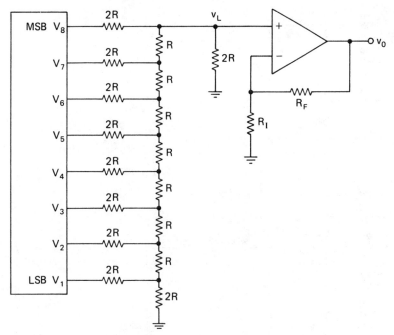

(a) R/2R Ladder DAC using noninverting amplifier

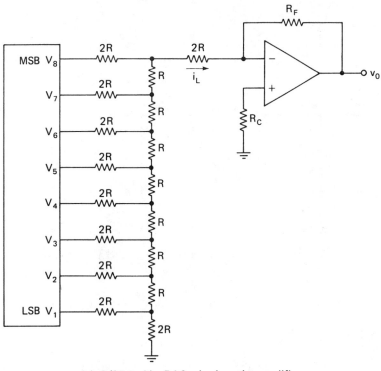

(b) R/2R Ladder DAC using inverting amplifier

Fig. 8-5 $R/2R$ ladder DAC.

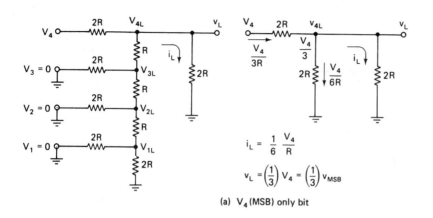

(a) V_4 (MSB) only bit

$$i_L = \frac{1}{6} \frac{V_4}{R}$$

$$v_L = \left(\frac{1}{3}\right) V_4 = \left(\frac{1}{3}\right) v_{MSB}$$

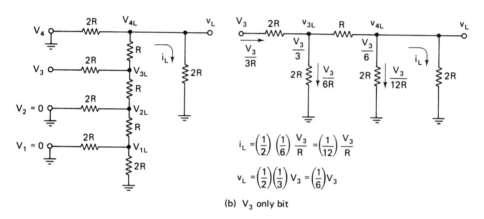

(b) V_3 only bit

$$i_L = \left(\frac{1}{2}\right)\left(\frac{1}{6}\right) \frac{V_3}{R} = \left(\frac{1}{12}\right) \frac{V_3}{R}$$

$$v_L = \left(\frac{1}{2}\right)\left(\frac{1}{3}\right) V_3 = \left(\frac{1}{6}\right) V_3$$

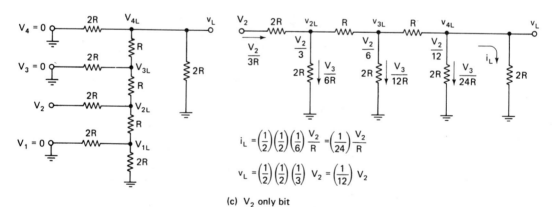

(c) V_2 only bit

$$i_L = \left(\frac{1}{2}\right)\left(\frac{1}{2}\right)\left(\frac{1}{6}\right) \frac{V_2}{R} = \left(\frac{1}{24}\right) \frac{V_2}{R}$$

$$v_L = \left(\frac{1}{2}\right)\left(\frac{1}{2}\right)\left(\frac{1}{3}\right) V_2 = \left(\frac{1}{12}\right) V_2$$

Fig. 8-6 Voltage and currents in 4-bit $R/2R$ ladder network.

EXAMPLE 8-6

a. Verify that the output voltage in Fig. 8-6c is correct.

b. Determine the output voltage v_L when $V_2 = +6$ V.

Solution

Using Eq. (8-7), we obtain

a. $v_L = (\frac{1}{2})^{(4-2)}(\frac{1}{3})V_1 = (\frac{1}{2})^2(\frac{1}{3})V_1 = \frac{1}{12}V_2$

b. $v_L = \frac{1}{12}(6 \text{ V}) = 0.5 \text{ V}$

The current in any $2R$ segment resistor is $(\frac{1}{3})(V_n/R)$ due to *its* V_n voltage only. As can be seen from Fig. 8-6, the current in each R resistor, adjacent to the segment producing the current, is one-half the value in the $2R$ segment resistor. The current in the $2R$ terminating resistor is no exception; this current is again diminished by half that in the previous segment. Thus the current in the $2R$ terminating resistor can be expressed by the equation

$$i_L = \left(\frac{1}{2}\right)^{(N-n)}\left(\frac{1}{6}\right)\frac{V_n}{R} \tag{8-8}$$

(for the inverting amplifier, see Fig. 8-5b) where N is the number of bits in the DAC and n is the segment number.

An example will help to illustrate the use of Eq. (8-8).

EXAMPLE 8-7

a. Verify that the output current equation indicated in Fig. 8-6b is correct.

b. If $V_3 = +6$ V and $R = 10$ kΩ, find the value of i_L.

Solution

Using Eq. (8-8), we obtain

a. $i_L = \left(\frac{1}{2}\right)^{(4-3)}\left(\frac{1}{6}\right)\frac{V_3}{R} = \left(\frac{1}{2}\right)^1\left(\frac{1}{6}\right)\frac{V_3}{R} = \frac{1}{12}\frac{V_3}{R}$

b. $i_L = \left(\frac{1}{12}\right)\frac{6 \text{ V}}{10 \text{ k}\Omega} = 50 \text{ }\mu\text{A}$

The combination of the effects due to various segment voltages and currents will *add* algebraically on the output of the DAC.

For the case where a noninverting amplifier is used on the ladder output, the output voltage is given by **Rule 3** as

$$v_O = \left(\frac{R_F + R_I}{R_I}\right)v_L$$

or

$$v_{O(n)} = \left(\frac{R_F + R_I}{R_I}\right)\left(\frac{1}{2}\right)^{(N-n)}\left(\frac{1}{3}\right)V_n \tag{8-9}$$

(for the noninverting amplifier, see Fig. 8-5a). Once the numerical value of $v_{O(\text{MSB})}$ is found from Eq. (8-9), the numerical value of the lesser-bit output voltages can be

Digital-to-Analog Converter   **215**

found from

$$v_{O(n)} = \frac{v_{o(\text{MSB})}}{2^{(N-n)}} \tag{8-10}$$

The amplifier output voltage at 2^N counts is V_N, which is twice the most-significant-bit output voltage, or $(A_{\text{VCL}})(\frac{1}{3}V_n)$. Thus the noninverting amplifier gain, by **Rule 3**, must be

$$A_{\text{VCL}} = \frac{R_F + R_I}{R_I} = (1.5)\frac{V_N}{V_n} \tag{8-11}$$

where V_N is on the amplifier output and V_n is a register output voltage.

An example will help to illustrate the use of Eq. (8-11).

EXAMPLE 8-8

An $R/2R$ ladder network is connected to a COS/MOS register with a supply voltage of $+5$ V. The amplifier output voltage (V_N) at 2^N counts should be $+12$ V. Determine the necessary gain of the noninverting amplifier.

Solution

The register output voltages (V_n) are $+5$ V, since the register output voltage of a COS/MOS register is virtually the value of the supply voltage. Using Eq. (8-11), the noninverting amplifier gain $= (1.5)(12\text{ V}/5\text{ V}) = 3.6$.

The noninverting amplifier used on the ladder output causes two simultaneous situations to occur which must be satisfied:

1. The ladder output voltage must be raised, using gain, to the desired value; this is seen in Eqs. (8-10) and (8-11).
2. The parallel combination of R_I and R_F must equal the source impedance (R) of the ladder network as seen by the amplifier for proper bias-current compensation. Thus

$$\frac{R_F R_I}{R_F + R_I} = \frac{2}{3}R \tag{8-12}$$

and by combining Eqs. (8-11) and (8-12), it is found that

$$R_F = \left(\frac{V_N}{V_n}\right)R \tag{8-13}$$

and that

$$R_I = \left(\frac{V_N}{\frac{3}{2}V_N - V_n}\right)R \tag{8-14}$$

where V_N is on the amplifier output and V_n is the register output voltage.

An example will help to clarify the use of Eqs. (8-12) to (8-14).

216   *Digital Applications*

EXAMPLE 8-9

The DAC in Example 8-8 uses a $R/2R$ network with an R of 10 kΩ. Determine the values of R_I and R_F.

Solution

$$R_F = \left(\frac{12\text{ V}}{5\text{ V}}\right)(10\text{ k}\Omega) = 24\text{ k}\Omega$$

$$R_I = \left[\frac{12\text{ V}}{\frac{3}{2}(12\text{ V}) - 5\text{ V}}\right](10\text{ k}\Omega) = \frac{12\text{ V}(10\text{ k}\Omega)}{18\text{ V} - 5\text{ V}} = 9.23\text{ k}\Omega$$

Select 23.7 kΩ and 9.09 kΩ from Appendix B1. The gain is then 3.6, and the parallel combination of R_I and R_F is 6.57 kΩ; thus Eq. (8-12) is nearly satisfied.

An example will help to illustrate the type of of problem encountered in a typical DAC.

EXAMPLE 8-10

An 8-bit DAC uses an $R/2R$ ladder network with $R = 20$ kΩ and a noninverting amplifier with a gain of 4 and register output voltages of $+6$ V.

a. Determine the output voltage for a register content of 1 0 0 1 0 0 1 1.
b. What is the maximum clock rate for the DAC if a μA741 op amp is used?

Solution

a. The segment output voltages present are

1	0	0	1	0	0	1	1
8	7	6	5	4	3	2	1

bits 8, 5, 2, and 1 are HI. The MSB output voltage (8) is given by Eq. (8-9)

$$v_{O(8)} = (4)\left(\frac{1}{2}\right)^{(8-8)}\left(\frac{1}{3}\right)(6\text{ V}) = +8\text{ V}$$

and from Eq. (8-10),

$$v_{O(5)} = \frac{8\text{ V}}{2^{8-5}} = \frac{8\text{ V}}{8} = 1\text{ V}$$

In a like manner, $v_{O(2)} = 0.125$ V and the LSB output voltage (1) is $v_{O(1)} = 0.0625$ V. Thus the output voltage is

$$(8 + 1 + 0.125 + 0.0625) = 9.188\text{ V}$$

b. The op amp slew rate is 0.5 V/μs and at a closed loop gain of 4, the op amp slew rate is still constant-current-limited, so the slew rate does not change. From Eq. (8-6), $V_N = 2(8\text{ V}) = 16$ V; then from Eq. (8-3), $V_{LSB} = V_N/2^N = 16$ V/256 = 62.5 mV. The time to slew by one step is 62.5 mV/(0.5 V/μs) = 125 ns. Each clock pulse represents one step; thus the clock rate is 1 cycle/125 ns = 8 MHz when time is only available for the op amp to *reach* the desired voltage on each step. If any time is *spent* on a step, the clock must run at a reduced rate; for 275 ns spent on each step, the clock must run at

$$\frac{1 \text{ pulse}}{125 \text{ ns} + 275 \text{ ns}} = 2.5 \text{ MHz}$$

The DAC spends 31% of the total step time *slewing to* the step voltage and 69% *on* the step.

If an inverting amplifier, as illustrated in Fig. 8-5b, is used on the output of the $R/2R$ ladder network, the current through the $2R$ terminating resistor is the *input current* to the inverting amplifier. The hypothetical output voltage at 2^N counts is given by

$$V_N = -\left(\frac{1}{3}\right)\left(\frac{R_F}{R}\right)V_n \tag{8-15}$$

and v_O due to any single segment voltage (V_n) is $-(i_L)R_F$ and is given by

$$v_{O(n)} = -\left(\frac{1}{2}\right)^{(N-n)}\left(\frac{1}{6}\right)\left(\frac{R_F}{R}\right)V_n \tag{8-16}$$

(for the inverting amplifier, see Fig. 8-5b) where N is the number of bits in the DAC and n is the segment number. Again, Eq. (8-10) can be used to determine the lesser segment output voltages which sum to form the amplifier output voltage.

The impedance seen from the inverting amplifier summing junction looking back into the ladder network is $3R$; thus the value of the bias-current compensation resistor (R_C) is

$$R_C = \frac{(3R)R_F}{(3R) + R_F} \tag{8-17}$$

An example will help to illustrate the use of Eqs. (8-15) to (8-17).

EXAMPLE 8-11

An inverting amplifier with $R_F = 20 \text{ k}\Omega$ is used on a 4-bit $R/2R$ ladder DAC.

   a. Determine the output voltage equation due to V_2.
   b. If $V_2 = +8$ V and $R = 10 \text{ k}\Omega$, determine the numerical value of the output voltage.
   c. Determine the numerical value of V_N, R_c, and V_{LSB}.

Solution

   a. $v_o = -\left(\frac{1}{2}\right)^{(4-2)}\left(\frac{1}{6}\right)(\frac{20 \text{ k}\Omega}{R})V_2 = -\left(\frac{833 \ \Omega}{R}\right)V_2$

   b. $v_o = -\left(\frac{0.833 \text{ k}\Omega}{10 \text{ k}\Omega}\right)(8 \text{ V}) = -0.666$ V

   c. $V_n = -\left(\frac{1}{3}\right)\left(\frac{20 \text{ k}\Omega}{10 \text{ k}\Omega}\right)(8 \text{ V}) = -5.33$ V

      $R_C = \frac{(30 \text{ k}\Omega)(20 \text{ k}\Omega)}{30 \text{ k}\Omega + 20 \text{ k}\Omega} = 12 \text{ k}\Omega$

      From Eq. (8-3),

$$V_{LSB} = -\left(\frac{5.33 \text{ V}}{16}\right) = -0.333 \text{ V}$$

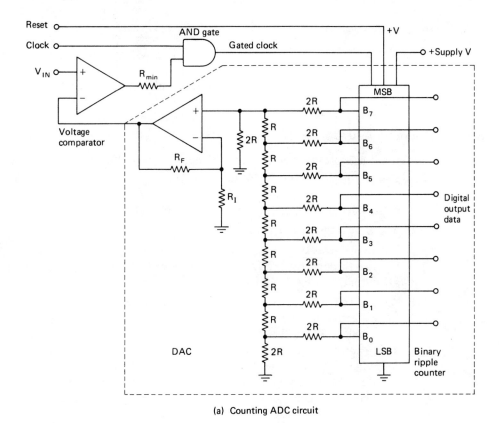

(a) Counting ADC circuit

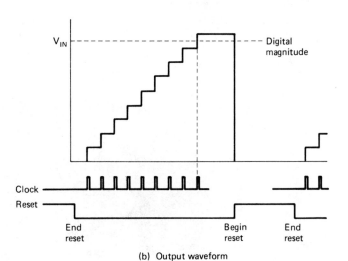

(b) Output waveform

Fig. 8-7 Counting A/D converter.

8.4 Analog-to-Digital Converter

An analog-to-digital (A/D) converter (ADC) is a system that accepts a dc voltage and causes a digital register to store a number that is proportional to the magnitude of the dc voltage. Many forms of ADCs are available, and several of these will be presented.

8.4.1 Counting A/D Converter

This circuit uses a DAC and a voltage comparator to perform the conversion and is illustrated in Fig. 8-7a. The circuit operates as follows (see Fig. 8-7b):

1. A dc voltage is applied to the voltage comparator.
2. The DAC register is RESET.
3. Upon the release of RESET, clock pulses are applied to the binary ripple counter through the AND gate which is enabled by the voltage comparator HI output.
4. As long as the output voltage from the DAC is *less* than the dc input voltage, the voltage comparator has a HI output and permits the AND gate to pass the clock pulses. When the staircase voltage passes the point where it is a partial step *greater* than the dc input voltage, the voltage comparator output changes to a LO state and the AND gate inhibits the clock from advancing the counter. The binary number contained in the counter register is then equivalent to the magnitude of the dc input voltage.

One disadvantage to this method is the time it takes the counter to advance to the desired count. But an overriding advantage is the number of bits in the DAC, which represents the number of discrete segments into which the input voltage can be divided. For an 8-bit DAC, the input voltage is divided into 256 segments; for a 10-bit DAC there are 1024 segments, etc.

Fig. 8-8 $3\frac{1}{2}$-Digit Panel meter. (Courtesy Datel-Intersil, Inc., Mansfield, Mass.)

8.4.2 Digital Readouts

It is probably well at this point to relate the ADC topic to digital panel meters and visual readouts. This will give the reader a familiar device with which to relate these somewhat abstract theories and help to place, in perspective, the number of bits required to digitize a voltage. A $3\frac{1}{2}$-digit digital panel meter (DPM) (see Fig. 8-8) will indicate voltages from 0.001 V to 1.999 V with ± 1 count accuracy; the ratio is 1999. This device requires an 11-bit (2048) ADC to perform this task, and the ratio nearly matches the bits required. A $4\frac{1}{2}$-digit digital panel meter will indicate voltages from 0.0001 V to 1.9999 V, a ratio of 19,999; this requires a 15-bit (32,768) ADC, as a 14-bit (16,384) ADC just falls short.

8.4.3 Tracking A/D Converter

The tracking ADC is similar to the counting ADC with the exception that the comparator output drives an up-down direction control on the counter rather than the clock; the clock is connected continuously to the counter. The circuit for a tracking ADC is shown in Fig. 8-9a. The counter, in this case, is an up-down counter rather than the binary ripple counter used in the counting ADC. This circuit will follow the changes in an applied input voltage illustrated in Fig. 8-9b, as long as the changes do not occur more rapidly than the ADC is able to follow (see Fig. 8-9c).

An example will help to illustrate the limitation shown in Fig. 8-9c.

EXAMPLE 8-12

A tracking ADC uses a 1 MHz clock and has a 10-bit (1024) counter. $V_N = 10$ V at 1024 counts. The input signal is given by equation $v_{in} = 6 + 3\sin(6280t)$V and shown in Fig. 8-9c. Determine the waveshape of the digital output information.

Solution

The ADC is able to slew from zero count to full count or vice versa in (1024 steps)/(10^6 steps/s) = 1.024 ms. The counter has an average "step rate" of 10 V/1.024 ms = 9.76 mV/μs. The input sine wave has a maximum slew rate given by Eq. (5-16), which is 18.8 mV/μs. This tracking ADC will appear to be "step-rate-limited" just as an op amp becomes slew-rate-limited, and the output waveform will be triangular with the equivalent of a +6 V dc average. The input signal frequency must be reduced to 500 Hz before the ADC can accurately follow the input waveform.

8.4.4 Dual-Slope A/D Converter

The dual-slope or integrating ADC is shown in Fig. 8-10a. This circuit operates on the principle that an integrator with a dc input voltage linearly "accumulates" charge on the capacitor in proportion to the product of the dc input voltage magnitude and time of integration while a counter *accumulates* a number that is proportional to the clock rate in pulses/second. Thus while the integrator output is changing by *volts/second*, the counter output is changing by *counts/second* and the two are *proportional* to each other. The circuit contains three major elements: an

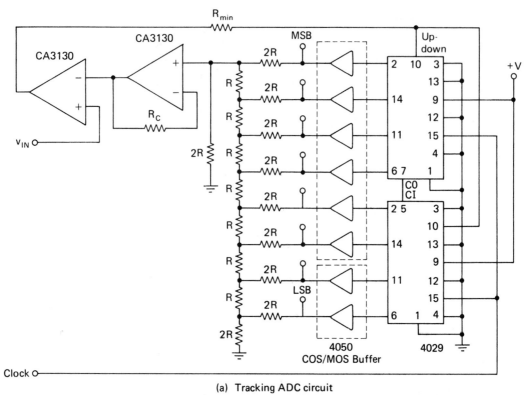

(a) Tracking ADC circuit

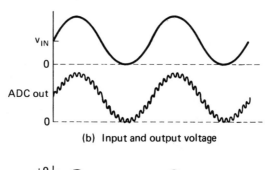

(b) Input and output voltage

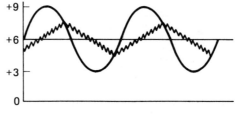

(c) Output step rate limited

Fig. 8-9 Tracking ADC.

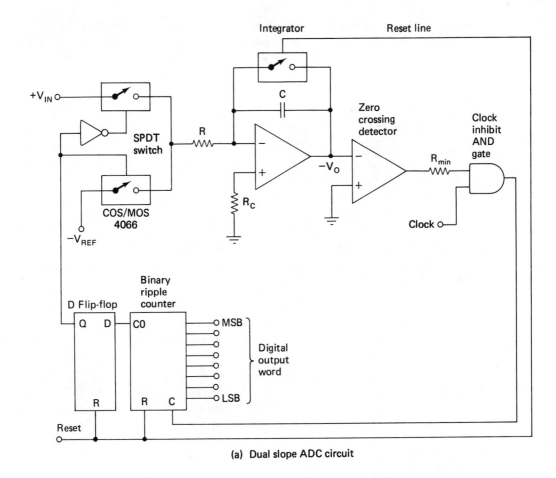

(a) Dual slope ADC circuit

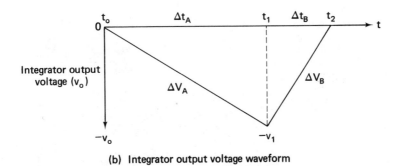

(b) Integrator output voltage waveform

Fig. 8-10 Dual-slope ADC.

analog integrator, a binary ripple counter, and a digital control system. The control system switches the input of the integrator between V_{IN} and $-V_R$, opens the reset switch on the integrating capacitor, senses the zero crossing of the integrator, and resets the binary ripple counter. An analysis of the circuit operation follows:

1. The integrator begins with its capacitor shorted (reset) and the counter begins at all 0's (reset).
2. The integrator is switched to V_{IN}; the integrator capacitor and counter resets are simultaneously removed. The integrator begins integrating and the counter begins counting at the same time.
3. They integrate and count, respectively, until the instant when the counter reaches all 0's (one count beyond full count where the carry out, CO, is high). The integrator input is switched to a reference voltage (V_R); V_R is equal in magnitude and opposite in polarity to the largest V_{IN} expected.
4. The integration then moves back toward zero volts and the counter continues to count up from zero again.
5. When the integrator output reaches zero volts, the zero crossing is detected and stops the counter.
6. The number contained in the counter is then proportional to V_{IN}.

This circuit is unusual in the respect that it does not employ a DAC in the A/D converter circuit. Remember, it is the DAC which is the limiting factor in the number of bits that an ADC can accurately digitize; this circuit does not contain a DAC. It is thus only limited in its ability to digitize an analog signal by the number of segments in the counter and the long-term stability of the offsets in the integrator. Thus this circuit is used for the A/D conversion in six-digit digital voltmeters (DVM) with reliable long-term accuracy. To place this in perspective, a six-digit DVM has 1 million counts or 20 bits; the accuracy required in a 20-bit DAC is beyond the current state of the art. The values for the integrator in the dual-slope ADC are found using the relationships defined in Fig. 8-10b.

$$\Delta t_A = t_1 - t_0 = \frac{2^N \text{ counts}}{\text{clock rate}} \tag{8-18}$$

and

$$\Delta t_B = t_2 - t_1 = \frac{\text{digital magnitude}}{\text{clock rate}} \tag{8-19}$$

and

$$\Delta v_A = -v_1 - 0 = -v_1 \qquad v_1 = -\Delta v_A$$

$$\Delta v_B = [0 - (-v_1)] = +v_1 \qquad v_1 = \Delta v_B$$

$$\Delta v_o = \left(-\frac{1}{RC}\right)(v_{IN})(\Delta t)$$

and

$$-\Delta v_A = v_1 = \left(-\frac{1}{RC}\right)(V_{IN})(\Delta t_A)$$

$$\text{Digital magnitude} = 2^N\left(\frac{V_{IN}}{-V_R}\right) \tag{8-20}$$

$$\Delta v_B = v_1 = \left(-\frac{1}{R_C}\right)(-V_R)(\Delta t_B) \tag{8-21}$$

Then for the integrator

$$RC = -\left(\frac{V_{IN}}{v_1}\right)\Delta t_A \tag{8-22}$$

An example will help to clarify the use of these equations.

EXAMPLE 8-13

A dual-slope ADC uses a 16-bit counter and a 4 MHz clock rate. The maximum input voltage is +10 V. The maximum integrator output voltage should be −8 V when the counter has cycled through 2^N counts. The integrator capacitor is 0.1 μF. Determine the value of R to use in the integrator.

Solution

From Eq. (8-18), $\Delta t_A = 65{,}536/4$ MHz = 16.38 ms. Then, from Eq. (8-22),

$$RC = -\left(\frac{10\text{ V}}{-8\text{ V}}\right)(16.38\text{ ms}) = 20.47\text{ ms}$$

and

$$R = \frac{20.47\text{ ms}}{0.1\ \mu\text{F}} = 204.7\text{ k}\Omega$$

use a 205 kΩ 1% from Appendix B1.

A subtle advantage to the integration time chosen in Example 8-13 is that it nearly matches one period of a 60 Hz sine wave. If any 60 Hz signal appears on the input voltage, it will be nearly canceled out at v_1 because of the integrating action of the integrator.

A second example will help to illustrate the type of op amp that must be used for the integrator.

EXAMPLE 8-14

For the dual-slope ADC in Example 8-13, determine the maximum long-term offset error of the integrator op amp. Also determine the open loop gain and slew rate of a comparator that drives the TTL logic gates.

Solution

Using Eq. (8-3), we obtain

$$V_{IN} = V_N = 10\text{ V}$$

$$V_{LSB} = \frac{10\text{ V}}{65{,}536} = 152\ \mu\text{V}$$

For the least significant bit to be valid, the long-term integrator output offset voltage should be held to 0 V within $\pm 15\ \mu V$. The zero-crossing detector must switch within $\pm 1/4\ V_{\mathrm{LSB}}$ or 38 μV. If the output must drive TTL gates, it must reach $+2.0$ V to guarantee the gate excitation. Thus the open loop gain must be $A_{\mathrm{VOL}} = 2.0\ \mathrm{V}/38\ \mu V = 52{,}632$ or 94.4 dB gain. The comparator must reach this $+2.0$ V in $\frac{1}{4}$ clock cycle or 62.5 ns; thus the slew rate must be

$$\frac{2.0\ \mathrm{V}}{62.5\ \mathrm{ns}} = 32\ V/\mu s$$

EXAMPLE 8-15

Using the circuit of Example 8-13, determine the digital number (in binary) for an input of $+5.162$ V.

Solution

Using Eq. (8-20), the digital magnitude is found to be

$$65{,}536\left[\frac{5.162\ \mathrm{V}}{-(-10\ \mathrm{V})}\right] = 33{,}830$$

which, when converted to the binary equivalent, is

1000010000100110

8.4.5 Successive-Approximation ADC

The first of two A/D converters which are very fast is presented here. It should be becoming apparent to the reader by now that as the number of bits increases, the time to complete an A/D conversion also increases. The successive approximation ADC is one that employs a DAC in the process of the A/D conversion, but, rather than a binary ripple counter, the DAC employs a digital processor, as illustrated in Fig. 8-11a. The circuit operates in the following manner:

Method 1

1. The outputs from the DAC are programmed to be all initially LO; the DAC is at "zero count."
2. The most significant DAC output bit is caused to go HI and the comparator is sensed for a state change. If the change occurs, the MSB output from the DAC is returned to LO, as the DAC output voltage was greater than the input voltage; if no change occurs the MSB output is left HI.
3. The next-lower DAC output bit is caused to go HI and the comparator is sensed for a state change. If a change occurs, the bit is returned to LO, as the new DAC output voltage was greater than the input voltage.
4. This process of changing the next lower DAC output bit and sensing the comparator for a change is continued through the least significant bit of the DAC.
5. When the process is complete, the final DAC output states represent the digital equivalent of the step just *below* the actual input voltage magnitude.
6. The whole process requires a maximum of only $2N$ pulses to complete the entire A/D conversion. More hardware is required than for a ripple counter type of DAC.

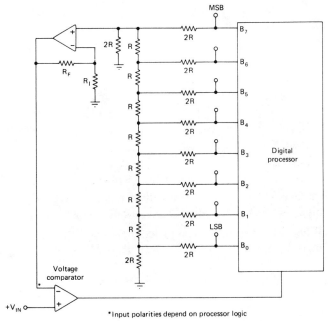

*Input polarities depend on processor logic

(a) Successive approximation ADC

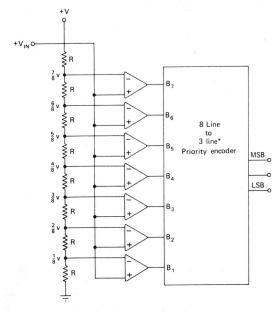

*This is an Octal-to Binary encoder and bit 0 (B_0) is redundant as all output are L0 when inputs are all L0.

(b) Simultaneous ADC

Fig. 8-11 Fast analog-to-digital converters.

Method 2

1. The outputs from the DAC are programmed to be all initially HI: the DAC is at "full count."

2. The most significant DAC output bit is caused to go LO and the comparator is sensed for a state change. If the change occurs, the MSB output from the DAC is returned to HI, as the DAC output voltage was less than the input voltage; if no change occurs, the bit is left LO, as the input voltage was greater than the DAC output.

3. The next lower DAC output bit is caused to go LO and the comparator sensed for a state change. If the change occurs, the bit is returned to HI, as the new DAC output was less than the input voltage; if no change occurs the bit is left LO, as the new DAC output was greater than the input voltage.

4. When the process is complete, the final DAC output states represent the digital equivalent of the step just *above* the actual voltage magnitude.

An example will help to illustrate these two methods.

EXAMPLE 8-16

A 4-bit successive approximation ADC uses a DAC with a V_N of $+10$ V. The DAC output, and an input voltage of 6.50 V, drive the voltage comparator (see Fig. 8-11a).

a. Determine the digital number that appears after the conversion is completed using Method 1.

b. Do the same for Method 2.

Solution

Bit	Output Voltage
4	5.000 V
3	2.500 V
2	1.250 V
1	0.625 V

a.

Bit HI	Comparator Change	Bits Forming CMPR Voltage			V_O at Compare	Bit after Compare
4	No	4			5.000 V	HI
3	Yes	4	3		7.500 V	LO
2	No	4	2		6.250 V	HI
1	Yes	4	2	1	6.875 V	LO
Digitized V_O is		4	2		6.250 V	

b.

Bit LO	Comparator Change	Bits Forming CMPR Voltage			V_O at Compare	Bit after Compare
4	Yes	3	2	1	4.375 V	HI
3	No	4	2	1	6.875 V	LO
2	Yes	4		1	5.625 V	HI
1	Yes	4	2		6.250 V	HI
Digitized V_O is		4	2	1	6.875 V	

From this example we see that in Method 1 the digitized output voltage is on the LO side of the input voltage, whereas in Method 2 the digitized output voltage is on the HI side of the input voltage. Knowing this characteristic, the method most suitable for a particular application may be chosen.

8.4.6 Simultaneous ADC

By far, the fastest conversion is obtained by using the simultaneous or parallel ADC*; the circuit is illustrated in Fig. 8-11b. Although fast, this circuit has a serious limitation in the number of bits that can be digitized to represent the analog input voltage. The circuit applies the input voltage simultaneously to a group of voltage comparators. The reference voltage to each voltage comparator is formed by connecting each one to a division point along a voltage divider with N resistors. The supply voltage (V_N) to the voltage divider is set to be

$$V_N = V_{IN_{max}} \tag{8-23}$$

The reference voltage to each comparator is formed by connecting each one to the junction of two divider resistors. This means that there are

$$n_{max} = 2^N - 1 \tag{8-24}$$

voltage comparators and each one has a reference voltage of

$$V_{R_n} = \left(\frac{V_N}{2^N}\right) n \tag{8-25}$$

where n represents the number of the voltage comparator, beginning at the lowest division point of the voltage divider. It can be seen from Eq. (8-24) that the number of comparators doubles for each bit added to the ADC. A 3-bit ADC requires 7 comparators; a 4-bit ADC requires 15 comparators, and so on. "This type of processing is called *bin conversion*, because the analog input is sorted into a given voltage range or "voltage bin" determined by the thresholds of two adjacent comparators."[†] An example will help to illustrate the use of Eqs. (8-23) to (8-25).

*Also known as a "flash encoder."
[†]Reference: Jacob Millman, *Micro-Electronics*: *Digital and Analog Circuits and Systems* (New York: McGraw-Hill Book Company, 1979), p. 613.

EXAMPLE 8-17

A simultaneous ADC must digitize an input voltage with a maximum value of -12 V to 4-bit accuracy.

   a. Determine the number of comparators and reference voltage for each.

   b. If the reference voltage is connected to the inverting input of each comparator, determine which comparator outputs are HI and which are LO for an input voltage of -5.5 V.

Solution

   a. $V_N = -12$ V from Eq. (8-23)

      $n_{max} = 2^4 - 1 = 15$ comparators from Eq. (8-25)

      $V_{R_1} = \left(\dfrac{-12\text{ V}}{16} \right)(1) = -0.75$ V,

      each succeeding V_R is 0.75 V more negative; thus V_{R_2} is -1.5 V, V_{R_3} is -2.25 V, and $V_{R_{15}}$ is -11.25 V.

   b. $V_{R_7} = -5.25$ and $V_8 = -6.0$ V; thus comparators 1 through 7 are LO and comparators 8 through 15 are HI.

TRW LSI Products manufactures two versions of integrated simultaneous ADCs; the 4-bit TDC-1021J and the 8-bit TDC-1007J. The 8-bit version is also sold, completely assembled in a circuit board, as their TDC-1007PCB. The conversion time for each is 33 ns.

PROBLEMS

1. A comparator using a μA 741 op amp has ± 12 V power supplies. It is driving a 7408 TTL logic AND gate.

   a. Find the minimum-size current-limiting resistor that can be used.

   b. Select a value from the 5% table that will suffice.

2. A 74S09 TTL open-collector-logic AND gate uses a 4.7 kΩ pull-up resistor. Find V_{OH} if the load to this gate draws 300 μA at a logic HI.

3. A COS/MOS CD4071 OR gate with ± 5 V supplies is preceded in a circuit by a comparator with ± 15 V supplies.

   a. Calculate the size of R_{min} for the circuit.

   b. Select an appropriate resistor value from the 5% table of Appendix B1.

4. A TTL binary counter uses a COS/MOS CD4050 buffer amplifier on its output as a switch. The COS/MOS buffer has an output impedance of 120 Ω in both the HI and LO states. If $+5$ V and common power supplies are used, find the HI output voltage when driving a 15 kΩ load resistor.

5. Repeat Problem 4 for:

   a. A 30 kΩ load resistor.
   b. A 60 kΩ load resistor.

6. A 4-bit inverting summer DAC has a MSB resistor of 20 kΩ.

   a. Find the ideal values of the other input resistors.
   b. Select values from the 1% table in Appendix B1.

7. The DAC of Problem 6 is driven by COS/MOS switches with ±12 V supplies. Bits 0 and 2 are HI, and bits 1 and 3 are LO.

   a. Find the current through the feedback resistor on the summing amplifier.
   b. Find the summer output voltage if $R_f = 10$ kΩ.

8. How many voltage levels can be obtained with an 8-bit DAC?

9. What resistor tolerance level must be used for the MSB in the DAC of Problem 8? The error must not exceed one-quarter step.

10. A 5-bit DAC has a hypothetical output voltage of -5 V at 32 counts.

    a. Find the least significant bit voltage.
    b. What is the actual full-count voltage?

11. A 7-bit inverting summer DAC has an output voltage of -15 V when only the MSB input voltage is HI (1000000).

    a. Find V_{LSB}.
    b. Find V_{FC}.
    c. Find the output voltage at count 1101100.

12. A 13-bit inverting summer DAC uses a LM 208 op amp. The LM 208 is specified as having a coefficient of input offset voltage of 5 μV/° C, and a coefficient of input offset current of 2.5 pA/° C. If $R_f = 6$ kΩ and $R_{MSB} = 5$ kΩ, find the change in temperature that causes a half-bit error to occur. The initial output offset is 0 V, and the initial temperature is 25° C.

13. The MSB input resistor (R_7) in Problem 11 is 0.1% lower than its ideal value. All other resistors have their correct values.

    a. Find the voltage at count 1110000 and 0001110.
    b. If R_8 is 0.5% high in value and all other resistors are within tolerance, which counts will be in error? How much error will each step have?
    c. Show the corresponding steps that have the same output voltage.
    d. Which steps in part (c) are the correct output voltage?

14. How much tolerance error does it take for the LSB resistor in Problem 11 to create a 1-bit error in the output voltage?

15. A 5-bit inverting summer DAC has a LSB output voltage of -200 mV. Find the values of the input resistors if $R_f = 11.5$ kΩ and the digital register has a $+5$ V supply voltage.

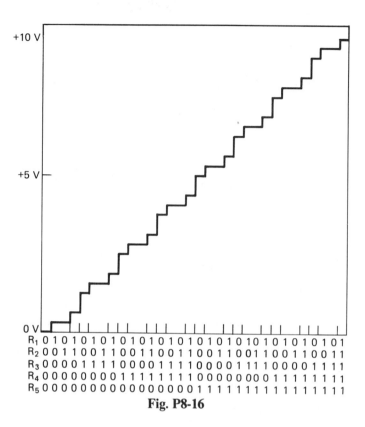

Fig. P8-16

16. The output voltage from a 5-bit DAC is shown in Fig. P8-16.

   a. Which input resistor is in error?

   b. Is it too large or too small?

   c. What is its percent error?

   d. If 1% resistors were used in this DAC, what is the most likely cause of the nonmonotonic staircase voltage?

17. A 7-bit $R/2R$ DAC has $R = 25$ kΩ. The register output voltage, V_N, is $+10$ V. Find the ladder output voltage, V_L, for each of the segment voltages: 1, 3, 5, and 7.

18. For Problem 17, find the current, I_L, in the $2R$ terminating resistor for the four segment voltages given.

19. For Problem 17, consider the case where all of the segment voltages are HI simultaneously.

   a. Find the ladder output voltage (V_L).

   b. Find the current (I_L) in the $2R$ terminating resistor.

20. An 8-bit $R/2R$ ladder DAC has $V_L = +5$ V when only the MSB segment voltage is HI. The ladder is connected to a noninverting amplifier with $R_f = 30$ kΩ and $R_I = 12$ kΩ.

   a. Find $V_{O(8)}$ for only the MSB segment voltage applied.

   b. Find $V_{O(6)}$.

21. A 12-bit $R/2R$ ladder DAC uses a register with a $+10$ V supply voltage. The MSB output voltage from the noninverting amplifier is 12 V.

    a. Find the gain in the noninverting amplifier.

    b. Find the output voltage when only the segment 5 register output voltage is active (HI).

22. In Problem 21, $R = 40$ kΩ for the $R/2R$ ladder.

    a. Find the value of R_1 and R_f for the noninverting amplifier.

    b. Select appropriate values from the 1% table in Appendix B1.

23. A 11-bit $R/2R$ ladder with $R = 10$ kΩ is driving an inverting amplifier. If the register supply voltage is $+5$ V, and the hypothetical output voltage at 2048 counts is $+10$ V:

    a. Find the value of the feedback resistor in the inverting amplifier.

    b. Find the current compensation resistor value.

    c. Find the value of the MSB and LSB output voltages.

24. A μA 741 op amp with a 0.5 V/μs slew rate is used on the op amp of Problem 23. Find the maximum allowable clock rate if the DAC must spend 75% of its available time on each step voltage (25% slewing to each step).

25. A counting A/D converter uses a 8-bit DAC. It has an MSB DAC output voltage of $+6$ V.

    a. If the input voltage is $+7.25$ V, what will be the $R/2R$ ladder output voltage (V_L) when the clock stops?

    b. How many clock pulses occur between the release of reset and the stopping of the clock?

26. The ADC in Problem 25 uses a 150 kHz clock. How long did it take to digitize 7.25 V?

27. Why does a digital panel meter (DPM) or digital voltmeter (DVM) specify a ± 1 count disparity between the actual input voltage and its indication?

28. A 10-bit tracking ADC uses a 1 MHz clock. $V_N = 8$ V at 2^n counts, and an input sine wave of $V_{in} = +4 + 4\sin(2\pi ft)$ V is applied. Find the highest frequency for the sine wave before any step rate limiting occurs in the tracking ADC.

29. A dual-slope ADC uses a 15-bit counter with a 2 MHz clock. The maximum input voltage is $+15$ V. The integrator output voltage at 2^n counts is -12 V.

    a. Find the size of the capacitor to use for the integrator if $R = 10$ kΩ.

    b. Select the nearest standard-value capacitor from Appendix B2. Calculate the appropriate resistor size to be selected from the 1% table in Appendix B1.

30. For the dual-slope ADC in Problem 29:

    a. Determine the maximum long-term offset error allowed in the integrator op amp.

    b. Determine the open loop gain and the slew rate of the comparator that drives the TTL logic gates.

31. The dual-slope ADC in Problem 29 has an input voltage of $+7.452$ V. Determine the digital number (in binary) which represents the count on the register.

32. A 5-bit successive approximation ADC uses a DAC with $V_N = 4$ V. The input voltage is 2.25 V.

  a. Find the digital number that appears when the conversion is complete through Method 1.

  b. Using Method 2, find the digital number that appears when the conversion is complete.

33. For Problem 32:

  a. How many clock cycles (pulses) were required to complete the conversion for part (a)?

  b. How many clock cycles were required to complete the conversion for part (b)?

34. An 8-bit simultaneous ADC must digitize a $+15$ V maximum input voltage.

  a. Determine the number of comparators and the reference voltages for each.

  b. Find which are HI and which are LO for an input voltage of $+11.32$ V. The reference voltages are connected to the inverting input of each comparator.

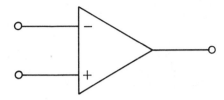

9 Active Filters

9.1 Introduction

The topic of active filters for students who have not had courses in complex variables and Laplace transforms is fraught with controversy. The author believes that an in-depth study of poles and zeros can be taught without involving the student with mathematics beyond simple algebraic computations. At the same time, the reader is provided with a deep understanding of the relationship between poles and zeros and the frequency-response characteristics of active filters. This chapter contains most of the mathematical relationships that describe filters and thus serves as a useful reference as well as a text. With these thoughts in mind, we begin the topic of active filters.

9.2 Definition of Terms

Many terms that are new to the reader will be introduced in this chapter. In order to effectively communicate the ideas presented in this chapter, the reader must understand the terms and concepts used so that the reader and author are together in thought.

9.2.1 s-Plane

A frequency-response plot, illustrated in Fig. 9-1a, is a plot of amplitude or gain versus frequency. The envelope of this frequency-response or frequency-spectrum plot represents the amplitude of each frequency where that particular frequency occurs. This familiar plot is the amplitude versus frequency along one axis of the s-plane. An s-plane with the *same* frequency response as in Fig. 9-1a plotted on it is

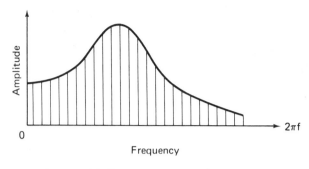

(a) Frequency response plot

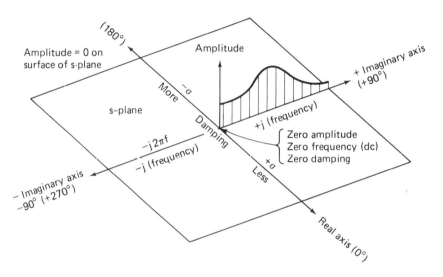

(b) Frequency response plot on S-plane

Fig. 9-1 Relationship of frequency response to *s*-plane.

illustrated in Fig. 9-1b, only this time it is shown on the *s*-plane. It can be seen from Fig. 9-1b that the frequency-response or spectrum plot is the amplitude spectrum along the *imaginary* or frequency axis of the *s*-plane. Any point along the frequency axis of this plot is labeled $+j\omega$, where the (j) is the *imaginary* operator, which just designates that frequencies appear along the $+90°$ axis. If one thinks of a right triangle where the X-axis is the real axis and the Y-axis is the imaginary axis and the axes are labeled (x) and (jy), it can be seen that the j operator just indicates that the Y-axis is at right angles to the X-axis; thus is the case for the damping and frequency axes on the *s*-plane. The damping is represented as values along the *real* axis and the frequency is represented as values along the *imaginary* axis. Another point to realize is that the damping everywhere along the j(frequency)-axis is *zero* and that the frequency everywhere along the damping-axis is *zero* (dc); the concept will be used later to simplify the mathematical relationships.

Figure 9-2a is a top view of the *s*-plane with the frequency and damping shown on figure as they occur in a time relationship. A careful analysis of the figure reveals

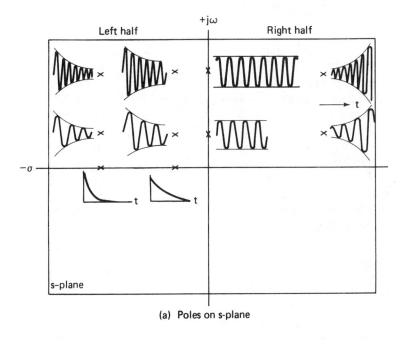

(a) Poles on s-plane

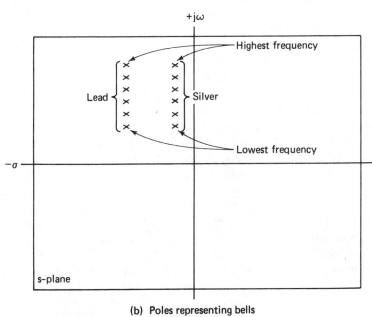

(b) Poles representing bells

Fig. 9-2 Frequency and damping on *s*-plane.

that as a point moves away from the damping axis ($f = 0$), the frequency increases; it also reveals that as a point moves away from the frequency axis to the left, the damping factor (σ) becomes more negative and a signal amplitude diminishes more rapidly with time. A damping factor to the right of the frequency axis indicates a positive damping factor, so that the signal amplitude increases with time.

Consider a set of 10 bells: 5 bells are of pure silver and 5 contain some percentage of lead. Each set of 5 bells has one of several ring frequencies, and each set of 5 bells has the same ring frequencies. The 5 silver bells will have points on the s-plane slightly to the left of the frequency axis, as the damping factor is very small and they ring for a long time after being struck; the other 5 bells have points farther to the left of the frequency axis, as the damping factor is greater and they ring for a shorter time after being struck. Each respective bell is the same distance from the damping axis, as illustrated in Fig. 9-2b. In all cases, the bells are to the left of the frequency axis as the amplitude of the ring diminishes with time. An air-raid siren that has just been turned on would appear to the right of the frequency axis, where the amplitude is increasing with time.

9.2.2 Poles and Zeros

A "pole" is a thin vertical line extending from a point on the surface of the s-plane to infinity. A "zero" is a point on the surface of the s-plane. Poles and zeros may occupy the same point on the s-plane, and multiple poles or multiple zeros may occupy the same position on the s-plane. To clarify the definition of a pole, each of the bells, both silver and lead, are represented as poles on the s-plane. The points, representing the bells in Fig. 9-2b, are the location of the poles of each bell. The distance of the pole from the frequency axis determines the damping, Q, or bandwidth (see Section 9.2.6) of the bell; these terms are all related and lead to the same conclusion. That is, the bell will not ring forever after being struck, unless the pole is exactly on the $j\omega$ axis of the s-plane where the damping is zero.

Poles that appear on the s-plane appear in *pairs*, one on each side of the damping axis, unless they are exactly on the damping axis. This may seem strange at first until the nature of the s-plane is understood. The s-plane is constructed as illustrated in Fig. 9-3, where it can be seen that a *mirror* extends along the damping axis and a translucent *window* extends along the frequency axis; this mirror and window, although shown limited in Fig. 9-3, have no limit in either height or length. Measurable frequencies are on the $+j\omega$ side of the mirror; those on the opposite side of the mirror ($-j\omega$) are not measurable, as they are only a mathematical reflection in the mirror. Thus any pole that appears on the $+j\omega$ side of the mirror (real pole) has an image pole on the reflected side of the mirror. This phenomenon is shown in Fig. 9-3. Similarly, any zero that appears on the $+j\omega$ side of the mirror will have a reflection on the $-j\omega$ side of the mirror. The only exception to this is for poles or zeros that lie directly on the σ-axis; they lie in the plane of the mirror and thus only one exists. The fact that the frequencies appear along the imaginary axis of the s-plane does not mean that they are "imaginary"; it only enhances the mathematical representation of the quantities.

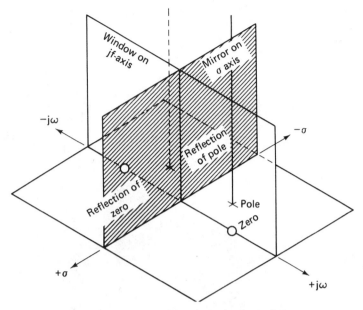

Fig. 9-3 *s*-Plane with mirror and window.

Notice that poles which appear on the $-\sigma$ side of the window have no reflected counterpart on the $+\sigma$ side of the window. Again, the same rules apply to zeros, and their counterparts do not appear on the $+\sigma$ side of the window. The reason for this is eloquently expressed by Benjamin C. Kuo in his book *Automatic Control Systems* [1, p. 155]: "In simple words, a system is defined as stable if the output response to any bounded input disturbance is finite. This implies that all the roots of the characteristic equation must be located in the left half of the *s*-plane. Roots that are in the right half of the plane give rise to transients which tend to diverge from the steady state and the system is said to be unstable." In Fig. 9-2 the complex *s*-plane is divided into two regions: the stable region, which is the left half of the *s*-plane, and the unstable region, which is the right half of the *s*-plane.

9.2.3 The Elastic Membrane

Consider now the *s*-plane with both poles and zeros on the left half of the plane, each one reflected in the mirror, so that there appears to be an equal number on each side of the mirror. A thin elastic membrane, which has a finite weight per unit area, hangs on the poles (it actually never touches the pole) and is attached to the *s*-plane at the zeros, as seen in Fig. 9-4. Except for its attachment to the *s*-plane at the zeros, no other part of the membrane touches the *s*-plane.

The membrane passes through the mirror and window as though they were not present. They do not really exist anyway; they are only a way to easily explain the phenomenon created by the mathematics of the functions of a complex variable which we are going to evade. The functions of a complex variable define the shape

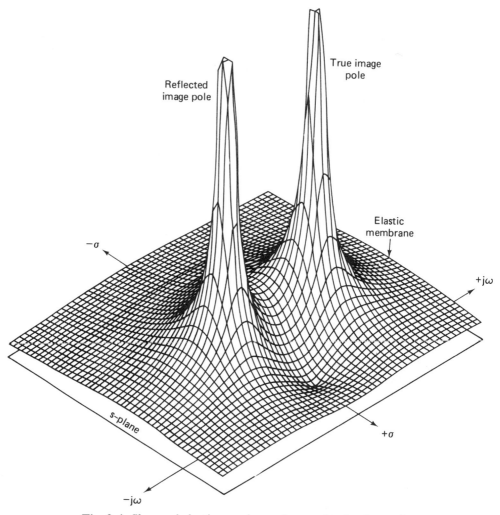

Fig. 9-4 Shape of elastic membrane for a pair of poles and a zero.

of the membrane. Poles exist at infinity which keep the membrane from touching the *s*-plane except at the zeros. These do not appear in the characteristic equation for the control function (filter response) and must be presumed to exist by the fact that the membrane maintains some elevation above the *s*-plane as it leaves the limited area, which we can see in the figure.

9.2.4 The East Face

As the membrane passes through the translucent window, it creates a thin line on the window which we can see. This thin line on the window represents the frequency response of the function defined by the poles and zeros to the left of this window. Thus it appears as though we were standing on the right side of the window looking through to the left side and seeing only the shadow of the membrane as it passes through the translucent window. The effect of this shadow, together with the poles

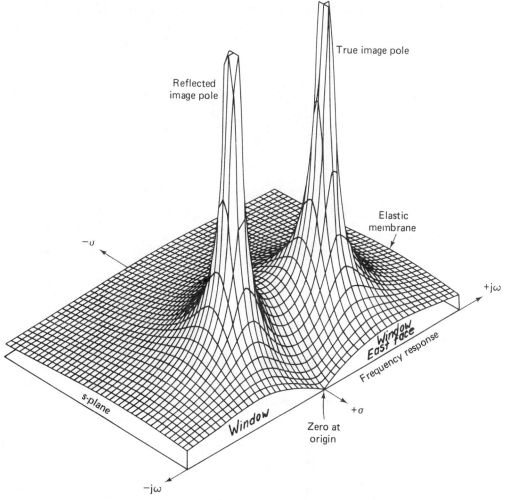

Fig. 9-5 Pole–zero diagram showing the east face.

and zeros behind the window, is illustrated in Fig. 9-5. Thus, when $s = j\omega$, we are in the plane of the window (frequency response).

The term "the *east face*" was coined to represent a mountain range with a vertical cliff at the window. Consider now that the membrane stops at the window, but this does not alter its shape. Now also consider that the window only exists between the membrane and the s-plane, so that a true cliff exists along the $j\omega$-axis looking in from the right half of the s-plane. The shape that the poles and zeros create at this frequency axis is the frequency response of the function formed by the poles and zeros to the left of the cliff, as illustrated in Fig. 9-5.

A mirror image of the frequency response on the $+j\omega$ side of the mirror also exists on the $-j\omega$ side of the mirror along the frequency axis. The image does not exist in reality because it is not measurable, but only exists so as to not disrupt the frequency response on the $+j\omega$ side of the mirror. Thus when a frequency response is transferred from the *east face* of the *s*-plane to a drawing of the frequency response, the mirror image part is omitted, as was done in Fig. 9-1a.

9.2.5 Frequency-Response Variations

As the poles and zeros move closer to the $j\omega$-axis, their effect on the frequency-response plot (*east face*) is more pronounced, as illustrated in Fig. 9-6a. As they

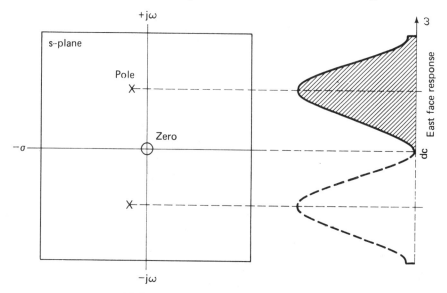

(a) Frequency response for pole close to jf axis

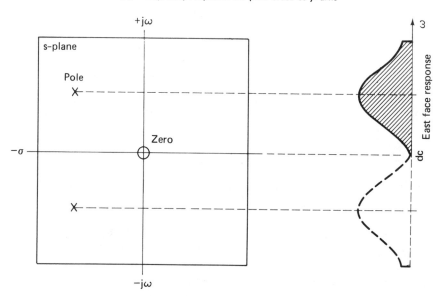

(b) Frequency response for pole far from jf axis

Fig. 9-6 Frequency response versus pole location.

move away from the *east face*, their effect is less pronounced on the frequency response (Fig. 9-6b). Let us now correlate this discovery with the damping, which is a measure of the distance away from the $j\omega$-axis.

Consider a single pair of poles, one on each side of the mirror, each one causing a rise in the frequency response to occur near its projection on the *east face*. As the poles move closer to the *east face*, the response becomes sharper and taller in amplitude, indicating that the damping is less; also, the Q is higher and the bandwidth is less, because of the sharpness of the *east face* frequency-response peak.

As the poles move farther away from the *east face*, their projection on the window is less pronounced. The amplitude is less and the bandwidth is greater, indicating that the damping has become greater resulting in a lower Q; and this agrees with the labels on the σ-axes.

As the pole or zero pairs move farther away from the σ-axis, the frequency of that pole or zero pair becomes greater. At the σ-axis, the frequency of the poles or zeros is 0 Hz (dc).

9.2.6 Mathematical Pole and Zero

The whole purpose of the preceding discussion was to give the reader a way to visualize the discussion in this section. Without this past work, the material that follows would be purely mathematical, with no way to relate the expressions to a visualization of what is really happening. As you have no doubt ascertained by now, the author wants the reader to have a deep understanding of the mathematical relationships without exceeding the reader's level of comprehension.

A mathematical function will "become undefined" or, in plainer terms, "go to infinity" if the denominator goes to zero. This is the method of creating a *pole*. A mathematical function will "go to zero" if the numerator becomes zero; this is how a *zero* is formed. Thus a pole is formed when the transfer function equation has the form

$$\boxed{\frac{v_o}{v_{in}} = \frac{1}{s + \sigma_p}} \tag{9-1}$$

and when $s + \sigma_p = 0$, Eq. (9-1) becomes infinite and a "pole" is created. A zero is formed when the transfer function has the form

$$\boxed{\frac{v_o}{v_{in}} = \frac{s + \sigma_z}{1}} \tag{9-2}$$

except that this time, when $s + \sigma_z = 0$, a "zero" is created. No relationship exists between Eqs. (9-1) and (9-2).

Figure 9-7a illustrates the location of this pole or zero as a point on the *s*-plane. Note that when $s + \sigma = 0$, then $s = -\sigma$,* and this is the point on the *s*-plane

---

*Because of the difficulty in understanding this topic at this level, σ (sigma), the real roots on the damping axis, and α (alpha), the real part of the complex roots, will both be represented by σ.

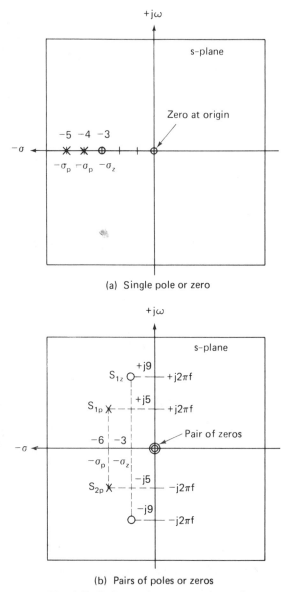

(a) Single pole or zero

(b) Pairs of poles or zeros

Fig. 9-7 Poles and zeros versus s-plane.

where the pole $(-\sigma_p)$ is indicated by an $\times$ and the zero $(-\sigma_z)$ is indicated by a $\bigcirc$. σ is called the *damping factor*. The fact that one is a pole and the other is a zero has no significance on their relative position on the s-plane (i.e., the location of the pole and zero could be reversed on the σ-axis).

When σ_p or σ_z is zero, the pole or zero appears at the origin of the s-plane. Thus the relationship, $1/s$ is a pole at the origin, while the relationship $s/1$ is a zero at the origin of the s-plane. This occurs because $s + 0 = 0$ results in a root of $s = 0$.

An example will help to illustrate this and the relationships in Eqs. (9-1) and (9-2).

EXAMPLE 9-1

a. A transfer function has the form

$$\frac{v_o}{v_{in}} = \frac{1}{s + 5}$$

Determine the numerical value of s at the pole location.

b. A transfer function has the form

$$\frac{v_o}{v_{in}} = s + 3$$

Determine the numerical value of s at the zero location.

c. A transfer function has the form

$$\frac{v_o}{v_{in}} = \frac{s}{s + 4}$$

Determine the numerical value of s at the pole and zero.

Solution

a. $s + 5 = 0$, $\sigma_p = 5$; thus $s = -\sigma_p = -5$, and the pole is plotted 5 units to the left of the origin on the σ-axis (see Fig. 9-7a).

b. $s + 3 = 0$, $\sigma_z = 3$; thus $s = -\sigma_z = -3$, and the zero is plotted 3 units to the left of the origin on the σ-axis.

c. $\sigma_z = 0$; $s + 4 = 0$, $\sigma_p = 4$; thus $s = -\sigma_p = -4$. In the plot, a zero appears at the origin of the s-plane, and a pole appears 4 units to the left of the origin on the σ-axis.

A pair of points is formed on the s-plane when the equation is a quadratic of the form

$$\boxed{s^2 + 2\sigma s + (\sigma^2 + \omega^2) = 0} \tag{9-3}$$

which when the roots are extracted, where $\omega = 2\pi f (\text{rad}/\text{s})$, becomes

$$\boxed{s_1 = -\sigma + j\omega} \qquad \text{true image point} \tag{9-4}$$

and

$$s_2 = -\sigma - j\omega \qquad \text{reflected image point} \tag{9-5}$$

which are illustrated on the s-plane in Fig. 9-7b. When Eq. (9-3) is in the denominator, a pair of "poles" is formed, and when Eq. (9-3) is in the numerator, a pair of "zeros" is created. Equation (9-5) represents the mirror image of the true image point. Again, if an (s^2) alone appears, it indicates a pair of points at the origin; they may be poles or zeros.

An example will help to illustrate how this equation is used to plot a pair of poles or zeros on the s-plane.

EXAMPLE 9-2

a. The transfer function of a circuit with a pair of poles is given by

$$\frac{v_o}{v_{in}} = \frac{1}{s^2 + 12s + 61}$$

Determine the values of σ_p and ω_p. Plot these on an s-plane.

b. The transfer function of a circuit with a pair of zeros is given by

$$\frac{v_o}{v_{in}} = s^2 + 6s + 90$$

Determine the values of σ_z and ω_z. Plot these on the s-plane.

c. The transfer function in part (a) has the same denominator, but the numerator is now s^2. The transfer function appears as

$$\frac{v_o}{v_{in}} = \frac{s^2}{s^2 + 12s + 61}$$

Determine the values of σ_z, ω_z, σ_p, and ω_p. Plot these on an s-plane.

Solution

a. The most formal way to solve the problem would be to let $s^2 + 12s + 61 = 0$ and solve using the quadratic formula. A much easier way to solve the problem is by comparison:
$s^2 + 2\sigma s + (\sigma^2 + \omega^2)$,
$s^2 + 12s + (61)$, $2\sigma = 12$, $\sigma = 6$, $\sigma^2 = 36$, $36 + \omega^2 = 61$, $\omega^2 = 25$, $\omega = 5$ rad/s.
Remember, now, that $\sigma_p = 6$ and $\omega_p = \pm j5$. Thus they are plotted as shown in Fig. 9-7b.

b. $s^2 + 6s + 90$; by comparison, $2\sigma = 6$, $\sigma = 3$, $\sigma^2 = 9$, $90 - 9 = \omega^2 = 81$, $\omega = 9$; then $\sigma_z = 3$ and $\omega_z = \pm j9$. Plot as before.

c. $\sigma_z = 0$, $\omega_z = 0$. σ_p and ω_p are as determined in part (a). The plot is the same as in part (a) except that now a pair of zeros appears at the origin, as illustrated in Fig. 9-7b.

Equations (9-3) to (9-5) were presented for the purpose of demonstrating the way a pole or zero is geometrically located on the s-plane. The actual equation for points on the s-plane is called the "characteristic equation" and is

$$\boxed{s^2 + 2\zeta\omega_0 s + \omega_0^2 = 0} \qquad \text{characteristic equation} \qquad (9\text{-}6)$$

where ζ (zeta) is the *damping ratio* and ω_0 is the *radian frequency* of the point on the s-plane. The roots of Eq. (9-6) yield the actual mathematical relationships for the roots; they are

$$\boxed{s_1 = -\zeta\omega_0 + j\omega_0\sqrt{1 - \zeta^2}} \qquad \text{true image point} \qquad (9\text{-}7)$$

and

$$s_2 = -\zeta\omega_0 - j\omega_0\sqrt{1 - \zeta^2} \qquad \text{reflected image point} \qquad (9\text{-}8)$$

where

$$\sigma = \zeta\omega_0 \qquad \text{rad/s} \qquad\qquad (9\text{-}9)$$

and

$$j\omega \neq j2\pi f = j\omega_0\sqrt{1 - \zeta^2} \qquad \text{rad/s} \qquad\qquad (9\text{-}10)$$

are the coordinates of the points on the s-plane.

An example to help illustrate the use of these equations is given next.

EXAMPLE 9-3

a. The characteristic equation for a transfer function having two zeros is $s^2 + 1000s + 10^8$. Determine the value of σ_z, ω_0, ζ, s_1, and s_2.

b. The characteristic equation for a transfer function having two poles is $s^2 + 10^4 s + 10^{12}$. Determine the value of σ_p, ω_0, ζ, s_1, and s_2.

Solution

a. By comparison,

$$s^2 + 2\zeta\omega_0 s + \omega_0^2$$
$$s^2 + 1000s + 10^8 \qquad\qquad \omega_0^2 = 10^8 \qquad\qquad \omega_0 = 10^4 \text{ rad/s}$$

$$2\zeta(10^4) = 1000 \qquad\qquad \zeta = 0.05.$$

$$\sigma_z = 500$$

$$s_1 = -500 + j10^4\sqrt{1 - 0.0025} = -500 + j9987.5 \text{ rad/s}$$

$$s_2 = -500 - j10^4\sqrt{1 - 0.0025} = -500 - j9987.5 \text{ rad/s}$$

b. By comparison, $s^2 + 10^4 s + 10^{12}$, $\omega_0^2 = 10^{12}$, $\omega_0 = 10^6$ rad/s, $2\zeta(10^6) = 10^4$, and $\zeta = 0.005$.

$$\sigma_p = 5000$$

$$s_1 = -5000 + j10^6\sqrt{1 - 25 \times 10^{-6}} = -5000 + j999{,}987 \text{ rad/s}$$

$$s_2 = -5000 - j10^6\sqrt{1 - 25 \times 10^{-6}} = -5000 - j999{,}987 \text{ rad/s}$$

One of the subtle points that Eqs. (9-9) and (9-10) reveals is that all along the *east face*, where the damping factor ($\sigma = \zeta\omega_0$) is zero, the value of s is given by

$$\boxed{s = j\omega = j2\pi f} \qquad \text{all along the } east\ face \qquad\qquad (9\text{-}11)$$

Thus when ever a frequency response is to be determined, s is set equal to $j\omega$. This will be used to advantage as the chapter develops. The student, at this point, should not be frustrated by the alternate use of f and ω for frequency; they just differ by 2π.

The frequency-response maximum on the *east face* is slightly different from the undamped resonant frequency (ω_0) by the factor $\sqrt{1 - \zeta^2}$.

The following example will illustrate this frequency shift.

EXAMPLE 9-4

Determine the actual resonant frequency for part (a) of Example 9-3.

Solution

Using Eq. (9-10), we obtain

$$j\omega = j10^4\sqrt{1 - 25 \times 10^{-4}} = j9987.5 \text{ rad/s}$$

Thus the actual resonant frequency on the *east face* is 12.5 rad/s below ω_0 (for zero damping) when $\zeta = 0.05$.

Most students have heard the term Q (quality factor) in relation to resonant circuits. ζ is related to Q, which is defined [2, p. 45] as

$$Q = 2\pi \frac{\left(\begin{array}{c} \text{energy stored in} \\ \text{resonant circuit} \end{array} \right)}{\left(\begin{array}{c} \text{energy dissipated in resonant} \\ \text{circuit during one cycle} \end{array} \right)} \tag{9-12}$$

or

$$Q = \frac{\omega_0 L}{R} \tag{9-13}$$

in a resonant circuit. Remember that the bells discussed in Section 9.2.1 are resonant circuits. When one is struck, energy is stored in the bell. As the bell rings, the energy diminishes and the amplitude of sound diminishes according to the relationship for Q given by Eq. (9-12). The damping ratio (ζ) is related to Q by

$$\boxed{Q = \frac{1}{2\zeta}} \tag{9-14}$$

An example will help to show the use of these equations.

EXAMPLE 9-5

The circuit generating the characteristic equation for Example 9-3(b) is a series resonant circuit with $C = 0.001 \ \mu\text{F}$ and $L = 1$ mH; the series circuit initially has 5 μJ (microjoules) of energy in the capacitor.

a. Calculate Q.

b. Calculate R.

c. Calculate the voltage across the capacitor initially (5 μJ in capacitor) and during the third cycle of energy in the capacitor. Remember from physics that $W_C = \frac{1}{2}CV_C^2$.

Solution

a. Using Eq. (9-14), we obtain

$$\zeta = 0.005 \qquad Q = \frac{1}{2\zeta} = 100$$

b. Using Eq. (9-13), we obtain

$$100 = \frac{10^6 10^{-3}}{R} \qquad \text{the series } R = 10 \; \Omega$$

c. V_c in the initial cycle is $5 \; \mu J = \frac{1}{2}(10^{-9} \; F)V_C^2$.

$$V_C = \sqrt{\frac{10 \; \mu J}{10^{-9} \; F}} = 100 \; V$$

The energy in the initial cycle is $5 \; \mu J$. The energy dissipated between the first (initial) and the second cycle is given by Eq. (9-12):

$$\text{energy dissipated} = \frac{2\pi(5 \; \mu J)}{100} = 0.314 \; \mu J$$

The energy in the capacitor during the second cycle is

$$5 - 0.314 \; \mu J = 4.686 \; \mu J, \; V_c = 96.81 \; V$$

The energy dissipated between the second and third cycles is

$$\text{energy dissipated} = \frac{2\pi(4.686 \; \mu J)}{100} = 0.294 \; \mu J$$

The energy during the third cycle is $4.686 - 0.294 \; \mu J = 4.391 \; \mu J$

$$V_C\sqrt{\frac{2(4.391 \; \mu J)}{10^{-9} \; F}} = 93.71 \; V \qquad \text{from the energy equation}$$

See Fig. 9-2a for resulting waveform.

The bandwidth (f_{BW}) of a resonant circuit, illustrated in Fig. 9-8a, is related to Q by

$$\boxed{f_{BW} = \frac{f_0}{Q}} \tag{9-15}$$

where the bandwidth is the frequency between the upper and lower -3 dB [0.707 of amplitude at resonance (f_0)]. Figure 9-8b and c show the effects of a pole moving farther away from the *east face* (frequency axis), where the frequency response appears. [*Note*: Eq. (9-15) is valid only for $Q > 10$.]

An example will show the use of Eq. (9-15).

EXAMPLE 9-6

Calculate the bandwidth of the frequency response projected on the *east face* due to the pair of poles in part (b) of Example 9-3.

Solution

$$Q = \frac{1}{2\zeta} = 100$$

$$f_0 = \frac{\omega_0}{2\pi} = \frac{10^6}{2\pi} = 159 \; kHz$$

From Eq. (9-15),

$$f_{BW} = \frac{f_0}{Q} = \frac{159 \; kHz}{100} = 1.59 \; kHz$$

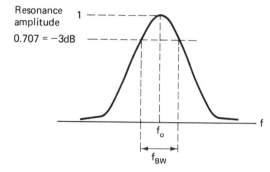

(a) Frequency response showing bandwidth

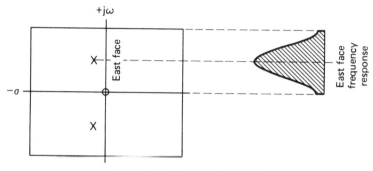

(b) Small bandwidth or high Q

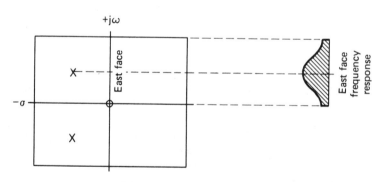

(c) Large bandwidth or low Q

Fig. 9-8 Bandwidth of resonant frequency-response curve.

Equations (9-13) to (9-15) permit us to relate Eqs. (9-7) and (9-8) in terms of Q as

$$s_1 = -\frac{\pi f_0}{Q} + j2\pi f_0\sqrt{1 - \frac{1}{4Q^2}} \qquad \text{true image} \qquad (9\text{-}16)$$

and

$$s_2 = -\frac{\pi f_0}{Q} - j2\pi f_0\sqrt{1 - \frac{1}{4Q^2}} \qquad \text{reflected image} \qquad (9\text{-}17)$$

and in terms of bandwidth for Q's > 10 as

$$s_1 = -\pi(f_{BW}) + j2\pi\sqrt{f_0^2 - \left(\frac{f_{BW}}{2}\right)^2} \qquad \text{true image} \qquad (9\text{-}18)$$

and

$$s_2 = -\pi(f_{BW}) - j2\pi\sqrt{f_0^2 - \left(\frac{f_{BW}}{2}\right)^2} \qquad \text{reflected image} \qquad (9\text{-}19)$$

Through these equations it becomes apparent that the unfamiliar damping ratio (ζ) is indeed related to the familiar terms Q and bandwidth. [Eqs. (9-18) and (9-19) are only valid for $Q > 10$.]

An example will help to illustrate the use of these equations.

EXAMPLE 9-7

a. A resonant circuit with a center frequency of 250 Hz has $Q = 50$. Determine the location of the pole pair on the s-plane.

b. A tuned circuit has an undamped resonance frequency of 3000 Hz. It has a bandwidth of 150 Hz.

   1. Locate the pole pair on the s-plane.

   2. Determine the actual frequency of the tuned circuit.

Solution

a.

$$s_1 = -\frac{250\pi}{50} + j500\pi\sqrt{1 - \frac{1}{4(2500)}}$$

$$s_2 = -\frac{250\pi}{50} - j500\pi\sqrt{1 - \frac{1}{4(2500)}}$$

$\sigma_p = 15.7$ rad/s; when $\sigma = 0$, $s = j\omega = 1570\sqrt{1 - 10^{-4}}$ rad/s $= 1569.2$ rad/s.

b.

   1. $s_1 = -\pi(150) + j2\pi\sqrt{9 \times 10^6 - \left(\frac{150}{2}\right)^2}$

     $\sigma_p = 471.2$ rad/s; when $\sigma = 0$, $s = j\omega = j2\pi(2999) = j(18.8 \times 10^3)$ rad/s.

   2. $jf = \frac{j\omega}{2\pi} = j2999$ Hz.

9.3   Single Op Amp Active Filter Sections

Active-filter representations of passive filters can be accomplished in many ways. One way of modeling passive filters, using op amps, involves the use of one op amp per single pole or one op amp per pole pair. The meaning of this will become more apparent as the section develops. If a multipole filter is desired, several appropriate single op amp filter sections are cascaded (connected in series) to form the required pole-zero plot for the desired composite frequency response. The low output impedance of the op amp circuit isolates one section from another so that they do not interact.

Passive filters fall into four general categories: low-pass, high-pass, band-pass and band-reject. The "passband" of a filter is defined as the range of frequencies where the input signal is passed without attenuation. The "rejection band" is defined as the range of frequencies where the input signal has a maximum of attenuation as it passes through the filter. The dividing line between the passband and the rejection band is the "cutoff" frequency. It has been shown [3, p. 404] that the high-pass, band-pass and band-reject filter characteristics can be determined using those of the low-pass filter and some simple transformations.

Passive-filter networks, in which all the components can be drawn without any two lines crossing, and connecting, have a transfer function called a "minimum phase" function [3, p. 346]. In a minimum phase function, it can be shown [3, p. 346] that "zeros of transmission," * not appearing directly in the transfer function, are either *on* the $\pm j\omega$ axis or in the *left half* of the s-plane. The importance of this situation is that the circuit is unconditionally stable and that all components are positive and real (no negative impedances or unrealizable components). A bridge circuit causes a "nonminimum phase" function to occur, as it has at least one line that crosses another (the circuit is not planar); this circuit will have zeros in the right half-plane, as will be seen in Section 9.4.4.

9.3.1 Low-Pass Filters

The passive low-pass filter passes input signals from dc ($j\omega = 0$) to some "cutoff" frequency where the output amplitude has reduced to 0.707 (-3 dB) of that at dc.

Figure 9-9a illustrates the circuit for a one-pole low-pass filter; its transfer-function equation is

$$\frac{v_o}{v_{\text{IN}}} = \frac{1/RC}{s + 1/RC} \tag{9-20}$$

where it is immediately apparent that it has one pole on the σ-axis of the s-plane; this pole is at $s = -\sigma_p = -1/RC$. The pole, plotted on the s-plane together with the *east face* frequency response, is shown in Fig. 9-9c. Take particular note of the form of the transfer function; it has only poles, no zeros. Also, the multiplier on the equation is σ_p. This is characteristic of the *low-pass* transfer function and should be remembered, as it is useful to be able to recognize particular filter types from the transfer function only. The cutoff frequency occurs at the point where the amplitude has reduced to 0.707 of that at dc and is given by

$$\boxed{f_c = \frac{1}{2\pi RC}} \tag{9-21}$$

The low-pass filter will pass dc, as can be seen by the circuit in Fig. 9-9a. An example will illustrate the use of Eqs. (9-20) and (9-21).

---

*A zero of transmission is formed when the reactance of a shunt capacitor goes to zero at $f = \infty$. A series inductor also forms a zero of transmission as its reactance goes to infinity at $f = \infty$. Also, it is the "impedance pole" of a series branch or the impedance zero of a shunt branch of the passive filter network.

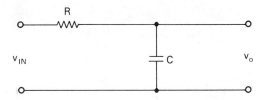

(a) Passive one pole low pass filter section

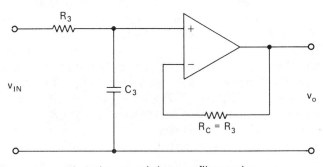

(b) Active one pole low pass filter section

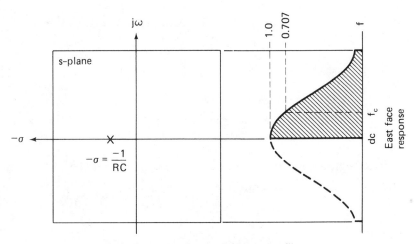

(c) Pole location for one pole low pass filter

Fig. 9-9 One-pole low-pass filter.

EXAMPLE 9-8

The transfer function of a single-pole low-pass filter is

$$\frac{v_o}{v_{\text{IN}}} = \frac{1256}{s + 1256}$$

Determine the location of the pole and the cutoff frequency for the low-pass filter.

$$\sigma_p = 1256 \text{ rad/s}$$

$$f_c = \frac{1256}{2\pi} = 200 \text{ Hz}$$

Above the cutoff frequency, the amplitude decreases at the rate of 20 dB/decade (6 dB/octave), as described in Chapter 5.

A zero of transmission exists at infinity, as the capacitive reactance is zero at $f = \infty$, which causes the elastic membrane to tend toward zero as the frequency increases. It is apparent that the transfer function has a lower power of s in the numerator than in the denominator and it goes to zero as s becomes infinity.

In Fig. 9-9b, it can be seen that the passive filter (Fig. 9-9a) and the active filter are identical except for a voltage follower which prevents the RC network from being loaded and provides a low output impedance for isolation from the following state. Equations (9-20) and (9-21) still apply and Fig. 9-9c also applies. The current-compensation resistor (R_C) is made equal to R.

A two-pole low-pass filter is composed of a series RLC circuit as shown in Fig. 9-10a, where the output voltage is taken across the capacitor. Again this filter will pass signal voltages from dc to the cutoff frequency with little attenuation. The transfer-function equation for the two-pole low-pass filter is

$$\frac{v_o}{v_{IN}} = \frac{\frac{1}{LC}}{s^2 + \frac{R_0}{L}s + \frac{1}{LC}} \tag{9-22}$$

where several features should be particularly noted:

1. Again, it has only poles (two) and no zeros.
2. The transfer-function equation has a multiplier of $1/LC$, which is ω_0^2 [see Eq. (9-6)], the constant term.

By comparing Eq. (9-22) with the characteristic equation (9-6), it can be seen that

$$\omega_0 = \sqrt{\frac{1}{LC}} \text{ rad/s} \tag{9-23}$$

and

$$2\zeta\omega_0 = \frac{R_0}{L} \tag{9-24}$$

Thus

$$\sigma_p = \zeta\omega_0 = \frac{R_0}{2L} \text{ rad/s} \tag{9-25}$$

Therefore,

$$\zeta = \frac{R_0}{2}\sqrt{\frac{C}{L}} \tag{9-26}$$

and

$$j\omega_p = j\sqrt{\frac{1}{LC}}\sqrt{1 - \frac{R_0^2 C}{4L}} \quad \text{rad/s} \tag{9-27}$$

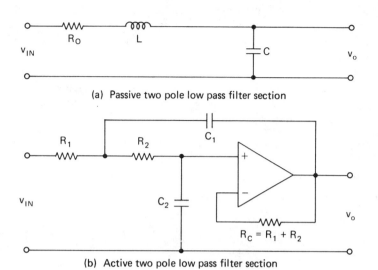

(a) Passive two pole low pass filter section

(b) Active two pole low pass filter section

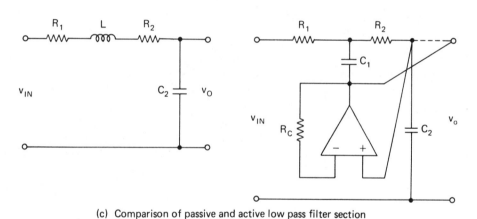

(c) Comparison of passive and active low pass filter section

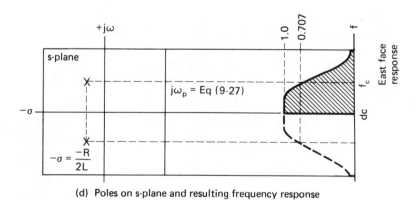

(d) Poles on s-plane and resulting frequency response

Fig. 9-10 Two-pole low-pass filter.

where σ_p and $j\omega_p$ determine the coordinates of the pair of poles on the s-plane. Figure 9-10d shows a plot of these poles and the resulting frequency response along the *east face*. The response is unity as it passes through dc and reduces to 0.707 of this amplitude at the pole frequency ($j\omega_p$), whereafter it drops at 40 dB/decade (12 dB/octave).

An example will illustrate the use of these equations.

EXAMPLE 9-9

A passive low-pass filter (Fig. 9-10a) has $R = 5\ \Omega$, $L = 4$ H, and $C = 40\ \mu$F.

    a. Determine the low-pass transfer-function equation.

    b. Calculate ω_0, ζ, σ_p, $j\omega_p$, and jf_p.

Solution

a. $\quad \dfrac{v_o}{v_{IN}} = (6250)\dfrac{1}{s^2 + \dfrac{5}{4}s + 6250}$

$\qquad\quad = (6250)\dfrac{1}{s^2 + 1.25s + 6250}$

b. $\quad \omega_0 = \sqrt{6250} = 79.05$ rad/s

$\qquad \zeta = \dfrac{5}{2}\sqrt{\dfrac{40 \times 10^{-6}}{4}} = (2.5)(3.162 \times 10^{-3}) = 0.0079$

$\qquad \sigma_p = \dfrac{R}{2L} = \dfrac{5}{8} = 0.625$ rad/s; when $\sigma = 0$,

$\qquad\qquad\qquad s = j\omega_p = (79.05)\sqrt{1 - \dfrac{25(40 \times 10^{-6})}{4(4)}}$

$\qquad\qquad\qquad\qquad = 79.05(0.999968) = 79.04$ rad/s

$\qquad\qquad\qquad jf_p = \dfrac{79.04}{2\pi} = 12.6$ Hz

The Sallen-Key active-filter version of the two-pole low-pass filter is illustrated in Fig. 9-10b. The inductor is replaced by a capacitor and voltage source (voltage follower). This can be seen by comparing the two circuits illustrated in Fig. 9-10c, where the resistor in the passive filter has been split into two parts, one on each side of the inductor. The transfer function for the low-pass filter is

$$\frac{v_o}{v_{in}} = \left[\frac{1}{(R_1 R_2 C_1)C_2}\right]\frac{1}{s^2 + \dfrac{R_1 + R_2}{R_1 R_2 C_1}s + \dfrac{1}{(R_1 R_2 C_1)C_2}} \qquad (9\text{-}28)$$

where by comparison with Eq. (9-22),

$$R_o = R_1 + R_2 \qquad\qquad \Omega \qquad\qquad (9\text{-}29)$$

and

$$L = R_1 R_2 C_1 \qquad\qquad \text{H} \qquad\qquad (9\text{-}30)$$

256

while

$$C = C_2 \qquad\qquad \text{F} \qquad\qquad (9\text{-}31)$$

By comparing Eq. (9-28) with the characteristic equation (9-6), it is found that

$$\omega_0 = \sqrt{\frac{1}{(R_1 R_2 C_1) C_2}} = \sqrt{\frac{1}{LC}} \qquad \text{rad/s} \qquad (9\text{-}32)$$

and

$$\sigma_p = \zeta\omega_0 = \frac{R_1 + R_2}{2(R_1 R_2 C_1)} = \frac{R_0}{2L} \qquad \text{rad/s} \qquad (9\text{-}33)$$

while

$$\zeta = \frac{R_1 + R_2}{2} \sqrt{\frac{C_2}{R_1 R_2 C_1}} = \frac{R_0}{2}\sqrt{\frac{C}{L}} \qquad\qquad (9\text{-}34)$$

An example will show how these equations are used.

EXAMPLE 9-10

A two-pole active low-pass filter has the following circuit components:

$$R_1 = R_2 = 10\ \text{k}\Omega$$
$$C_1 = 0.2\ \mu\text{F}$$
$$C_2 = 0.1\ \mu\text{F}$$

Determine the numerical value of R_o, L, C, ω_0, σ_p, and ζ.

Solution

$$R_o = 2(10\ \text{k}\Omega) = 20\ \text{k}\Omega$$

$$L = (10\ \text{k}\Omega)^2 (0.2\ \mu\text{F}) = 20\ \text{H}$$

$$C = 0.1\ \mu\text{F}$$

$$\omega_0 = \sqrt{\frac{1}{(20\ \text{H})(0.1\ \mu\text{F})}} = 707\ \text{rad/s}$$

$$\sigma_p = \frac{20\ \text{k}\Omega}{2(20\ \text{H})} = 500\ \text{rad/s}$$

$$\zeta = \frac{\sigma_p}{\omega_0} = \frac{500}{707} = 0.707$$

or

$$\zeta = \frac{20\ \text{k}\Omega}{2}\sqrt{\frac{0.1\ \mu\text{F}}{20\ \text{H}}} = 0.707$$

9.3.2 Normalization of Circuit Components

It would be virtually an impossible task to develop the component values for filters if each one were an individual case. A method to reduce the calculations is presented in Franklin F. Kuo's excellent book, *Network Analysis and Synthesis*, [3, pp. 402–410] where he outlines a method to normalize the component values to

$R_n = 1 \; \Omega$ and $\omega_n = 1$ rad/s. Equation (9-20) would be written as

$$\boxed{\frac{v_o}{v_{\text{IN}}} = \frac{1/C_n}{s + 1/C_n}} \tag{9-35}$$

where the subscript n indicates a normalized capacitance value.

For the two-pole low-pass transfer function, Eq. (9-22) would be rewritten, when normalized, to be

$$\boxed{\frac{v_o}{v_{\text{IN}}} = \frac{1/L_n C_n}{s^2 + \dfrac{1}{L_n}s + \dfrac{1}{L_n C_n}}} \tag{9-36}$$

where L_n is the normalized inductance (nearly 1 H) and C_n is the normalized capacitance (nearly 1 F).

A normalized characteristic equation would look like those listed in Table 9-1, where the undamped resonant radian frequency (ω_0) and the characteristic impedance (R_0) are unity (1).

The most important skill to be developed in this section is the ability to instantly recognize L_n and C_n from the normalized characteristic equations (9-35) and (9-36). It is not difficult and the work in the balance of the chapter depends on this ability. It is best shown by example.

EXAMPLE 9-11

Determine the L_n and C_n values for the five-pole normalized Butterworth low-pass filter. The problem must be solved in two parts:

    a. The one-pole filter section.

    b. Two individual sections of two-pole filters.

Solution

The characteristic equation for the five-pole Butterworth filter from Table 9-1 is

$$(s + 1) \qquad (s^2 + 0.6180s + 1) \qquad (s^2 + 1.6180s + 1)$$
$$\text{1 pole} \qquad\qquad \text{2 poles} \qquad\qquad\qquad \text{2 poles}$$

where the three sections are clearly delineated.

    a. Comparison of $s + 1$ with $s + 1/C_n$, shows that $C_n = 1$ F.

    b. Comparison of $s^2 + 0.6180s + 1$ with $s^2 + (1/L_n)s + 1/L_n C_n$ shows that $1/L_n C_n = 1$; thus $C_n = 1/L_n = 0.6180$ F and $L_n = 1/0.6180 = 1.6180$ H for the first two-pole filter section. For the second section the values can be determined by inspection. $s^2 + 1.6180s + 1$, $C_n = 1.6180$ F, and $L_n = 0.6180$ H.

The pattern of poles for the Butterworth filter characteristic is an equally spaced set of points on a circle of radius ($\zeta\omega_0$) in the left half of the s-plane; the angle between poles is $180°$/poles, as illustrated in Fig. 9-11. The Butterworth filter is

Table 9-1   Normalized Low-Pass Butterworth Polynomials (Factored Form)

Poles	Characteristic Equation (Normalized)
1	$(s + 1)$
2	$(s^2 + \sqrt{2}\,s + 1)$
3	$(s + 1)(s^2 + s + 1)$
4	$(s^2 + 0.76536s + 1)(s^2 + 1.8477s + 1)$
5	$(s + 1)(s^2 + 0.6180s + 1)(s^2 + 1.6180s + 1)$
6	$(s^2 + 0.5176s + 1)(s^2 + \sqrt{2}\,s + 1)(s^2 + 1.9318s + 1)$

"maximally flat" in the passband, as the location of the poles causes the membrane to be stretched out in a straight line as it passes through the *east face* window in the area of the passband. The steepness of the attenuation "skirt" increases as the number of poles increases; the beginning of this trend can be observed by comparing Figs. 9-9c and 9-10d.

9.3.3 Active Low-Pass Frequency and Impedance Transformations

Three active low-pass filter sections are shown in Fig. 9-12. The component values are found by using the normalized polynomials (Trade 9-1, for Butterworth) and a transformation equation. For the one-pole low-pass filter section

$$R_o = R_3 \qquad (9\text{-}37)$$

and

$$C_3 = \frac{C_n}{\omega_0 R_3} \qquad (9\text{-}38)$$

An example will help to illustrate the use of these equations.

EXAMPLE 9-12

A one-pole Butterworth low-pass filter has a cutoff frequency of 200 Hz and should use a resistor (R_3) between 5 kΩ and 10 kΩ. Determine the appropriate capacitor to use.

Solution

From Eq. (9-35),

$$s + \frac{1}{C_n} = s + 1 \qquad C_n = 1$$

Then from Eq. (9-38),

$$\text{If } R_3 = 5\,\text{k}\Omega: \quad C_3 = \frac{1}{2\pi(200)(5\,\text{k}\Omega)} = 0.16\,\mu\text{F}$$

$$\text{If } R_3 = 10\,\text{k}\Omega: \quad C_3 = \frac{1}{2\pi(200)(10\,\text{k}\Omega)} = 0.08\,\mu\text{F}$$

Thus the capacitor should range between 0.08 and 0.16 μF. Choose the 0.1 μF capacitor from

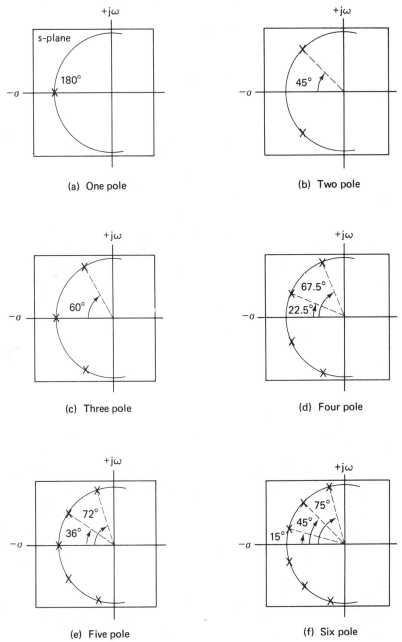

(a) One pole

(b) Two pole

(c) Three pole

(d) Four pole

(e) Five pole

(f) Six pole

Fig. 9-11 Butterworth poles on *s*-plane.

Appendix B2; then using Eq. (9-38)

$$R_3 = \frac{C_n}{\omega_0 C_3} = \frac{1}{2\pi(200)(0.1\ \mu F)} = 7.96\ k\Omega$$

Use the 7.87 kΩ resistor from Appendix B1 for R_3 as it is the one nearest to the ideal value.

For the two-pole low-pass filter, two possible methods are available for finding the components:

1. Choose the capacitor values and calculate the resistor values.
2. Choose the resistor values and calculate the capacitor values.

Method 1 (Better). The capacitor values are chosen initially and they must:

1. Satisfy the inequality

$$\boxed{C_1 \geq 4\frac{L_n}{C_n}C_2} \tag{9-39}$$

2. The capacitive reactance should be between 2 and 20 kΩ and film capacitors from Appendix B2 should be used.

Then

$$\boxed{R_1 = \frac{C_n}{2C_2\omega_0}\left(1 + \sqrt{1 - 4\frac{L_n C_2}{C_n C_1}}\right)} \tag{9-40}$$

[for conditions of Eq. (9-39)]

$$\boxed{R_2 = \frac{C_n}{2C_2\omega_0}\left(1 - \sqrt{1 - 4\frac{L_n C_2}{C_n C_1}}\right)} \tag{9-41}$$

are valid; either R_1 or R_2 may have the larger value. For those with a programmable electronic calculator, the quadratic equation for R_1 and R_2 is

$$R^2 - \frac{C_n}{\omega_0 C_2}R + \frac{L_n C_n}{\omega_0^2 C_1 C_2} = 0 \tag{9-42}$$

where the inequality in Eq. (9-39) must be satisfied before using the quadratic equation. The two roots are R_1 and R_2, where it is unimportant which root is larger.

When the inequality Eq. (9-39) is equal, or $C_1 = 4(L_n/C_n)C_2$,

$$R_1 = R_2 = \frac{C_n}{2C_2\omega_0} \qquad \left(\text{for } C_1 = 4\frac{L_n}{C_n}C_2\right) \tag{9-43}$$

This is not usually a desirable situation, as it results in at least one capacitor with an irrational value (see Example 9-14). It is best to select capacitors with standard values that just meet the inequality conditions and then use resistors with irrational values which are more easily obtained.

An example will illustrate the use of Eqs. (9-39) to (9-41).

EXAMPLE 9-13

A two-pole active low-pass filter is to operate at $\omega_0 = 1000$ rad/s; the characteristic equation is $s^2 + 1.414s + 1$.

Method 1   Select C_1 and C_2 from Appendix B2 (film) and calculate R_1 and R_2; select the nearest values of R_1 and R_2 from Appendix B1 (1%).

Solution (Method 1)

$C_n = 1.414$, $L_n = 0.707$. Using Eq. (9-39), we obtain $C_1 > 4(0.707/1.414)C_2$; thus C_1 must be slightly greater than $2C_2$. From the experience gained in Example 9-12, the range of capacitor values should be between 0.04 and 0.14 μF. Knowing this, choose $C_2 = 0.047\ \mu$F; then if $C_1 = 0.1\ \mu$F, it is slightly more than twice C_2. The conditions of Eq. (9-39) have been met.

$$R_1 = \frac{1.414}{2(0.047\ \mu\mathrm{F})1000}\left(1 + \sqrt{1 - 4\frac{0.707(0.047\ \mu\mathrm{F})}{1.414(0.10\ \mu\mathrm{F})}}\right)$$

$$R_1 = 15{,}042(1 + \sqrt{0.06})$$

$$R_1 = 15{,}042(1 + 0.245) = 18.73\ \mathrm{k}\Omega,$$

use 18.7 kΩ 1%; then

$$R_2 = 15042(1 - 0.245) = 11.36\ \mathrm{k}\Omega,$$

use 11.3 kΩ 1% from Appendix B1.

Method 2. The resistor values (R_1 and R_2) may be equal or have a variety of ratios where R_1 may be greater than or less than R_2. Typical values for the two resistors range between 2 and 20 kΩ; a good beginning value is 10 kΩ. The capacitor values are then found using

$$C_1 = \frac{L_n}{\omega_0\left(\dfrac{R_1R_2}{R_1 + R_2}\right)} \qquad (\text{for } R_1 \neq R_2) \qquad (9\text{-}44\mathrm{a})$$

and

$$C_2 = \frac{C_n}{\omega_0(R_1 + R_2)} \qquad (\text{for } R_1 \neq R_2) \qquad (9\text{-}45\mathrm{a})$$

If $R_1 = R_2 = R$, then

$$C_1 = \frac{2L_n}{\omega_0 R} \qquad (\text{for } R_1 = R_2) \qquad (9\text{-}44\mathrm{b})$$

and

$$C_2 = \frac{C_n}{2\omega_0 R} \qquad (\text{for } R_1 = R_2) \qquad (9\text{-}45\mathrm{b})$$

If $R = R_1 = 2R_2$ or if $R = R_2 = 2R$, then

$$C_1 = \frac{3}{2} \frac{L_n}{\omega_0 R} \qquad \text{(for } R_1 = 2R_2 \text{ or } R_2 = 2R_1) \qquad (9\text{-}44c)$$

and

$$C_2 = \frac{1}{3} \frac{C_n}{\omega_0 R} \qquad \text{(for } R_1 = 2R_2 \text{ or } R_2 = 2R_1) \qquad (9\text{-}45c)$$

An example of the use of Method 2 equations follows.

EXAMPLE 9-14

A two-pole low-pass filter has the resistor values shown below. $L_n = 0.707$, $C_n = 1.414$, and $\omega_0 = 1000$ rad/s.

a. $R_1 = 15$ kΩ, $R_2 = 20$ kΩ.
b. $R_1 = 10$ kΩ, $R_2 = 10$ kΩ.
c. $R_1 = 10$ kΩ, $R_2 = 20$ kΩ.

Determine the values of C_1 and C_2 for all cases.

Solution (Method 2)

a. $C_1 = \dfrac{0.707}{(1000)\left(\dfrac{15 \times 20}{35}\text{ k}\Omega\right)} = 0.0825 \ \mu\text{F}$

$C_2 = \dfrac{1.414}{(1000)(35\text{ k}\Omega)} = 0.0404 \ \mu\text{F}$

b. $C_1 = \dfrac{2(0.707)}{(1000)(10\text{ k}\Omega)} = 0.1414 \ \mu\text{F}$

$C_2 = \dfrac{1.414}{2(1000)(10\text{ k}\Omega)} = 0.0707 \ \mu\text{F}$

c. $C_1 = 1.5\left[\dfrac{0.707}{1000(10\text{ k}\Omega)}\right] = 0.1061 \ \mu\text{F}$

$C_2 = \dfrac{1}{3}\left[\dfrac{1.414}{1000(10\text{ k}\Omega)}\right] = 0.0471 \ \mu\text{F}$

The complete five-pole active low-pass filter is shown in Fig. 9-12. The order of the sections is not important. Notice that dc coupling is maintained throughout, as is expected for a low-pass filter. The amplifier-voltage offsets will have to be adjusted to zero (see Chapter 3).

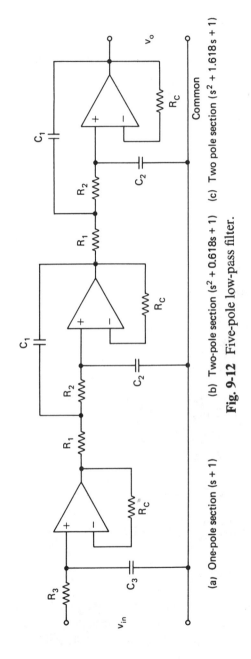

(a) One-pole section $(s + 1)$ (b) Two-pole section $(s^2 + 0.618s + 1)$ (c) Two pole section $(s^2 + 1.618s + 1)$

Fig. 9-12 Five-pole low-pass filter.

9.3.4 Characteristic Equations for Filters

Table 9-1 listed the normalized characteristic equations for the Butterworth filter approximation. This is only the first approximation for filters shapes, although it is the most familiar. It produces the frequency rolloff found in most amplifier circuits and was the filter shape discussed in Chapter 5. It is by no means the best filter shape. Each filter approximation was developed to overcome a particular disadvantage in other approximations. With each particular advantage in an approximation, several disadvantages manifest themselves. The user must decide which of the disadvantages can be tolerated to justify the advantage of that particular filter approximation.

Six filter approximations are relevant.

1. Butterworth (maximally flat magnitude).
2. Chebyshev (Tchebycheff or equal ripple).
3. Bessel (Thompson or maximally flat delay).
4. Optimal (Legendre or monotonic L).
5. Parabolic (P)
6. Elliptic (Cauer or biquadratic).

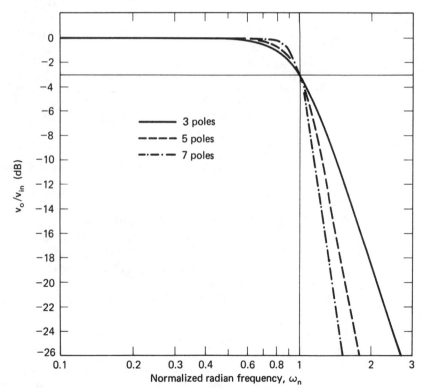

Fig. 9-13 Response curves for Butterworth filter. (From *Network Analysis and Synthesis*, Franklin F. Kuo Copyright © 1966: Reprinted by permission of John Wiley & Sons, Inc.)

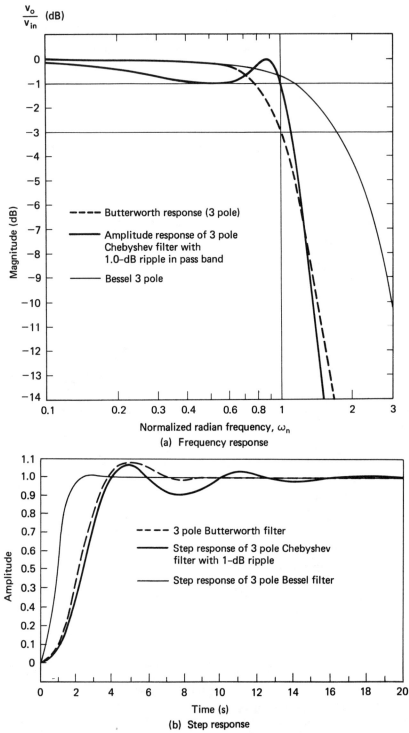

Fig. 9-14 Comparison of three-pole low-pass filters. (From *Network Analysis and Synthesis*, Franklin F. Kuo; Copyright © 1966: Reprinted by permission of John Wiley & Sons, Inc.)

The Butterworth filter has the characteristics of being "maximally flat" in the passband; the slope is as flat as possible at dc [3, p. 369]. This means that the frequency response is level out to some frequency just before cutoff, where the response begins to diminish. The poles of a Butterworth filter are located on a half-circle of radius $\zeta\omega_0$ in the left half-plane. Regardless of how many poles exist, they all lie, equally spaced, on this half-circle, as illustrated in Fig. 9-11. The response of the Butterworth filter for various numbers of poles is illustrated in Fig. 9-13. The frequency response of this filter is monotonic (i.e., it is always decreasing, even though the rate of decrease changes as the frequency increases).

The Chebyshev filter approximation was developed in an attempt to improve on the steepness of the skirt beyond cutoff. To accomplish this, the poles were placed on an ellipse instead of a circle. As the poles become closer to the $j\omega$-axis, their proximity to the *east face* can be seen as humps in the passband near the pole locations. The Chebyshev poles are located in such a manner that the peaks and valleys of these humps are equal in height between dc and cutoff frequencies for the low-pass filter. These humps are called "ripple" in the passband; the Chebyshev filter is *not* monotonic because of the ripple, although it is monotonic beyond cutoff. Figure 9-14a illustrates the comparison between rolloffs of Butterworth and Chebyshev filters of the third order (three poles).

The amount of ripple in the passband of a Chebyshev filter is specified in dB, where each amount of ripple creates a different set of normalized characteristic equations. Three sets of characteristic equations plus a method of finding those for other amounts of ripple will be given. Table 9-2 lists the characteristic equations for 0.25 dB, 0.5 dB, and 1 dB passband ripple for a Chebyshev filter.

The circuit for the low-pass Chebyshev filter is identical to that for the Butterworth; only the component values change. The variables, L_n and C_n, which describe the pole locations are different. This is evident by observing the coefficients of the terms in the polynomial equations. The methods for determining L_n and C_n are identical to those used in Example 9-11 except that the constant term is not unity.

An example will help illustrate the way to determine L_n and C_n for the Chebyshev low-pass filter.

EXAMPLE 9-15

Determine L_n and C_n for the three-pole low-pass Chebyshev Filter with 1 dB passband ripple.

Solution

From Table 9-2, the characteristic equation is

$$(s + 0.494)(s^2 + 0.494s + 0.994)$$

For the one-pole low-pass filter, Eq. (9-35) gives C_n from

$$s + \frac{1}{C_n} = s + 0.494 \qquad \text{as } C_n = \frac{1}{0.494} = 2.023\ F$$

Table 9-2   Normalized Low-Pass Chebyshev Polynomials (Factored Form)*

Poles	(0.25 dB Ripple) Characteristic Equation (Normalized)
1	$(s + 4.10811)$
2	$(s^2 + 1.7967s + 2.1140)$
3	$(s + 0.76722)(s^2 + 0.7672s + 1.3386)$
4	$(s^2 + 0.4250s + 1.1619)(s^2 + 1.0261s + 0.4548)$
5	$(s + 0.43695)(s^2 + 0.2700s + 1.0954)(s^2 + 0.70700s + 0.5364)$

Poles	(0.5 dB Ripple) Characteristic Equation (Normalized)
1	$(s + 2.8628)$
2	$(s^2 + 1.4256s + 1.5162)$
3	$(s + 0.6264)(s^2 + 0.6264s + 1.1424)$
4	$(s^2 + 0.3507s + 1.0635)(s^2 + 0.8467s + 0.3564)$
5	$(s + 0.36233)(s^2 + 0.5862s + 0.4767)(s^2 + 0.2239s + 1.0358)$

Poles	(1.0 dB Ripple) Characteristic Equation (Normalized)
1	$(s + 1.9652)$
2	$(s^2 + 1.0977s + 1.1025)$
3	$(s + 0.4942)(s^2 + 0.4942s + 0.9942)$
4	$(s^2 + 0.2791s + 0.9865)(s^2 + 0.6737s + 0.2794)$
5	$(s + 0.2895)(s^2 + 0.1789s + 0.9883)(s^2 + 0.4684s + 0.4293)$

*From *Principles of Active Network Synthesis and Design*, Gobind Daryanani; Copyright © 1976: Reprinted by permission of John Wiley & Sons, Inc.

For the two-pole low-pass section, C_n and L_n are found from Eq. (9-36) by comparison:

$$s^2 + 0.4942s + 0.9942 = s^2 + \frac{1}{L_n}s + \frac{1}{L_n C_n}$$

where it can be seen that $L_n = 1/0.4942 = 2.023$ H and $1/(2.023 \text{ H})(C_n) = 0.9942$, $C_n = 0.4972$ F.

Since this function has an even number of poles, the response at dc should be -1 dB (0.8921) from that given by the low pass equation with 0.9942 in the numerator. The actual numerator constant should be $(0.9942)(0.8921) = 0.8861$ and the divider ratio of the preceding voltage divider is 0.8921.

$$R_A = 10 \text{ k}\Omega \left[\frac{(1 - 0.8921)}{0.8921} \right] = 12.2 \text{ k}\Omega$$

Use a 12.2 kΩ from Appendix B1.

The normalized characteristic polynomials for the Chebyshev filter can be determined from the Butterworth polynomials using the following conversion scheme: [3, pp. 376–379]

1. The individual Butterworth polynomials are separated into one root or two roots. For one root, $s = -\sigma$; for two roots, $s_1 = -\sigma + j\omega$ and $s_2 = -\sigma - j\omega$. This separation of two roots is done using Eqs. (9-1), (9-3), (9-4), and (9-5) and the methods of Examples 9-1 and 9-2.

2. A term β_k is found from the equation

$$\beta_k = \left(\frac{1}{\text{poles}}\right)\sinh^{-1}\left(\frac{1}{\delta}\right) \qquad (9\text{-}46)$$

and δ is found from

$$\delta = \sqrt{10^{\text{dB}/10} - 1} \qquad (9\text{-}47)$$

where dB is the amount of ripple in the passband. If $\sinh^{-1}$ is not a function on your electronic calculator, then the relationship

$$\sinh^{-1}\left(\frac{1}{\delta}\right) = \ln\left(\frac{1}{\delta} + \sqrt{\left(\frac{1}{\delta}\right)^2 + 1}\right) \qquad (9\text{-}48)$$

can be used to find $\sinh^{-1}(1/\delta)$.

3. The real parts (σ) of the roots are multiplied by the hyperbolic tangent of β_k, or $\tanh \beta_k$.

4. Both the new real and old imaginary parts of the roots are multiplied by the hyperbolic cosine of β_k, or $\cosh \beta_k$. [*Note:* These functions are some times specified as HYP COS and HYP TAN on electronic calculators or use the Euler* expressions, $\cosh x = (\varepsilon^x + \varepsilon^{-x})/2$ and $\tanh x = (\varepsilon^x - \varepsilon^{-x})/(\varepsilon^x + \varepsilon^{-x})$.]

5. The polynomials are then reassembled from the new roots to form the new normalized characteristic equations for the Chebyshev filter. An example will help to illustrate the use of these steps.

EXAMPLE 9-16

Determine the normalized Chebyshev polynomials for the three pole filter with 1 dB passband ripple using the Butterworth polynomials and the five steps.

Solution

From Table 9-1, the Butterworth polynomials are $(s + 1)(s^2 + s + 1)$.

1. One pole: $\sigma = 1$
   Two poles: $(s^2 + 1s + 1)$
   $\qquad\qquad\quad (s^2 + 2\sigma s + \sigma^2 + \omega^2)$
   $\qquad\qquad\quad s_1 = -0.5 + j0.8660$
   $\qquad\qquad\quad s_2 = -0.5 - j0.8660$
2. For 1 dB, $\delta = \sqrt{10^{(0.1)} - 1} = \sqrt{0.2589} = 0.5088$.

$$\sinh^{-1}\left(\frac{1}{0.5088}\right) = \ln(4.17025) = 1.4279$$

$$\beta_k = \frac{1.4279}{3} = 0.4760 \qquad \text{from Eq. (9-46)}$$

*ε is the base of the natural logarithm (2.718282).

3. Tanh(0.4760) = 0.443

One pole: σ_{new} = 1.0(0.443) = 0.443
Two poles: σ_{new} = 0.5(0.443) = 0.2215

$s_1 = -0.2215 + j0.8660$

$s_2 = -0.2215 - j0.8660$

4. Cosh(0.4760) = 1.11544

One pole: $s = -0.443(1.11544) = -0.4942$
Two poles: $s_1 = (-0.2215 + j0.866)(1.11544)$
$= -0.2471 + j0.966$
$s_2 = -0.2471 - j0.966$

5. The reassembly of the two-pole equation is as follows:

$$2\sigma = 2(0.2471) = 0.4942$$

$$\sigma^2 + \omega^2 = (0.2471)^2 + (0.966)^2 = 0.9942$$

The new polynomial is

$$(s + 0.4942)(s^2 + 0.4942s + 0.9942)$$

which agrees with that in Table 9-2.

The overshoot for a step input is less in a Chebyshev than in the Butterworth filter; the undershoot is virtually zero for the Butterworth, while the Chebyshev has nearly as much undershoot as overshoot, as illustrated in Fig. 9-14b.

The Bessel filter was developed to yield the fastest pulse rise time. It is the worst filter for frequency-response characteristics but has the best step response, as can be seen in Fig. 9-14b. The overshoot for a Bessel filter is less than 1% for any number of poles in the filter [4, p. 94]. Table 9-3 lists the normalized characteristic polynomials for the low-pass Bessel filter.

The optimal and parabolic filters fall in a midrange between the Bessel, which has poor frequency response but the best step response, and the Butterworth and Chebyshev, which have good frequency response but poor step response. Fig. 9-15a illustrates the frequency response for the five filter shapes, each with three poles, while Fig. 9-15b illustrates the step response for the same filters. The optimal filter has the sharpest frequency response while still being monotonic. Various other advantages and disadvantages can be observed on the two figures where they are plotted on the same coordinates. The frequency response is normalized to 1 Hz

Table 9-3   Normalized Low-Pass Bessel Polynomials (Factored Form)*

Poles	Characteristic Equation (Normalized)
1	$(s + 1)$
2	$(s^2 + 3s + 3)$
3	$(s + 2.322)(s^2 + 3.678s + 6.6460)$
4	$(s^2 + 4.208s + 11.488)(s^2 + 5.792s + 9.140)$
5	$(s + 3.647)(s^2 + 4.649s + 18.156)(s^2 + 6.704s + 14.272)$

*From *Principles and Design of Linear Active Circuits*, Mohammed Ghausi; Copyright © McGraw-Hill, Inc. 1965. All rights reserved.

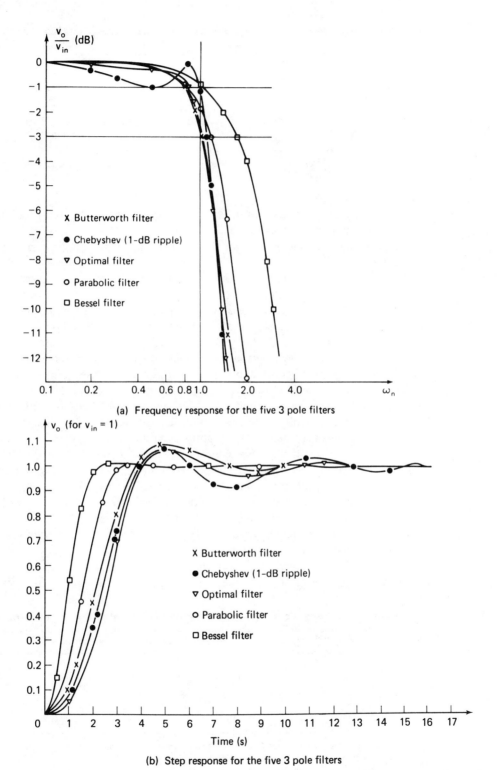

(a) Frequency response for the five 3 pole filters

(b) Step response for the five 3 pole filters

Fig. 9-15 Comparison of the five three-pole low-pass filters. (From *Principles and Design of Linear Active Circuits*, Mohammed Ghausi: Copyright © McGraw-Hill, Inc. 1965. All rights reserved.)

Table 9-4   Normalized Low-Pass Optimal Polynomials (Factored Form)*

Poles	Characteristic Equation (Normalized)
1	$(s + 1)$
2	$(s^2 + \sqrt{2}s + 1)$
3	$(s + 0.620)(s^2 + 0.690s + 0.929)$
4	$(s^2 + 0.464s + 0.949)(s^2 + 1.100s + 0.430)$
5	$(s + 0.486)(s^2 + 0.308s + 0.959)(s^2 + 0.776s + 0.496)$

while the step response is normalized to 1 s, which is the response for the normalized characteristic polynomials.

The normalized characteristics for the optimal and parabolic filters are given in Tables 9-4 and 9-5.

Figure 9-16 illustrates the pole locations for the five three-pole low-pass filters. By comparing the distance of the poles from the $j\omega$-axis (*east face*), it can be seen that the frequency-response skirt beyond cutoff becomes steeper for poles closer to the $j\omega$-axis, while the step response has more overshoot. As the poles move farther away from the *east face*, their overall effect on the frequency response is reduced, resulting in a more shallow slope on the response skirt beyond cutoff. The advantage gained is better pulse response and more linear phase response.

The ideal low-pass filter has a phase response that changes *linearly* as the frequency shifts over the passband. The Butterworth and Chebyshev filters do not have linear phase responses. The Bessel filter has the most linear phase response.

All of the previously defined filter approximations except the Chebyshev have one characteristic in common. The constant terms in all the individual polynomial terms must be multiplied together to form the multiplier of the transfer function. An example of the use of this multiplier was presented in Example 9-15. If the multiplier is less than unity, a voltage divider buffered by a voltage follower is used in the series string of circuits (see Fig. 9-12). If the multiplier is greater than unity, a noninverting amplifier must be used to maintain the proper phase relationship while increasing the gain of the total circuit.

The multiplier obtained in this fashion is correct for all filters except the Chebyshev and Cauer. For these filter approximations, it is true for filters with an

Table 9-5   Normalized Low-Pass Parabolic Polynomials (Factored Form)*

Poles	Characteristic Equation (Normalized)
1	$(s + 1)$
2	$(s^2 + 1.818s + 1.656)$
3	$(s + 1)(s^2 + 1.582s + 2.507)$
4	$(s^2 + 1.382s + 3.266)(s^2 + 1.958s + 1.124)$
5	$(s + 1)(s^2 + 1.218s + 3.880)(s^2 + 1.894s + 1.368)$

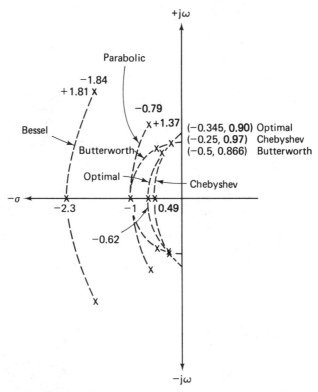

Fig. 9-16 Normalized pole locations for the five three-pole low-pass filters.

odd number of poles. For filters with an *even* number of poles, the multiplier is less than the actual value by the amount of the ripple. Thus for any filter with ripple *and* having an even number of poles, the amount of ripple (in dB) is converted to a ratio (Example 2-16) and used as a factor to *increase* the multiplier to the actual value (see Example 9-17).

The sixth filter characteristic is the elliptic filter. The word "elliptic" comes from the fact that both the poles and zeros produce equal ripple as in a Chebyshev filter. This is developed from the complex theory of elliptic functions. The elliptic filter is generally called a Cauer filter approximation and has a biquadratic transfer function, as the low-pass filter contains *zeros*, not at the origin, as well as poles; thus it is sometimes called a "biquadratic" filter. This is a new concept, as the five previous low-pass filters contained only poles. These zeros are all *on* the $j\omega$-axis; thus their effect is extremely visible in the *east face* frequency response, as seen in Fig. 9-17. Because there are zeros on the $j\omega$-axis to draw the response down to "zero," the poles can be placed very close to the $j\omega$-axis for tight control on the shape of the frequency response. This filter permits the most versatile filter shape. It is also the most widely used filter approximation at the time of this writing. The filter appears to be like a Chebyshev, with ripple appearing in the passband due to the close proximity of the poles to the *east face*. The frequency response drops suddenly from the pole nearest the cutoff frequency to zero at the "zero" nearest cutoff in the

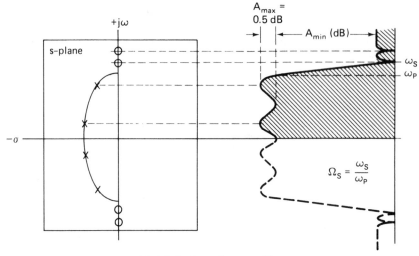

(a) 4 Pole Cauer low pass filter

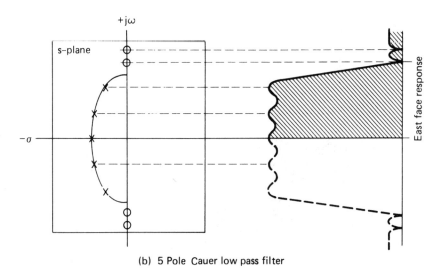

(b) 5 Pole Cauer low pass filter

Fig. 9-17 Multipole low-pass Cauer filters.

plane of the *east face*. This pole and zero can be made as close to each other as desired, but other undesirable effects begin to appear; one such effect is the reduction of Amin (see Fig. 9-17a).

The study of Cauer filters is very extensive and complex. A book of tables including the design of these filters has been compiled by Christian and Eisenmann [5].

Table 9-6 lists the normalized characteristic polynomials for several versions of the Cauer filter approximation. The term Ω in the tables is the ratio of the

Table 9-6 Normalized Low-Pass Elliptic (Cauer) Polynomials (A_{max} = 0.5 dB)*

Poles	Denominator Multiplier	Numerator (Zeros)	Denominator (Poles)	A_{min} (dB)
			$\Omega_S = 1.5$	
2	0.38540	$(s^2 + 3.92705)$	$(s^2 + 1.03153s + 1.60319)$	8.3
3	0.31410	$(s^2 + 2.80601)$	$(s^2 + 0.45286s + 1.14917)(s + 0.766952)$	21.9
4	0.015397	$(s^2 + 2.53555)(s^2 + 12.09931)$	$(s^2 + 0.25496s + 1.06044)(s^2 + 0.92001s + 0.47183)$	36.3
5	0.019197	$(s^2 + 2.42551)(s^2 + 5.43764)$	$(s^2 + 0.16346s + 1.03189)(s^2 + 0.57023s + 0.57601)(s + 0.42597)$	50.6
			$\Omega_S = 2.0$	
2	0.20133	$(s^2 + 7.4641)$	$(s^2 + 1.24504s + 1.59179)$	13.9
3	0.15424	$(s^2 + 5.15321)$	$(s^2 + 0.53787s + 1.14849)(s + 0.69212)$	31.2
4	0.0036987	$(s^2 + 4.59326)(s^2 + 24.22720)$	$(s^2 + 0.30116s + 1.06258)(s^2 + 0.88456s + 0.41032)$	48.6
5	0.0046205	$(s^2 + 4.36495)(s^2 + 10.56773)$	$(s^2 + 0.19255s + 1.03402)(s^2 + 0.59054s + 0.52500)(s + 0.392612)$	66.1
			$\Omega_S = 3.0$	
2	0.083974	$(s^2 + 17.48528)$	$(s^2 + 1.35715s + 1.55532)$	21.5
3	0.063211	$(s^2 + 11.82781)$	$(s^2 + 0.58942s + 1.14559)(s + 0.65263)$	42.8
4	0.00062046	$(s^2 + 10.4554)(s^2 + 58.471)$	$(s^2 + 0.32979s + 1.063281)(s^2 + 0.86258s + 0.37787)$	64.1
5	0.00077547	$(s^2 + 9.8955)(s^2 + 25.0769)$	$(s^2 + 0.21066s + 1.0351)(s^2 + 0.58441s + 0.496388)(s + 0.37452)$	85.5

*From *Principles of Active Network Synthesis and Design*, Gobind Daryanani; Copyright © 1976: Reprinted by permission of John Wiley & Sons, Inc.

"stopband" frequency to the "passband" frequency and is given by

$$\Omega_S = \frac{\omega_S}{\omega_P} \qquad (9\text{-}49)$$

where the location of ω_S and ω_P is defined in Fig. 9-17a. The circuit to implement a Cauer low-pass filter is biquadratic and will be presented in Section 9.4.3. The method of finding L_{np} and C_{np} for the poles is identical to that already presented for the case where only poles exist. By observing the nature of the transfer functions in Table 9-6, it becomes apparent that the zero polynomials, in the numerator, do not have a coefficient for the single power of s. This indicates that σ is zero and any points determined through the equation lie directly on the $j\omega$-axis in the plane of the *east face*. The biquadratic transfer function polynomial appears as

$$\frac{v_o}{v_{in}} = (\text{multiplier})\frac{s^2 + 1/L_{zn}C_{zn}}{s^2 + \dfrac{1}{L_{pn}}s + \dfrac{1}{L_{pn}C_{pn}}} \qquad (9\text{-}50)$$

To determine the product of L_{nz} and C_{nz} for the particular normalized polynomial, one needs only to use the numerator function and compare it with the form of Eq. (9-50).

This technique is best shown by example.

EXAMPLE 9-17

a. Determine the value of ω_0 ad the product of L_{nz} and C_{nz} for the normalized Cauer polynomial having three poles with $\Omega_S = 2.0$ in Table 9-6.

b. Verify the numerator multiplier value for $\Omega_S = 2.0$ and four poles.

Solution

a. The zero polynomial, (numerator), is $(s^2 + 5.15321)$; by comparing the numerator of Eq. (9-50) with Eq. (9-3), it becomes apparent that $\sigma = 0$. Thus the form of the equation is $(s^2 + \omega^2)$ and $\omega_0 = \omega$ since $\zeta = 0$. By comparison, $\omega_0 = \sqrt{5.15321} = 2.2700$, as $1/L_{nz}C_{nz} = 5.15321$.

b. $A_{max} = 0.5$ dB, $0.5/20 = 0.025$, $10^{0.025} = 1.059254$;

$$\text{multiplier} = \frac{(1.05924)(0.0036987)(4.59326)(24.22720)}{(1.06258)(0.41032)} = 0.999975$$

which is sufficiently close to 1.0000 to verify the multiplier value.

9.3.5 High-Pass Filters

The passive high-pass filter passes input signals with unity gain from infinity (very high) to the "cutoff" frequency, where the output amplitude has reduced to 0.707 (-3 dB) of that at high frequencies. Figure 9-18a illustrates the circuit for a one-pole high-pass filter; its transfer function is

$$\frac{v_o}{v_{in}} = \frac{s}{s + 1/RC} \qquad (9\text{-}51)$$

where it is immediately apparent that it has one zero at the origin and one pole at

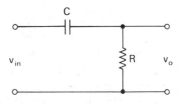

(a) One pole passive high pass filter section

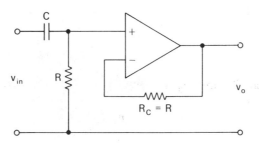

(b) One pole active high pass filter section

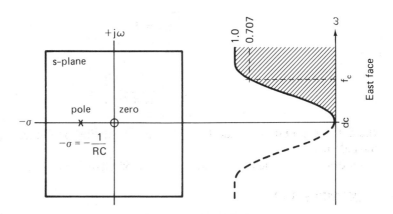

(c) Pole and zero locations for one pole high pass filter

Fig. 9-18 One-pole high-pass filter.

$s = -\sigma = -1/RC$. The zero and pole plotted on the s-plane, together with the *east face* response, are shown in Fig. 9-18c. Take particular note of the form of the transfer function; it has one zero at the origin and one pole. There is *no* multiplier on the transfer function, as the gain is unity at frequencies well above cutoff. This is typical of *high-pass* transfer function and should be remembered, as it is useful to be able to recognize this filter type from the transfer function only. The cutoff frequency occurs at the point where the output amplitude has reduced to 0.707 of that at high frequencies and is given by

$$f_c = \frac{1}{2\pi RC} \qquad (9\text{-}52)$$

The active high-pass filter is the passive circuit followed by a voltage follower to prevent loading of the circuit and to provide a low output impedance for stage isolation; this circuit is shown in Fig. 9-18b. A pole exists at infinity which holds the membrane at a constant level at frequencies above cutoff. This is apparent from the transfer function as the order of s is equal in the numerator and denominator.

The transfer-function equation for the single-pole high-pass filter in terms of the normalized capacitance is given by

$$\frac{v_o}{v_{in}} = \frac{s}{s + \dfrac{1}{R_o\left(\dfrac{1}{R_0\omega_0 C_n}\right)}} \tag{9-53}$$

which, when R_o and ω_0 are unity (1), yields the normalized high-pass transfer function

$$\frac{v_o}{v_{in}} = \frac{s}{s + C_n} \tag{9-54}$$

By comparing Eq. (9-53) with (9-51), we find that

$$\boxed{R_3 = R_o} \tag{9-55}$$

and

$$\boxed{C_3 = \frac{1}{R_3\omega_0 C_n}} \tag{9-56}$$

where C_n is found from the one-pole characteristic polynomial, again using the low-pass equation, (9-35).

An example will help to illustrate the use of these equations.

EXAMPLE 9-18

A one-pole high-pass Chebyshev filter with 1 dB of passband ripple has a cutoff frequency of 625 Hz. The capacitor that is available is 0.022 μF (film).

a. Calculate the value of R_3.

b. Explain how the plot would appear on the s-plane.

Solution

a. $\omega_0 = 2\pi(625$ Hz$) = 3927$ rad/s. For the one-pole Chebyshev low-pass filter, the normalized characteristic equation is $(s + 1.9652)$. From Eq. (9-35), $C_n = 0.5088$; thus the normalized high-pass transfer function, (9-54), is $v_o/v_{in} = s/s + 0.5088$. Rearranging Eq. (956), we get

$$R_o = R_3 = \frac{1}{C_3\omega_0 C_n} = \frac{1}{(0.022\ \mu\text{F})(3927)(0.5088)} = 22.75\ \text{k}\Omega$$

Use a 22.5 kΩ resistor from Appendix B1.

b. By letting $s = j\omega$ from Eq. (9-11) in Eq. (9-53), the shape of the elastic membrane along the *east face* is the desired frequency response. It appears as plotted in Fig. 9-18c, where $f_c = 625$ Hz.

The two-pole high-pass passive filter is composed of a series RLC circuit where the output voltage is taken across the inductor as shown in Fig. 9-19a. Again, this high-pass filter has unity gain at frequencies well above cutoff. The transfer function

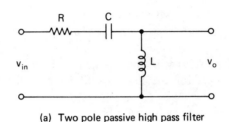

(a) Two pole passive high pass filter

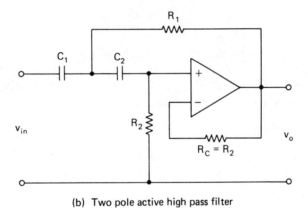

(b) Two pole active high pass filter

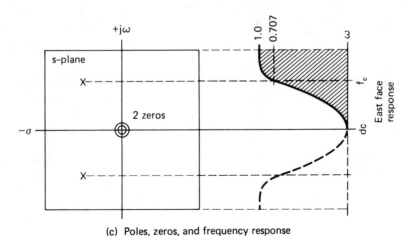

(c) Poles, zeros, and frequency response

Fig. 9-19 Two-pole high-pass filter.

is given by

$$\frac{v_o}{v_{in}} = \frac{s^2}{s^2 + \frac{R_o}{L}s + \frac{1}{LC}} \tag{9-57}$$

where the properties of a high-pass filter are again present:

1. It has two zeros at the origin of the s-plane (the numerator is the s^2 term in the denominator).
2. It has no multiplier on the equation.
3. When $s = j\omega$ and $\omega \gg \omega_c$ (cutoff) [see Eq. (9-6)], the gain becomes unity.

By comparing Eq. (9-57) with the characteristic equation (9-6), it can be seen that Eqs. (9-23) to (9-27) for the low-pass filter are still valid even though this is a high-pass filter.

The normalized transfer function is

$$\frac{v_o}{v_{in}} = \frac{s^2}{s^2 + C_n s + L_n C_n} \tag{9-58}$$

where it can be seen that the normalized polynomial is now $(s^2 + C_n s + L_n C_n)$. L_n and C_n are found by comparing the low-pass equation (9-36) with the normalized polynomials. The values of L_n and C_n are then inserted in Eq. (9-58).

An example will help reinforce this technique.

EXAMPLE 9-19

Determine the values of L_n and C_n for the two-pole Chebyshev high-pass filter with 0.25 dB passband ripple. Find the normalized transfer function.

Solution

From Table 9-2 the normalized characteristic polynomial is $(s^2 + 1.7967s + 2.1140)$. Compare with the low-pass equation

$$s^2 + \frac{1}{L_n}s + \frac{1}{L_n C_n}$$

which by inspection shows that $1/L_n = 1.7967$ or $L_n = 0.5566$. Since $1/C_n(0.5566) = 2.1140$, it follows that $C_n = 0.8499$.

$$\frac{v_o}{v_{in}} = \frac{s^2}{s^2 + 0.8499s + 0.4730}$$

The transfer function for the Sallen-Key high-pass active filter shown [6, p. 285] in Fig. 9-19b is

$$\frac{v_o}{v_{in}} = \frac{s^2}{s^2 + \frac{R_1}{R_1 R_2 \left(\frac{C_1 C_2}{C_1 + C_2}\right)}s + \frac{1}{\left[R_1 R_2 \left(\frac{C_1 C_2}{C_1 + C_2}\right)\right](C_1 + C_2)}} \tag{9-59}$$

where by comparison with Eq. (9-57) it can be seen that

$$R_o = R_1 \tag{9-60}$$

and

$$L = R_1 R_2 \left(\frac{C_1 C_2}{C_1 + C_2} \right) \qquad \text{Henries} \tag{9-61}$$

while

$$C = C_1 + C_2 \qquad \text{Farads} \tag{9-62}$$

By comparing Eq. (9-59) with the characteristic equation (9-6), it is found that

$$\omega_0 = \frac{1}{\sqrt{R_1 R_2 C_1 C_2}} \qquad \text{rad/s} \tag{9-63}$$

and

$$\sigma_p = \frac{C_1 + C_2}{2 R_2 C_1 C_2} = \zeta \omega_0 \qquad \text{rad/s} \tag{9-64}$$

while

$$\zeta = \frac{C_1 + C_2}{2} \sqrt{\frac{R_1}{R_2 C_1 C_2}} \tag{9-65}$$

Two methods are possible for finding the component values.

Method I. Select appropriate (standard) capacitor sizes from Appendix B2 (film); the capacitive reactance at the cutoff frequency should be between 2 and 20 kΩ. Either capacitor may be larger. Then

$$\boxed{R_1 = R_o = \frac{1}{(C_1 + C_2) \omega_0 L_n}} \tag{9-66}$$

and

$$\boxed{R_2 = \frac{1}{\omega_0 \left(\dfrac{C_1 C_2}{C_1 + C_2} \right) C_n}} \tag{9-67}$$

The approximate capacitor size should be

$$\boxed{C \simeq \frac{1}{2 \pi X_C f_c}} \tag{9-68}$$

where X_C is the capacitive reactance at the cutoff frequency.
   An example will illustrate the use of these equations.

EXAMPLE 9-20

Determine the component values for a two-pole high-pass Chebyshev filter with a cutoff frequency of 425 Hz. Use the L_n and C_n values from Example 9-19 for 0.25 dB passband ripple.

Single Op Amp Active Filter Sections   **281**

Solution

To determine the range of capacitor sizes, use Eq. (9-68) between 2 and 20 kΩ.

$$C(\text{for 2 k}\Omega) = \frac{1}{2\pi(2\ k\Omega)425\ \text{Hz}} = 0.18\ \mu F$$

$$C(\text{for 20 k}\Omega) = 0.019\ \mu F$$

Thus the capacitors should range between 0.02 and 0.2 μF; use two 0.05 μF capacitors from Appendix B2.

$$R_1 = \frac{1}{(0.1\ \mu F)2\pi 425(0.5566)} = 6.73\ k\Omega$$

Use a 6.81 kΩ resistor from Appendix B1.

$$R_2 = \frac{1}{2\pi 425(0.025\ \mu F)(0.8499)} = 17.62\ k\Omega$$

Use a 17.8 kΩ resistor from Appendix B1. The circuit is shown in Fig. 9-19c.

Method 2. Choose the resistor values first; the capacitor values are determined using the following procedure:

1. The resistors must have the relationship given by the inequality

$$R_2 \geq 4\frac{L_n}{C_n}R_1 \tag{9-69}$$

2. The capacitor values are given by the quadratic equation

$$C^2 - \frac{1}{\omega_0 R_1 L_n}C + \frac{1}{\omega_0^2 L_n C_n R_1 R_2} = 0 \tag{9-70}$$

where the inequality, (9-69) must be satisfied before using the quadratic with an electronic calculator. Either solution to the quadratic may be assigned to C_1 or C_2.

3. The solution to the quadratic gives the two equations

$$C_1 = \frac{1}{2\omega_0 L_n R_1}\left[1 + \sqrt{1 - \frac{4L_n R_1}{C_n R_2}}\right] \tag{9-71}$$

[and for conditions of Eq. (9-68)]

$$C_2 = \frac{1}{2\omega_0 L_n R_1}\left[1 - \sqrt{1 - \frac{4L_n R_1}{C_n R_2}}\right] \tag{9-72}$$

Either capacitor may have the larger value. This method usually produces capacitors with an irrational value; thus it is the less desirable method for solution of component values.

The pole locations on the s-plane for the high-pass filter are found by using the high-pass form of the normalized equation,

$$\frac{v_0}{v_{in}} = \frac{s}{(s + C_n)} \frac{s^2}{(s^2 + C_n s + L_n C_n)} \frac{s^2}{(s^2 + C_n s + L_n C_n)} \qquad (9\text{-}73)$$

The equation shown is for the five-pole filter; any number of poles may be used by omitting the inappropriate terms. The values of L_n and C_n are determined using Eqs. (9-35) for the one-pole and (9-36) for the two-pole sections; these values are then inserted into Eq. (9-73) to find the high-pass pole locations.

An example will help to illustrate this technique.

EXAMPLE 9-21

Determine the pole locations for the five-pole high-pass optimal filter.

Solution

The low-pass characteristic equation is

$$(s + 0.486)(s^2 + 0.308s + 0.959)(s^2 + 0.776s + 0.496)$$

One-pole: $C_n = \dfrac{1}{0.486} = 2.057$

First two-pole:

$$\frac{1}{L_n} = 0.308 \qquad L_n = 3.247$$

$$\frac{1}{L_n C_n} = 0.959 \qquad C_n = \frac{1}{(0.959)(3.247)} = 0.321 \qquad L_n C_n = 1.043$$

Second two-pole:

$$\frac{1}{L_n} = 0.776 \qquad L_n = 1.289$$

$$\frac{1}{L_n C_n} = 0.496 \qquad C_n = 1.564 \qquad L_n C_n = 2.016$$

The normalized high-pass transfer-function equation for pole and zero locations is

$$\frac{s^5}{(s + 2.057)(s^2 + 0.321s + 1.043)(s^2 + 1.564s + 2.016)}$$

Using Eqs. (9-1) and (9-3) and the methods of Examples 9-1 and 9-2, the pole locations are:
One pole:

$$s = -\sigma = -2.06$$

First two-pole:

$$s_1 = -0.16 + j1.01$$
$$s_2 = -0.16 - j1.01$$

Second two-pole:

$$s_1 = -0.78 + j1.18$$
$$s_2 = -0.78 - j1.18$$

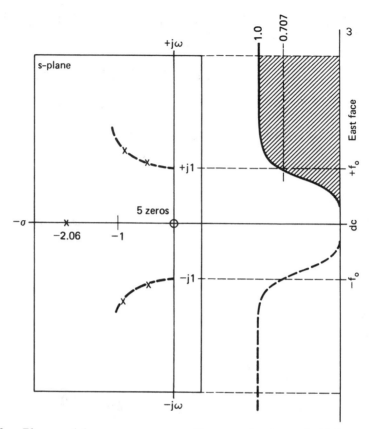

Fig. 9-20 *s*-Plane and frequency-response diagrams for five-pole high-pass optimal filter (Example 9-21).

The single high-pass pole is now more negative than $-\sigma = -1$, and the complex poles are now outside (greater value) the $\pm j\omega = \pm j1$ points on the *s*-plane. Five zeros appear at the origin and impedance poles appear at $\pm j$ infinity. The pole-zero plot and resulting *east face* response is as shown in Fig. 9-20. The pole–zero for a high-pass filter has two characteristics, which are related to the low-pass diagram. The *s*-plane "folds" on three hinges;

1. Low-pass poles on the $-\sigma$ axis fold on the -1 point where the distance of the high-pass poles out from the -1 hinge are the reciprocal of the distance in from -1 for the low-pass poles.

2. The low-pass poles that are inside the $+j1$ point fold on the $+j1$ hinge and become high-pass poles outside the $+j1$ hinge line or vice versa. No simple reciprocal relates the two sets now. The methods of Example 9-21 must be used to find their position.

The shape of the high-pass frequency response is the exact inverse of the low-pass response, folding on the cutoff frequency (f_c).

9.3.6 Band-Pass Filter

The band-pass filter passes frequencies over a narrow range and rejects all others. Unlike the low-pass and high-pass filters, which passed frequencies in the passband with unity gain, the band-pass filter has a gain in the passband. This gain is related to the Q of the resonant circuit response being simulated by the active filter. The shape of the frequency response of a band-pass filter is identical to the true-image and reflected-image low-pass filter considered as one continuous shape. Figure 9-21a illustrates a low-pass filter shape and its poles. Figure 9-21b illustrates a band-pass

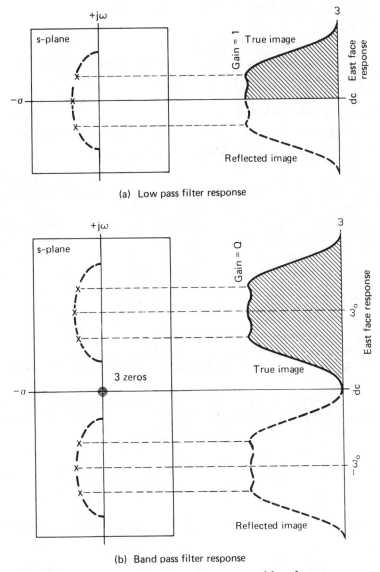

(a) Low pass filter response

(b) Band pass filter response

Fig. 9-21 Comparison between low-pass and band-pass response.

filter and its respective poles. The only difference between the two response shapes is the position on the frequency axis; the low-pass filter response is centered at dc ($f = 0$) while the band-pass response is centered at $f = f_0$.

The transfer-function equation for a band-pass filter is

$$\frac{v_0}{v_{\text{in}}} = (G)\frac{\frac{1}{RC}s}{s^2 + \frac{1}{QRC}s + \left(\frac{1}{RC}\right)^2} \qquad (9\text{-}74)$$

where, again, the form of the equation is distinctive and has the following properties:

1. The denominator has the same configuration as the low- and high-pass filters: namely, two poles.
2. The numerator has *one* zero at the origin, the second-order high-pass filter had two [see Eq. (9-57)].
3. The numerator has a coefficient term of 2σ or $1/RC$.
4. The gain (G) at maximum is related to Q.

The nature of the transfer function should be remembered, as it will be useful to recognize a band-pass filter from the transfer function only. The frequency response rises from zero at the origin to a peak near the pole frequency; then a zero of transmission (s in numerator, s^2 in denominator), at infinity, draws the response back down to zero as the frequency becomes higher, as illustrated in Fig. 9-22a. The shape of the curve is one of pure resonance, as illustrated in Fig. 9-22b, where the bandwidth is between the lower and upper -3-dB (0.707) points.

The development of a band-pass circuit is illustrated in Fig. 9-23, where a low-pass filter is shown in Fig. 9-23a and each element as it converts to the band-pass equivalent is shown on Fig. 9-23b. The circuit in Fig. 9-23b has four poles and two zeros. This circuit is not realizable in an active equivalent; thus only half of the circuit is used for the active equivalent. This half is illustrated [6, p. 286] in Fig. 9-23c.

The Sallen–Key basic circuit for the active band-pass filter circuit is illustrated in Fig. 9-24. The transfer-function equation for the active band-pass filter is

$$\frac{v_o}{v_{\text{in}}} = (G)\frac{\frac{1}{RC}s}{s^2 + \frac{4 - G}{RC}s + \left(\frac{\sqrt{2}}{RC}\right)^2} \qquad (9\text{-}75)$$

where G is the noninverting gain of the amplifier *(Rule 3)* and is given by

$$\boxed{G = \frac{R_F + R_I}{R_I}} \qquad (9\text{-}76)$$

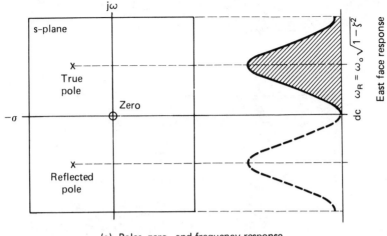

(a) Poles, zero, and frequency response

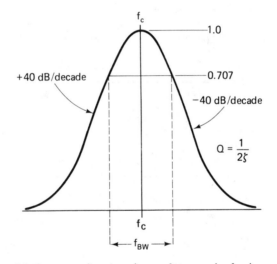

(b) Resonance (band pass) curve for one pair of poles

Fig. 9-22 Poles, zero, and frequency response for band-pass filter section.

By comparing Eq. (9-75) with the characteristic equation (9-6), it can be seen that

$$\boxed{f_0 = \frac{1}{\sqrt{2}\,\pi RC}} \tag{9-77}$$

and that

$$Q = \frac{\sqrt{2}}{4 - G} \tag{9-78}$$

while the stage gain at resonance is given by

$$\frac{v_o}{v_{in}} = \frac{G}{4 - G} \tag{9-79}$$

Single Op Amp Active Filter Sections

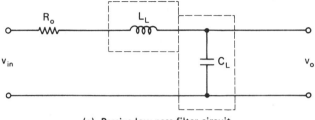

(a) Passive low pass filter circuit

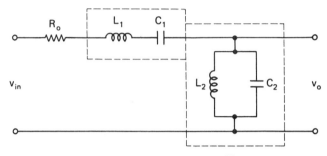

(b) Transformation to band pass filter

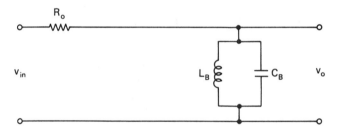

(c) Portion of passive filter used for active band pass filter

Fig. 9-23 Low-pass to band-pass transformation.

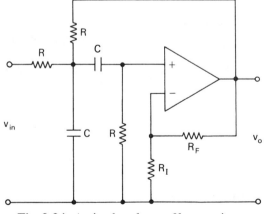

Fig. 9-24 Active band-pass filter section.

which causes the noninverting gain (G) to be

$$\boxed{\frac{R_F}{R_I} + 1 = 4 - \frac{\sqrt{2}}{Q}} \qquad \text{(for } Q > 0.4714\text{)} \qquad (9\text{-}80)$$

where Q must be greater than 0.4714, as the gain must not go below unity. The noninverting amplifier gain ranges from 4 for a large Q to a gain of 1 for a $Q = 0.4714$. This change in gain has no detrimental effect on the filtering action and may be altered if desired by placing a voltage divider and voltage follower somewhere between stages of the filter. The Bessel filter approximation may not be used with this circuit, as the Q of all poles falls below the usable range.

Each section (circuit of Fig. 9-24) of the filter creates one true-image pole and one reflected-image pole; we are only interested in the former. To construct a three-pole filter, we need three true-image poles thus three filter sections (circuits). They are connected in cascade (series), where the order is not of any concern. The relative position of poles on the s-plane is the *same* for a normalized characteristic equation as it is for one at the desired frequency. Thus the constants found in the normalized equation may be used to find the location of the poles on the s-plane for an actual filter section. The zeros of a filter, one for each section, all appear at the origin and do not enter into any calculations. The poles appear in exactly the same arrangement as for the low-pass filter; the only difference is their displacement due to the change in frequency. The equation for comparison with the characteristic equation is given by a combination of Eqs. (9-1) and (9-3), or the characteristic equation appears as

$$\boxed{(s + \sigma_1)\left(s^2 + 2\sigma_2 s + \sigma_2^2 + \omega_2^2\right)\left(s^2 + 2\sigma_3 s + \sigma_3^2 + \omega_3^2\right)} \qquad (9\text{-}81)$$

The process of finding the three pole frequencies and respective Q's is called the Geffe* algorithm which is applied as follows:

1. The resonant frequency and bandwidth of the desired filter is determined using the relationship,

$$Q_{BW} = \frac{f_0}{BW} \qquad (9\text{-}82)$$

where BW is the frequency difference between the low and high cutoff frequencies (-3 dB frequencies for Butterworth).

2. The normalized low pass characteristic equation for the filter shape (three poles are desired) is selected and appears as

$$(S + \sigma_1)\left(S^2 + 2\sigma_2 S + \sigma_2^2 + \omega_2^2\right) \qquad (9\text{-}83)$$

---

*M. E. Van Valkenberg, "Analog Filter Design," Holt, Rinehart, and Winston, 1982. Originally from P. R. Geffe, "Designers' Guide to Active band pass Filters," EDN, pp. 46–52, Apr. 5, 1974. The name is pronounced "geffe," with the g sounded as in "golley."

3. The Geffe algorithm is begun with the single pole part of Eq. (9-82) by finding the single pole band pass Q, designated Q_1.

$$Q_1 = \frac{Q_{BW}}{\sigma_1} \qquad (9\text{-}84)$$

This pole is at the center frequency of the band pass filter.

4. The Geffe algorithm is now applied to the 2-pole low pass characteristic equation using the following sequence

$$C = (\sigma_2^2 + \omega_2^2) \quad \text{(constant term of 2 pole function)}$$

$$D = \frac{(2\sigma_2)}{Q_{BW}}$$

$$E = 4 + \frac{(\sigma_2^2 + \omega_2^2)}{Q_{BW}^2} = 4 + \frac{C}{Q_{BW}^2}$$

$$G = \sqrt{E^2 - 4D^2}$$

$$Q_2 = \frac{1}{D}\sqrt{\frac{1}{2}(E + G)} \qquad (9\text{-}85)$$

$$K = \frac{\sigma_2 Q_2}{Q_{BW}}$$

$$\omega_n = K + \sqrt{K^2 - 1} \qquad (9\text{-}86)$$

5. Now, the actual pole frequencies for the low and high frequency poles can be determined as follows:

$$f_{pL} = \frac{f_0}{\omega_n} \qquad (9\text{-}87)$$

$$f_{pH} = f_0(\omega_n) \qquad (9\text{-}88)$$

EXAMPLE 9-22

Determine the circuit values for a three-pole Chebyshev band pass filter with 0.5 dB ripple. The center frequency (f_0) is 400 Hz and the bandwidth is 150 Hz.

Solution

Three filter sections are required, as each creates one true-image pole. The normalized characteristic equation is

$$(s + 0.6264)(s^2 + 0.6264s + 1.1424)$$

$$(s + \sigma_1)(s^2 + 2\sigma_2 s + \sigma_2^2 + \omega_2^2)$$

Using Eq. (9-81),

$$Q_{BW} = \frac{400}{150} = 2.667$$

The Geffe algorithm is begun with the single pole Q;

$$Q_1 = 2.667/0.6264 = 4.257.$$

The two pole procedure begins with the constant term of the quadratic:

$$C = 1.1424$$

$$D = 0.6264/2.667 = 0.2348$$

$$E = 4 + 1.1424/2.667^2 = 4.160$$

$$G = \sqrt{4.16^2 - (4)0.2348^2} = 4.107$$

$$Q_2 = \frac{1}{0.2348} \sqrt{\frac{1}{2}(4.160 + 4.107)} = 8.659$$

$$K = (0.3132)(8.659)/2.667 = 8.659$$

$$\omega_n = 1.02 + \sqrt{(1.017)^2 - 1} = 1.201$$

$$f_{pL} = 400/1.201 = 333 \text{ Hz}, \ Q_2 = 8.659$$

$$f_0 = 400 \text{ Hz}, \ Q_1 = 4.257$$

$$f_{pH} = 400(1.201) = 2480 \text{ Hz}, \ Q_2 = 8.659$$

$$R_{(p1)} = \frac{1}{\sqrt{2}\,\pi(333)(0.047\,\mu F)} = 14.4 \text{ k}\Omega$$

$$R_{(0)} = \frac{1}{\sqrt{2}\,\pi(400)(0.047\,\mu F)} = 11.9 \text{ k}\Omega$$

$$R_{(ph)} = \frac{1}{\sqrt{2}\,\pi(480)(0.047\,\mu F)} = 9.98 \text{ k}\Omega$$

Determine the gain for the center frequency pole using Eq. (9-80);

$$G_1 = 4 - \frac{\sqrt{2}}{4.257} = 3.668 \qquad \frac{R_F}{R_I} = 2.668$$

Use $R_I = 10 \text{ k}\Omega$, $R_F = 26.7 \text{ k}\Omega$. For the two outer poles,

$$G_2 = 4 - \frac{\sqrt{2}}{8.659} = 3.837 \qquad \frac{R_F}{R_I} = 2.837$$

Use $R_I = 10 \text{ k}\Omega$, $R_F = 28.7 \text{ k}\Omega$.

The gain of the complete system (three sections) is given by Eq. (9-79) as

$$\left(\frac{3.668}{4 - 3.668}\right)\left(\frac{3.837}{4 - 3.837}\right)\left(\frac{3.837}{4 - 3.837}\right) = (11)(23.5)(23.5) = 6100$$

The actual gain is less than this, as the two outer sections are operating off their resonance frequency. Precede the first stage with a voltage divider and a voltage follower having a gain of 1/6100. Thus, for this amplifier stage, $R_A = 619 \text{ k}\Omega$ and $R_B = 102 \text{ }\Omega$, see Eq. (7-7). A better solution would be to use three divider stages spread over the three filter sections. The complete system now has unity gain in the pass band and the filter will not saturate at large signal levels. The circuit for the complete filter is shown in Fig. 9-25.

The "shape factor" of a filter is a measure of the skirt slope; it is usually specified by the letter (y). Since the skirt shape for a band-pass filter is identical to the skirt shape for the low-pass filter, the low-pass filter-response curves can be used to determine the shape factor for the band-pass filter. The shape factor is the ratio of the width of the passband, in frequency, at 20 dB down from the response at f_0 to the width of the passband at 3 dB down from f_0 and is given by

$$y = \frac{(f_h - f_l) \text{ at 20 dB down}}{(f_h - f_l) \text{ at 3 dB down}} \qquad (9-89)$$

For example:

EXAMPLE 9-23

The three-pole Butterworth low-pass filter response curves have the shape illustrated in Fig. 9-13. Determine the shape factor between 3 and 20 dB down if this curve is used as a band-pass filter with a bandwidth of 100 Hz at $f_0 = 1000$ Hz.

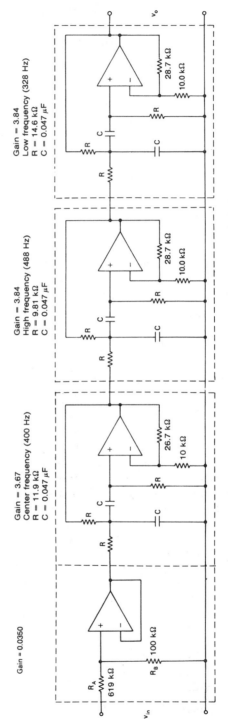

Fig. 9-25 Three-pole Chebyshev band-pass filter (Example 9-22).

Solution

Using Eqs. (9-85) and (9-86), the frequencies at 3 dB down are

$$f_{ch} = 1000 + 50 = 1050$$
$$f_{cl} = 1000 - 50 = 950$$

The frequencies at 20 dB down are twice and half, respectively; thus at -20 dB,

$$f_h = 2(1050) = 2100 \quad f_l = \frac{950}{2} = 475 \quad y = \frac{2100 - 475}{1050 - 950} = 16.25$$

9.4 Biquadratic Filters

The term "biquadratic," in its most general sense, means any equation that has all or parts of a quadratic equation in both the numerator and denominator. This would place all the transfer-function equations for the low-pass, high-pass, and band-pass filters in the biquadratic category; they are biquadratic equations. For the purposes of this chapter though, only equations that have both the s^2 term and the constant term of the quadratic equation in the numerator and the full quadratic equation in the denominator will be labeled biquadratic equations. This distinguishes equations with zeros at the origin from those with zeros not at the origin; only those with zeros not at the origin will be called "biquadratic equations." A broad range of filter sections fall into this category. There are five specific filter functions which the biquadratic transfer function represents:

1. Notch filter.
2. Band-reject filter.
3. Low-pass notch filter.
4. High-pass notch filter.
5. All-pass equalizer.

The three notch filters have zeros that are on the $j\omega$-axis in the plane of the *east face*, as the damping term (σ) in the numerator is zero. The band-reject filter has a zero slightly to the left of the $j\omega$-axis, as it has a small damping term in the numerator. The "all-pass" equalizer has a non-minimum-phase transfer function; thus it has zeros in the right half of the s-plane, as the circuit is of the bridge type and has at least one lead crossing another without touching.

9.4.1 Notch Filter

The notch filter is used to eliminate or notch out one frequency from a band of frequencies. An example of its use is the nulling of a 60-Hz noise from a signal comprised of frequencies both below and above 60 Hz. Specific frequencies are nulled out on a telephone line so that someone listening on the line will not hear these "control" frequencies. For many years the most commonly used notch filter has been the *twin-tee RC* network illustrated in Fig. 9-26a. This circuit is not now widely used because of the extremely close matching of components required for a deep notch. Historically though, it is a good circuit with which to be familiar, as it

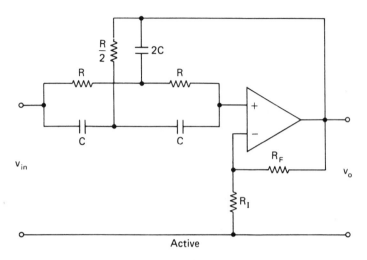

(a) Active twin-tee RC notch filter

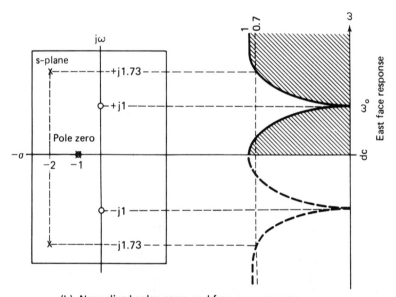

(b) Normalized poles, zeros, and frequency response

Fig. 9-26 Twin-tee notch filter.

appears in much of the literature on filters. It has a biquadratic transfer function of

$$\frac{v_o}{v_{in}} \quad \frac{s^2 + \dfrac{1}{R^2C^2}}{s^2 + \dfrac{4}{RC}s + \dfrac{1}{R^2C^2}} \tag{9-90}$$

where the resonant frequency is given by $f_0 = 1/2\,\pi RC$. The notch is at maximum depth when the component ratios shown in Fig. 9-26a are maintained. Notch depths of 60 dB are easily obtained with standard 1% components. Deeper notches require significantly closer matching of components. The filter is only good for low frequencies, below 1 kHz, as stray capacitances begin to alter the circuit performance as the capacitors become smaller. The *twin-tee* network has unity gain when far off resonance.

In its most general form, the arrangement of poles for a band-reject filter is a combination of low-pass poles and high-pass poles with a zero between them at the resonant frequency. The normalized pole-zero diagram for the *twin-tee RC* network is illustrated in Fig. 9-26b, where $1/R^2C^2 = 1$. The normalized transfer function is then

$$\frac{v_o}{v_{in}} = \frac{(s^2 + 1)(s+1)}{(s^2 + 4s + 1)(s+1)} \tag{9-91}$$

where the two $(s + 1)$ terms do not appear in Eq. (9-90), as they cancel [6, p. 289] (pole–zero cancellation) each other. Note that only the high-pass poles (higher frequency than $\pm j1$) appear in Fig. 9-26b; the low-pass poles do not exist.

An example will help to illustrate the use of Eq. (9-90).

EXAMPLE 9-24

Determine the component values for a twin-tee notch filter operating at a resonant frequency of 360 Hz with a gain, off resonance, of 3.

Solution

The only capacitors in Appendix B2 that have one value twice the other are 0.05 μF and 0.1 μF; if the capacitive reactance is to be between 2 and 20 kΩ, let $C = 0.05$ μF and $2C = 0.1$ μF. From Eq. (9-90),

$$R = \frac{1}{2\pi(360)(0.05\ \mu F)} = 8.84\ k\Omega$$

Use 8.87 kΩ. Then $R/2 = 4.42$ kΩ, which is a standard 1% value. Let $R_F = 20$ kΩ and $R_I = 10$ kΩ, which creates a noninverting gain of 3.

The reader is referred to *Operational Amplifiers and Linear Integrated Circuits* by Coughlin and Driscoll, published by Prentice-Hall, Inc., for an excellent example of an active notch filter in Fig. 12-12.

9.4.2 Band-Reject Filter

The band-reject filter has zeros as well as poles, but the zeros are not on the $j\omega$-axis, as in the notch filter, but to the left of it, as the notch depth does not reach zero response at resonance. Thus the numerator polynomial has a damping term $(2\sigma s)$. Although this filter response may be created by altering the shunt component ratios in the *twin-tee* notch circuit, this does not produce predictable results. The circuit that is most predictable is a combination of a band-pass filter and a subtractor circuit (Section 2.4.2), as illustrated in Fig. 9-27. This circuit effectively subtracts the band-pass shaped response from the original input signal; thus when the band-pass circuit has a near-unity output signal level, this subtracts from the original signal to produce a rejection of that frequency at the subtractor output. The gain in the band-pass amplifier is adjusted to yield the desired notch depth at the subtractor output; it should always be slightly less than unity. Through this means, a band-reject filter having any desired shape, Chebyshev, optimal, two-pole, three-pole, and

Biquadratic Filters

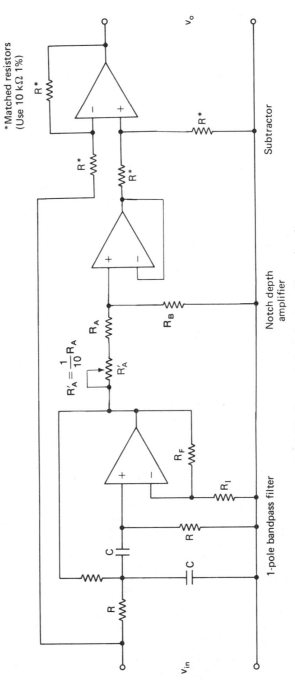

Fig. 9-27 One-pole band-reject filter circuit.

so on, may be constructed by first constructing the desired band-pass filter using the methods of Section 9.3.6.

An example will help to illustrate the use of this process.

EXAMPLE 9-25

Convert the three-pole Chebyshev band-pass filter from Example 9-22 into a band-reject filter with a notch depth of 50 dB.

Solution

The gain in the band-pass filter must be reduced 1 k from unity in the passband to slightly less than unity in the circuit of Fig. 9-25. Insert a potentiometer in series with the 27.4 kΩ (R_A) resistor in the input voltage divider. Since a notch depth of 50 dB is required, the gain in the passband should be: gain ratio for -50 dB is $10^{-2.5} = 0.00316$ and $1 - 0.00316 =$

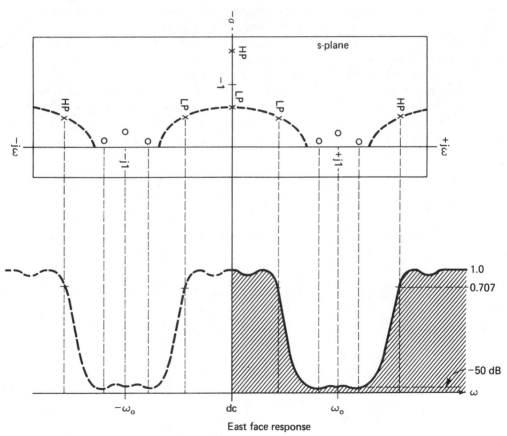

East face response

Fig. 9-28 Pole–zero plot and frequency response for Example 9-25.

0.9964. Thus the gain of the input voltage divider should be $(0.9964)(0.035) = 0.03487$. The total value of the 27.4 kΩ and the 1 kΩ potentiometer from Eq. (7-7) is $R_A = 1$ kΩ $(1 - 0.03487)/0.03487 = 27.67$ kΩ. Thus the potentiometer is adjusted to 200 Ω. This three-pole band-pass filter is followed by a subtractor circuit, as shown in Fig. 9-27. The "notch depth" amplifier, shown in Fig. 9-27, is now actually the band-pass input voltage divider shown in Fig. 9-25.

The pole–zero plot for the band-reject amplifier is shown in Fig. 9-28.

9.4.3 State-Variable Biquad Filters

In Chapter 6, an analog-computer circuit was presented and the term "state variable" was identified. A state variable is one way of representing the effect of an energy-storing element in any physical system. The electrical method is to implement an analog-computer simulation circuit (Fig. 6-18) where the integrator output voltages represent the state variables. For instance, in Chapter 6, an analog computer represented the mechanical spring–mass–damper system where both the mass and the spring are energy-storing elements. It requires one integrator to represent each energy-storing element. Thus one can count the integrators and know how many energy-storing elements are in the system being represented. Because the circuit, to be discussed in this section, is a form of analog computer and has a biquadratic transfer function, it is called a "state-variable biquad" [7, p. 558] circuit, or a "biquad" [6, p. 339] for short.

Several circuits are available for implementing the biquadratic equations. The first of these is the KHN state-variable biquad circuit [7, p. 557], named after Kerwin, Huelsman, and Newcomb, which is illustrated in Fig. 9-29a. This circuit can simultaneously represent the regular low-pass, high-pass, and band-pass filters at three different output points. The equations for these functions are

$$v_{LP} = \left(\frac{2R_3}{R_2 + R_3} \right) \left[\frac{1/R^2C^2}{s^2 + \dfrac{1}{\left(\dfrac{R_2 + R_3}{2R_2} \right)RC}s + \dfrac{1}{R^2C^2}} \right] v_{in} \qquad (9\text{-}92)$$

$$v_{HP} = \left(\frac{2R_3}{R_2 + R_3} \right) \left[\frac{s^2}{s^2 + \dfrac{1}{\left(\dfrac{R_2 + R_3}{2R_2} \right)RC}s + \dfrac{1}{R^2C^2}} \right] v_{in} \qquad (9\text{-}93)$$

$$v_{BP} = \left(\frac{2R_3}{R_2 + R_3} \right) \left[\frac{(1/RC)s}{s^2 + \dfrac{1}{\left(\dfrac{R_2 + R_3}{2R_2} \right)RC}s + \dfrac{1}{R^2C^2}} \right] v_{in} \qquad (9\text{-}94)$$

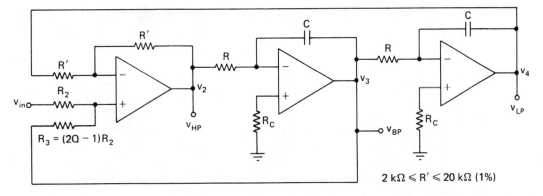

(a) KHN state-variable biquad circuit

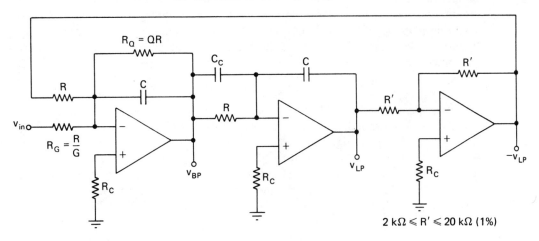

(b) Tow-Thomas state-variable biquad circuit

Fig. 9-29 Original state-variable biquads. (From *Filter Theory and Design: Active and Passive*, A. S. Sedra and P. O. Brackett; Matrix Publishers, 30 NW 23rd Place, Portland, OR 97210.)

where for all three equations, the denominator is in the form of Eq. (9-74), and

$$G = \frac{2R_3}{R_2 + R_3} \tag{9-95}$$

$$Q = \frac{R_2 + R_3}{2R_2} \tag{9-96}$$

$$\omega_0 = \frac{1}{RC} \tag{9-97}$$

where R is any value between 2 and 20 kΩ. This circuit has the first amplifier ungrounded, as a feedback loop enters at the noninverting input lead. The circuit also suffers from a change in Q as the op amp excess loop gain changes with frequency; this is called the Q-enhancement effect.

Biquadratic Filters

The second circuit, shown in Fig. 9-29b, is called the Tow-Thomas [7, p. 559] (T-T) state-variable biquad circuit. It is able to directly represent the low-pass and the band-pass filter functions. Because of the change in circuit structure, it is unable to represent the high-pass filter function. This is not a problem, though, as its primary use is for representing low- and high-pass notch filters. These are implemented through the use of an external summing amplifier which cancels out the numerator term, incorporating the single power of s; this leaves only the s^2 and the ω_z^2 terms in the numerator, which define zeros on the $j\omega$-axis. Although this circuit utilizes a passive compensation capacitor (C_c) to reduce the Q-enhancement effect, it still is somewhat dependent on both temperature and power-supply variations.

The circuit shown in Fig. 9-30a is called the Akerberg–Mossberg [7, p. 568] (A-M) state-variable biquad. Its operation is identical to the T-T biquad. The A-M biquad utilizes the Antoniou active compensated noninverting integrator; this circuit operates in a fashion identical to that of the integrator [7, p. 569] and inverting amplifier in the T-T biquad. Note that this compensated integrator (Fig. 9-30b) has the summing junction on the noninverting op amp input. This is because the output of this integrator amplifier is fed back to the integrating capacitor through an inverting amplifier. The result is a significant reduction of the undesirable Q-enhancement [7, p. 563] effect while suppressing the temperature and power-supply-dependence effects. In short, this circuit yields the best of both worlds. Its only drawback, when compared to the KHN biquad, is, again, the lack of a realization for the high-pass function. The use of matched [7, p. 571] op amps is extremely important in this circuit for the suppression of Q-enhancement, temperature, and power-supply variation effects. Thus a quad monolithic op amp should be used in the A-M biquad circuit for the necessary matching. The equation for the low-pass output voltage in the A-M biquad (Fig. 9-30a) is

$$v_{LP} = G \left(\frac{\dfrac{1}{R^2 C^2}}{s^2 + \dfrac{1}{QRC}s + \dfrac{1}{R^2 C^2}} \right) v_{in} \tag{9-98}$$

while that for the band-pass output voltage is

$$v_{BP} = -G \left(\frac{(1/RC)s}{s^2 + \dfrac{1}{QRC}s + \dfrac{1}{R^2 C^2}} \right) v_{in} \tag{9-99}$$

The T-T and A-M biquad circuits are seldom used for a low-pass or band-pass filter, as those functions can be obtained with single op amp circuits. Their primary function is to realize the low-pass and high-pass notch filters which are required to implement the Cauer (elliptic) approximations. The circuit for the low-pass notch filter is shown in Fig. 9-31a, while the change in connection at the inverting summer to produce the high-pass notch filter is illustrated in Fig. 9-31b. Both use either a T-T or A-M biquad. The inverting summer permits cancellation of the coefficient for the single power of the s term in the numerator through the selection of appropriate resistor ratios. It also permits the adjustment of Q, independent of frequency, for either connection through the selection of summing resistor values.

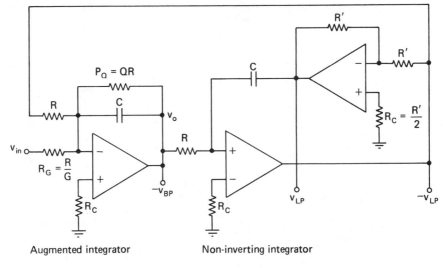

Augmented integrator   Non-inverting integrator

(a) Akerberg-Mossberg state-variable biquad

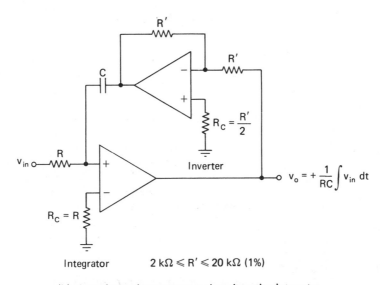

Integrator $2 \text{ k}\Omega \leqslant R' \leqslant 20 \text{ k}\Omega \ (1\%)$

(b) Antoniou active compensated noninverting integrator

Fig. 9-30 Modern state-variable biquad. (From *Filter Theory and Design: Active and Passive*, A. S. Sedra and P. O. Brackett; Matrix Publishers, 30 NW 23rd Place, Portland, OR 97210.)

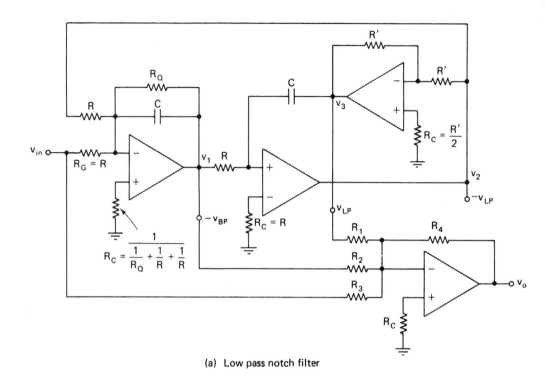

(a) Low pass notch filter

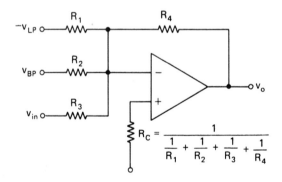

(b) Summer connection for high pass notch filter

Fig. 9-31 Biquad notch filters. (From *Filter Theory and Design: Active and Passive*, A. S. Sedra and P. O. Brackett; Matrix Publishers, 30 NW 23rd Place, Portland, OR 97210.)

The transfer function for the low-pass notch filter is

$$\frac{v_o}{v_{in}} = -G\left[\frac{\dfrac{R_4}{R_3}s^2 + \left(\dfrac{R_4}{QR_3} - \dfrac{R_4}{R_2}\right)\left(\dfrac{1}{RC}\right)s + \left(\dfrac{R_4}{R_3} + \dfrac{R_4}{R_1}\right)\left(\dfrac{1}{RC}\right)^2}{s^2 + \dfrac{1}{QRC}s + \left(\dfrac{1}{RC}\right)^2}\right] \quad (9\text{-}100)$$

If

$$R_3 = R_4$$
$$R_G = R$$
$$(9\text{-}101)$$

and

$$R_Q = QR \quad (9\text{-}102a)$$
$$R_2 = QR_3 \quad (9\text{-}102b)$$

the s term in the numerator drops out and the equation for the low-pass notch filter becomes

$$\frac{v_0}{v_{in}} = -\left[\frac{s^2 + (1 + R_4/R_1)(1/RC)^2}{s^2 + \dfrac{1}{QRC}s + \left(\dfrac{1}{RC}\right)^2}\right] \quad (9\text{-}103)$$

where the coefficient of $(1/RC)^2$ in the numerator is

$$M = 1 + \frac{R_4}{R_1}$$

or

$$R_1 = \frac{R_4}{M - 1} \quad (9\text{-}104)$$

The low-pass notch filter has six significant frequencies, three of which are used to determine resistor values. They are:

1. ω_0 is the resonant frequency of the filter.
2. ω_z is the frequency of the zero.
3. ω_m is the frequency of the peak amplitude.
4. ω_p is the frequency of the true-image pole.
5. ω_p is the frequency where the response crosses the amplitude at dc on its way from the peak amplitude to the zero; this frequency, together with the frequency of the zero ($\omega_z = \omega_S$), determines Ω_S, for Eq. (9-49).
6. The last frequency, which helps locate the others but is not useful for component selection, is the frequency corresponding to $+j1$ (normalized).

These frequencies are illustrated in Fig. 9-32a for the low-pass notch filter; for the high-pass notch filter, they are shown in Fig. 9-32b.

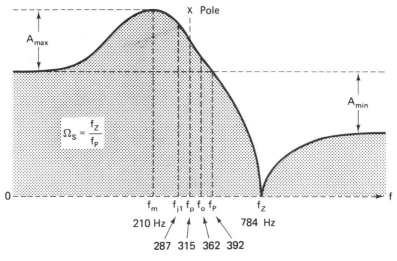

(a) Low pass notch filter response (Example 9-27)

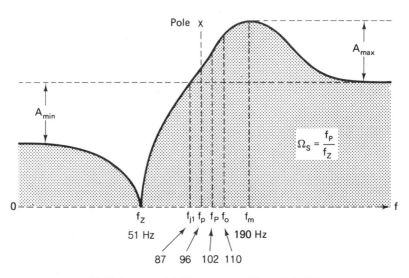

(b) High pass notch filter response (Example 9-29)

Fig. 9-32 Low- and high-pass notch filter responses.

The following series of steps are used to determine the component values and significant frequencies for the low-pass notch filter:

1. Find the normalized low-pass frequencies and coefficients ω_{0n}, ω_{zn}, Q, and M by comparing the normalized polynomial from Table 9-6 or a similar table of Cauer polynomials.

$$\frac{s^2 + (\omega_{zn})^2}{s^2 + \dfrac{\omega_{0n}}{Q}s + (\omega_{0n})^2} \tag{9-105}$$

where it can be seen that:

a. $(\omega_{0n})^2$ is the constant term in the denominator.
b. ω_{0n}/Q is the coefficient of s in the denominator.
c. $(\omega_{zn})^2$ is the constant term in the numerator.
d. $(\omega_{zn})^2 = M(\omega_{0n})^2$ by observation of Eq. (9-103); thus

$$M = \frac{(\omega_{zn})^2}{(\omega_{0n})^2} \tag{9-106}$$

2. The normalized radian frequency of the peak amplitude is given by [7, Table 1.1, p. 29]

$$\omega_{mn} = \sqrt{\left[\frac{M(1 - 1/2Q^2) - 1}{M + 1/2Q^2 - 1}\right]}(\omega_{0n}) \tag{9-107}$$

where $M > 1$ for the low-pass notch filter and

$$Q \geq \frac{M}{2(M - 1)} \tag{9-108}$$

for ω_{mn} to be real (not imaginary).

3. The peak amplitude is most likely to be the frequency around which the filter is developed. The resonant frequency is then determined using the peak amplitude frequency (ω_m) and

$$\omega_0 = \frac{1}{RC} = \omega_m\left(\frac{\omega_{0n}}{\omega_{mn}}\right) \tag{9-109a}$$

$$f_0 = f_m\left(\frac{\omega_{0n}}{\omega_{mn}}\right) \tag{9-109b}$$

The frequency of the zero is given by

$$f_z = f_m\left(\frac{\omega_{zn}}{\omega_{mn}}\right) \tag{9-110}$$

and the pole frequency is given by

$$f_p = f_m\left(\frac{\omega_{pn}}{\omega_{mn}}\right) \tag{9-111}$$

The value of ω_p in Eq. (9-49) is found by first determining the normalized value of ω_{Pn} from

$$\omega_{Pn} = \frac{\omega_{Sn}}{\Omega_S} = \frac{\omega_{zn}}{\Omega_S} \tag{9-112}$$

as ω_S is located at the radian frequency of the zero. The frequency of f_P is then determined from

$$f_P = \frac{f_m}{\Omega_S}\left(\frac{\omega_{zn}}{\omega_{mn}}\right) = \frac{f_z}{\Omega_S} \tag{9-113}$$

Choose a standard value for C from Appendix B2 which has a capacitive reactance near 20 kΩ. This causes R to be high in value, which is necessary, as the values of R_1 and R_2 are always less than R, sometimes significantly.

Biquadratic Filters

4. Since $R_3 = R_4$ from Eq. (9-101), they may be any value; a convenient value is the same as R.

5. R_1 controls the value of M, which establishes the zero frequency, given by Eq. (9-104), as

$$R_1 = \frac{R}{M - 1} \qquad \text{low-pass notch filter}$$

if $R_4 = R$.

6. R_2 and R_Q set the value of Q, independent of frequency, and their values are given by Eq. (9-102) as

$$R_Q = QR$$

and

$$R_2 = QR$$

if $R_3 = R$.

The use of these six steps will be explained through the use of two examples. The first example will illustrate steps 1 and 2; the second example will illustrate steps 3 through 6.

EXAMPLE 9-26

Determine the value of ω_{zn}, ω_{0n}, Q, M, ω_{mn}, σ_{pn}, and ω_{pn} for the normalized Cauer polynomial with two poles, $A_{max} = 0.5$ dB, and $\Omega_S = 2.0$. Construct a diagram showing this low-pass notch filter.

Solution

The normalized low-pass Cauer polynomial from Table 9-6 is

$$(4.967)\left(\frac{s^2 + 7.464}{s^2 + 1.245s + 1.592} \right)$$

Step 1:

a. $(\omega_{0n})^2 = 1.592$, $\omega_{0n} = 1.262$.

b. $\dfrac{\omega_{0n}}{Q} = 1.245 = \dfrac{1.262}{Q}$, $Q = \dfrac{1.262}{1.245} = 1.013$.

c. $(\omega_{zn})^2 = 7.464 = M(1.592)$, $M = \dfrac{7.464}{1.592} = 4.688$, $\omega_{zn} = 2.732$.

Step 2: $Q = 1.013$, $Q^2 = 1.026$, $1/2Q^2 = 0.487$; then using Eq. (9-107),

$$\omega_{mn} = \sqrt{\left[\frac{(4.688)(1 - 0.487) - 1}{4.688 + 0.487 - 1} \right]} (1.262) = 0.7321$$

Check the validity of Q using Eq. (9-108):

$$Q \geq \frac{4.688}{2(4.688 - 1)} = 0.6355$$

$$1.013 \geq 0.6355$$

Thus Q is a valid number.

$$\sigma_{pn} = \frac{1.245}{2} = 0.622 \qquad \sigma_{pn}^2 = 0.387$$

$$\omega_{pn}^2 = 1.592 - 0.387 = 1.205$$

$$\omega_{pn} = 1.097$$

The normalized low-pass notch filter is shown in Fig. 9-32a.

EXAMPLE 9-27

Using the normalized low-pass constants determined in Example 9-26, calculate the values of R, C, R_1, R_2, R_3, and R_4 for a low-pass notch filter with a peak amplitude (f_m) at 210 Hz.

Solution

Step 3: $\omega_0 = 1/RC = 2\pi(210)(1.262/0.732) = 2275$ rad/s. C (at 20 kΩ) = 0.02 μF. Choose the 0.022 μF that makes $R = 1/(2275)(0.022\ \mu\text{F}) = 20.0$ kΩ; use $R = 20.0$ kΩ from Appendix B1.

Step 4: Let $R_4 = R = 20.0$ k$\Omega = R_3$.

Step 5: From Example 9-26, $M = 4.688$. $R_1 = 20.0$ k$\Omega/(4.688 - 1) = 5.42$ kΩ; use 5.36 kΩ from Appendix B1.

Step 6: From Example 9-26, $Q = 1.013$. $R_2 = (1.013)(20.0$ k$\Omega) = 20.3$ kΩ; use a 20.5 kΩ. $R_Q = 20.5$ kΩ. From Appendix B1. Then, from Step 3,

$$f_0 = 210\left(\frac{1.262}{0.732}\right) = 362 \text{ Hz}$$

$$f_z = 210\left(\frac{2.732}{0.732}\right) = 784 \text{ Hz} = f_S$$

$$f_p = 210\left(\frac{1.097}{0.732}\right) = 315 \text{ Hz}$$

$$f_P = \frac{784}{2} = 392 \text{ Hz} \qquad \Omega_S = 2$$

$$f_{j1} = 210\left(\frac{1}{0.732}\right) = 287 \text{ Hz}$$

These values are plotted as shown in Fig. 9-32a for the normalized low-pass notch filter response curve.

The high-pass notch filter is developed in much the same way as for the low-pass notch filter. The first significant difference is the change in the numerator of the high pass normalized polynomial equation. Equation (9-50) is converted to the high-pass form,

$$\frac{s^2 + \left(1/L_{np}C_{np}\right)^2/\left(1/L_{nz}C_{nz}\right)}{s^2 + \left(1/L_{np}\right)s + \left(1/L_{np}C_{np}\right)} = \frac{s^2 + (1/M)\left(1/L_{np}C_{np}\right)}{s^2 + \left(1/L_{np}\right)s + \left(1/L_{np}C_{np}\right)} \qquad (9\text{-}114)$$

where the new M is the reciprocal of the low-pass M and the new $(\omega_{zn})^2$ is the new M times the $(\omega_{0n})^2$ term from the denominator. The denominator remains the same as used for the low-pass notch filter, see Eq. (9-50). Eq. (9-100) remains the same except that a minus sign separates the two parts of the coefficient of $(1/RC)^2$ in the numerator. Example 9-21 illustrates the design steps in detail. After the low-pass Cauer polynomial has been converted to the high-pass form, it is then in the form of Eq. (9-105) and is treated just as the low-pass notch filter with three exceptions.

Biquadratic Filters

Step 2A: Equation (9-107) is valid for the high-pass notch filter if $M < 1$ and ω_{mn} is real (not imaginary) if

$$Q \geq \frac{1}{2(1 - M)} \tag{9-108a}$$

Step 3A: The value of ω_{Pn} is now found from

$$\omega_{Pn} = \Omega_s \omega_{Sn} = \Omega_s \omega_{zn} \tag{9-115}$$

Step 4A: The value of M is now less than unity and R_1 is found using

$$R_1 = \frac{R}{1 - M} \tag{9-116}$$

if $R_4 = R$.

Again, two examples will be used to illustrate the use of these steps.

EXAMPLE 9-28

Determine the functions ω_{zn}, ω_{0n}, Q, M, ω_{mn}, σ_{pn}, and ω_{pn} for the normalized high-pass notch filter using the same Cauer polynomial as in Example 9-26.

Solution

$(\omega_{zn})^2$ for the normalized high-pass polynomial is found by taking the reciprocal of M for the low-pass polynomial and multiplying it by $(\omega_{0n})^2$. The denominator remains the same as before. The procedure for producing the high-pass polynomial is then low-pass $M = 4.688$, high-pass $M = \frac{1}{4.688} = 0.2133$, $(\omega_{zn})^2 = (0.2133)(1.592) = 0.3396$,

$$\frac{s^2 + 0.3396}{s^2 + 1.245s + 1.592}$$

$$(\omega_{zn})^2 = 0.3396 \qquad\qquad\qquad \omega_{zn} = 0.583$$

$$(\omega_{0n})^2 = 1.592 \qquad\qquad\qquad \omega_{0n} = 1.262$$

$$Q = \frac{1.262}{1.245} = 1.013 \qquad 2Q^2 = 2.052 \qquad \frac{1}{2Q^2} = 0.487$$

Then using Eq. (9-108a) gives us

$$Q \geq \frac{1}{2(1 - 0.2133)} = 0.635$$

$1.013 \geq 0.635$, so Q is valid. Equation (9-106) yields

$$M = \frac{(\omega_{zn})^2}{(\omega_{0n})^2} = \frac{0.3396}{1.592} = 0.2133$$

$$\omega_{mn} = \sqrt{\left[\frac{(0.2133)(1 - 0.487) - 1}{0.2133 + 0.487 - 1}\right](1.262)} = 2.176$$

$$\sigma_{pn} = \frac{1.245}{2} = 0.622 \qquad \sigma_{pn}^2 = 0.387$$

$$\omega_{pn}^2 = 1.592 - 0.387 = 1.205$$

$$\omega_{pn} = 1.097$$

$$\omega_{Pn} = 2(\omega_{zn}) = 2(0.583) = 1.166$$

EXAMPLE 9-29

Determine the circuit component values and the significant frequencies for a high-pass notch filter with a peak amplitude frequency of 190 Hz and the normalized functions of Example 9-28.

Solution

$$\omega_0 = \frac{1}{RC} = 2\pi(190 \text{ Hz}) = 1194 \text{ rad/s}$$

$$C(\text{at } 20 \text{ k}\Omega) = 0.042 \ \mu\text{F}$$

Choose $C = 0.05 \ \mu\text{F}$, so that R will be fairly large.

$$R = \frac{1}{(1194)(0.05 \ \mu\text{F})} = 16.75 \text{ k}\Omega$$

Use 16.9 kΩ from Appendix B1. Let $R_3 = R_4 = R = 16.9$ kΩ. Using Eq. (9-116), we obtain

$$R_1 = \frac{16.75 \text{ k}\Omega}{0.7867} = 21.3 \text{ k}\Omega \qquad \text{use } 21.5 \text{ k}\Omega \text{ from Appendix B1}$$

$$R_2 = (1.013)(16.75 \text{ k}\Omega) = 16.96 \text{ k}\Omega \qquad \text{use } 16.9 \text{ k}\Omega$$

$$R_Q = 16.9 \text{ k}\Omega$$

$$f_0 = 190\left(\frac{12.62}{2.176}\right) = 110 \text{ Hz}$$

$$f_{j1} = 190\left(\frac{1}{1.262}\right) = 87.3 \text{ Hz}$$

$$f_z = 190\left(\frac{0.583}{1.262}\right) = 51 \text{ Hz}$$

$$f_P = 2(51 \text{ Hz}) = 102 \text{ Hz}$$

$$f_p = 190\left(\frac{1.097}{1.262}\right) = 95.8 \text{ Hz}$$

These frequencies are plotted in Fig. 9-32b.

The main advantage of the low-pass and high-pass notch filters is their use in the construction of band-pass filters with small shape factors. The usual case involves the cascading of three filter sections; a band pass with a resonant frequency at the center of the passband, a low-pass notch filter with its peak amplitude frequency slightly higher than the band-pass resonant frequency, and a high-pass notch filter with its peak amplitude frequency slightly lower than the band-pass resonant frequency. The development of this filter is illustrated in Fig. 9-33, where it can be seen that the composite Cauer band-pass filter has an excellent shape factor and a virtually linear passband amplitude. The phase within the passband approaches the ideal case of linear phase change with frequency. This means that waveforms composed of many harmonics, within the passband, are passed without any appre-

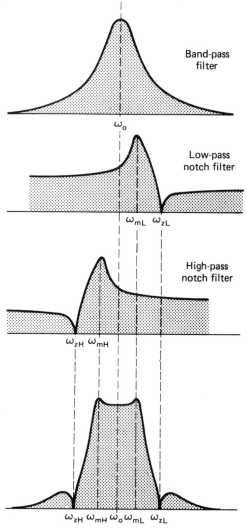

Fig. 9-33 Development of Cauer band-pass filter.

ciable distortion; they only have a time delay. The fact that zeros exist outside the passband can be further used to notch out adjacent channel frequencies. Figure 9-34 illustrates the use of Tow-Thomas state-variable biquad filter sections. Each of the three stacked circuit boards has space for two filter sections, including the inverting summer (Fig. 9-31); the top circuit board has the components installed for only one of those sections. The whole filter (three circuit boards) contains a single-section band-pass filter, a single-section low-pass filter, and a three-section Cauer band-pass filter like that illustrated in Fig. 9-33. This filter is one of many at contiguous frequencies. Each Cauer band-pass filter has its two zeros strategically located to be at the adjacent channel's resonant frequency, for minimum interchannel interference.

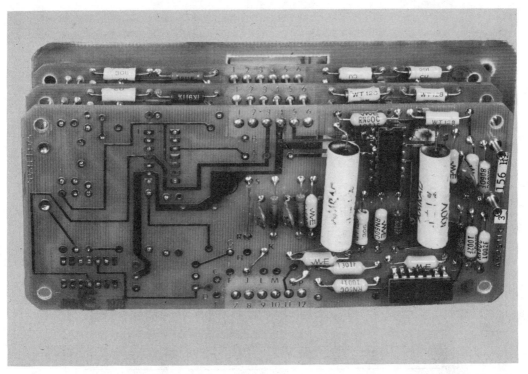

Fig. 9-34 Five-section Tow–Thomas biquad filter. (Courtesy of Safetran Systems Corporation, Louisville, KY.)

9.4.4 All-Pass Equalizers

The all-pass equalizer is a circuit that has unity gain at all frequencies and exhibits only a phase change as the input signal frequency varies. These circuits have many uses, in the phase and delay compensation of filters, as well as in the phase change of any signal. The all-pass equalizer is a non-minimum-phase system (i.e., it has zeros in the right half of the *s*-plane). Two all-pass equalizers will be presented in this section: the one-pole and the two-pole circuit.

The circuits for a one-pole all-pass equalizer are shown in Fig. 9-35. It is primarily a capacitor and resistor connected together with the output signal at the junction. One of the two components is fed out of phase from the other. The transfer function of the all-pass equalizer is

$$\frac{v_o}{v_{\text{in}}} = \pm \frac{s - 1/RC}{s + 1/RC} \tag{9-117}$$

where the transfer function is $(+)$ if the inverter is on the capacitor as shown in Fig. 9-35b and the transfer function is $(-)$ if the inverter is on the resistor as shown in Fig. 9-35c. The only significance of this sign is a 180° phase reversal in output signal phase. Aside from its use for delay equalization in filters, the primary use of the single-pole all-pass equalizer is to phase-shift a signal frequency a specified

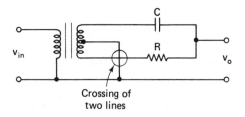

(a) Passive one pole all-pass equalizer

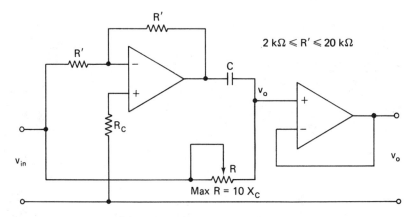

$2 \text{ k}\Omega \leqslant R' \leqslant 20 \text{ k}\Omega$

Max R = 10 X_C

(b) Positive phase active equalizer

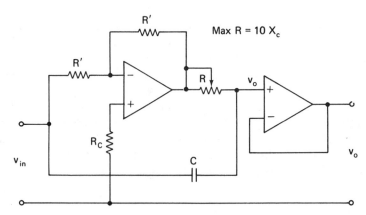

Max R = 10 X_c

(c) Negative phase active equalizer

Fig. 9-35 One-pole all-pass equalizer.

amount. The frequency should be constant as the phase will vary with frequency. When the capacitive reactance (X_C) equals the resistance (R), the phase shift is $90°$. It is $-90°$ if the inverter is on the resistor and $+90°$ if the inverter is on the capacitor. The phase can be varied by making R a potentiometer, and as the resistance becomes smaller, the output phase approaches that on the input of R. Similarly, as the resistance becomes larger, the output phase approaches the phase on the input to the capacitor. The maximum phase change, as the resistor value varies, is $0°$ to $-180°$ or $+180°$ to $0°$, depending on the connection. The later value of phase can only be obtained with an infinite resistance, which is not practical. The actual phase variation is usually $0°$ to $-174.3°$ or $+180°$ to $+5.7°$, as the potentiometer should have a value such that $R_{pot} = 10X_c$. In this circuit's most general use, the pot is adjusted to produce a fixed phase shift; in this case the pot value should be chosen such that the desired phase shift occurs at the pot's center. The phase shift of the circuit can be calculated by letting $s = j\omega$ in Eq. (9-117), in which case

$$\boxed{\theta^+ = 2\tan^{-1}\omega RC} \tag{9-118a}$$

or

$$\boxed{\theta^- + 180° = 2\tan^{-1}\omega RC} \tag{9-118b}$$

which rearranged becomes

$$\boxed{\omega RC = \tan\left(\frac{\theta^+}{2}\right)} \tag{9-119a}$$

or

$$\boxed{\omega RC = \tan\left(\frac{\theta^-}{2} + 90°\right)} \tag{9-119b}$$

where θ^+ is the case for positive phase shift (inverter on capacitor) and θ^- is the case for negative phase shift (inverter on resistor). Figure 9-36a illustrates the position of the pole and zero; they are equidistant from the $j\omega$-axis, producing a constant amplitude at all frequencies. Figure 9-36b illustrates the phase shift for the two circuits as the resistance (R) is varied from zero to infinity ohms. The value of X_C should be about 20 kΩ at the frequency being phase-shifted. This permits R to vary between 2 and 200 kΩ for the phase variation indicated in Fig. 9-36b.

An example will help to illustrate the use of Eq. (9-119b).

EXAMPLE 9-30

A phase shift of $-120°$ is required for a signal-processing controller that operates at a fixed frequency of 400 Hz.

  a. Select the circuit type (i.e., resistor or capacitor) on the inverter.
  b. Calculate the capacitor size if X_C is approximately 20 kΩ. Select the nearest standard value from Appendix B2 (film).

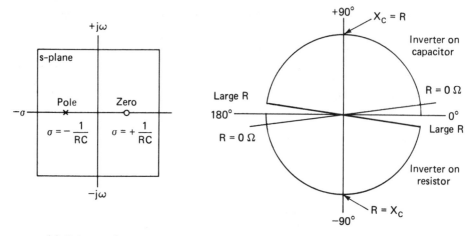

(a) Pole zero diagram

(b) Phase shift as R varies

Fig. 9-36 Pole–zero diagram and phase shift for one-pole all-pass equalizer.

c. Using the actual capacitance value selected, calculate the resistance value to produce $-120°$. Select a potentiometer from among 10 kΩ, 20 kΩ, or 50 kΩ that will produce the phase shift most near the midrange of the pot.

Solution

a. Since the phase is negative, select the circuit with the inverter on the resistor.

b. $C = \dfrac{1}{2\pi(400)(20\text{ k}\Omega)} = 0.02\ \mu\text{F}$; select the 0.022 μF capacitor from Appendix B2.

c. $X_C = \dfrac{1}{2\pi(400)(0.022\ \mu\text{F})} = 18.08\text{ k}\Omega = \dfrac{1}{\omega C}$ Using Eq. (9-119b), we obtain

$$\omega RC = \tan\left(-\frac{120°}{2} + 90°\right) = \tan(+30°) = 0.577$$

$$R = \frac{0.577}{\omega C} = (0.577)(18.08\text{ k}\Omega) = 10.4\text{ k}\Omega$$

Use the 20 kΩ potentiometer.

The two-pole all-pass equalizer is primarily used as a delay equalizer in active-filter circuits. The passive version of this circuit is illustrated in Fig. 9-37a, where the "line crossing" can clearly be seen. The active version of the all-pass equalizer can be implemented in many ways. One convenient way is to use the T-T or A-M state-variable biquad circuit and connect the summer in such a manner that the coefficient for the single power of s remains and is negative. The transfer function for the two-pole all-pass equalizer is

$$\frac{v_o}{v_{in}} = \pm\frac{s^2 - \dfrac{1}{QRC}s + \left(\dfrac{1}{RC}\right)^2}{s^2 + \dfrac{1}{QRC}s + \left(\dfrac{1}{RC}\right)^2} \tag{9-120}$$

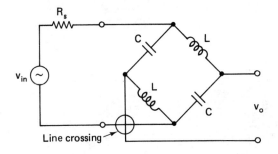

(a) Passive two pole all-pass equalizer

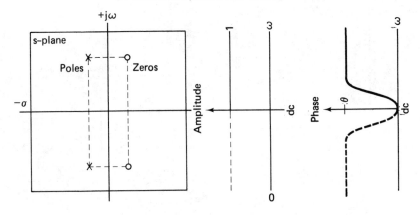

(b) Pole-zero diagram, amplitude, phase

Fig. 9-37 Two-pole all-pass equalizer.

where the zeros are in the right half of the s-plane. The zeros are always placed the same distance from the $j\omega$-axis as the poles as shown in Fig. 9-37b. This means that the pole–zero diagram on the s-plane has symmetry about both axes. The amount of phase or delay that the circuit yields is dependent upon the Q and the resonant frequency of the all-pass network with respect to the filter being equalized.

Figure 9-37c illustrates the A-M biquad connected as an all-pass equalizer. The transfer function of this circuit is Eq. (9-100) where $R_1 = \infty$, thus

$$\frac{v_o}{v_{in}} = -\left[\frac{\dfrac{R_4}{R_3}s^2 + \left(\dfrac{R_4}{QR_3} - \dfrac{R_4}{R_2}\right)\left(\dfrac{1}{RC}\right)s + \left(\dfrac{R_4}{R_3}\right)\left(\dfrac{1}{RC}\right)^2}{s^2 + \dfrac{1}{QRC}s + \left(\dfrac{1}{RC}\right)^2}\right] \tag{9-121}$$

where if

$$\boxed{R_3 = R_4 = R} \tag{9-122}$$

and

$$\boxed{R_Q = QR} \tag{9-123}$$

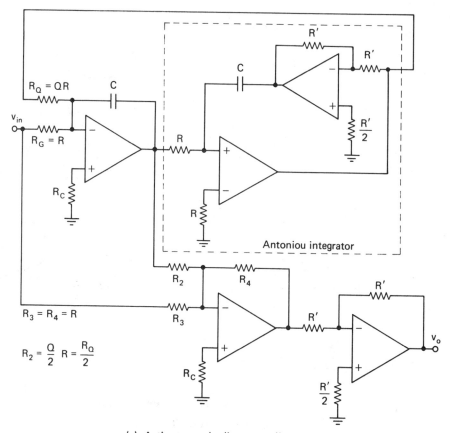

(c) Active two pole all-pass equalizer

Fig. 9-37 (*cont.*)

while

$$R_2 = \frac{QR}{2} \tag{9-124}$$

Equation (9-121) becomes equal to the negative of Eq. (9-120). The negative sign must be canceled by following the equalizer circuit by a unity-gain inverting amplifier, as shown in Fig. 9-37c. The equation for the phase shift in the circuit is

$$\theta = 2\tan^{-1}\left[\frac{\omega RC}{Q[(\omega RC)^2 - 1]}\right] \tag{9-125}$$

while the equation for the ωRC is

$$\omega RC = \frac{1}{2Q\tan(\theta/2)}\left[1 \pm \sqrt{1 + 4Q^2\tan^2\frac{\theta}{2}}\right] \tag{9-126}$$

Two-pole all-pass equalizers are used in series with other active-filter circuits to compensate for undesirable phase characteristics. Since the all-pass equalizer has unity gain at all frequencies, it does not alter the response shape of the filter being corrected. Time-delay characteristics, such as those illustrated in Fig. 9-15b, may also be modified through the use of an all-pass equalizer in series with the filter circuit. The selection of components in an all-pass equalizer circuit to alter phase or delay characteristics in an active filter is beyond the scope of this text.

9.5   System Concepts

The transfer admittance of a passive two-port network placed in a feedback loop will appear to be the inverse of the function *(Rule 6)* when observed as a system in the op amp circuit. This concept was first presented in Section 6.3, where the "two-port network" was a voltage divider. Equation (6-5) for the inverting amplifier illustrated this effect. Nowhere does this effect become more evident than in active filters. The transfer admittance of the effective network in the feedback loop is the inverse of the actual circuit transfer admittance; for this purpose, the input component can be presumed to be a resistor. Figure 9-38a illustrates a low-pass filter. Figure 9-38b illustrates the effective "feedback resistance" of this filter. Figure 9-38c places the low-pass response on the open-loop gain response (see Chapter 5) and illustrates where the excess loop gain is lost. Figure 9-38d illustrates the actual low-pass characteristic that can be expected from this active low-pass filter.

The same phenomenon is true for all filter functions. The response shape of the filter response is inverted and placed on the op amp open-loop-gain characteristic curve. Where the inverted filter function and the open-loop-gain curve intersect, no excess loop gain is available for control and the response begins to rise. Figures 9-39 to 9-41 illustrate the limitations of the high-pass, band-pass, and band-reject filters. The worst cases exist for the low-pass and band-pass filters. In these situations, op amps with the largest gain-bandwidth possible should be used. Feedforward compensation does not help for an active filter, as the open loop gain has returned to the dominate pole compensation rolloff slope, just where more excess loop gain is needed most. Thus only an increase in gain–bandwidth product will improve the high-frequency rejection characteristics of the low-pass and band-pass filters. There is no free lunch.

The effective closed loop gain (A_{VEF}) can be determined by presuming that a resistor in on the input to an inverting amplifier and that it has the inverse of the filter response in the feedback loop as a two-port network. This way the amount of loss in the filter response, in ratio, can be represented as an (R_F/R_I) closed-loop-gain ratio. A filter response is usually specified in dB/decade loss, and this must be converted to a loss ratio; thus the equation for effective closed loop gain can be written as

$$A_{\text{VEF}} = 10^{(\text{dB loss}/20)} \tag{9-127}$$

which represents the effective closed loop gain at any point along the response loss curve.

The use of Eq. (9-127) to determine the highest frequency that a low-pass or band-pass filter may have is illustrated by an example.

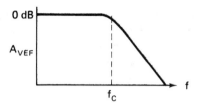

(a) Low pass filter response

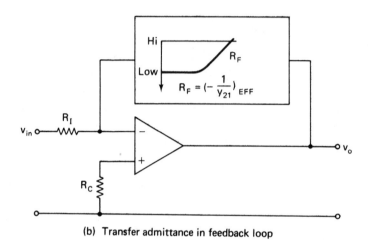

(b) Transfer admittance in feedback loop

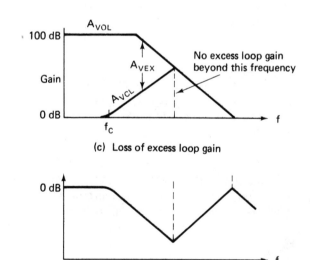

(c) Loss of excess loop gain

(d) Actual filter frequency response

Fig. 9-38 Active low-pass-filter frequency limitations.

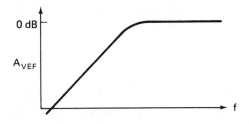

(a) High pass filter response

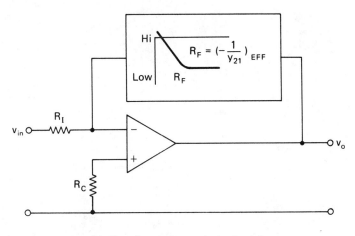

(b) Transfer admittance in feedback loop

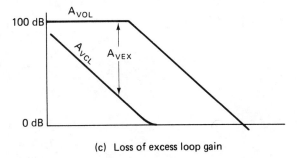

(c) Loss of excess loop gain

Fig. 9-39 Active high-pass-filter frequency limitations.

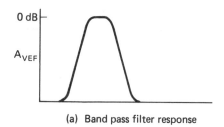

(a) Band pass filter response

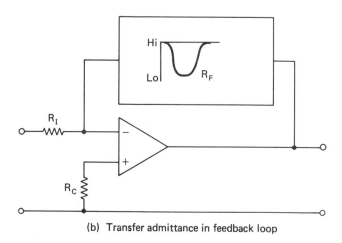

(b) Transfer admittance in feedback loop

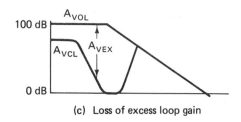

(c) Loss of excess loop gain

(d) Actual filter frequency response

Fig. 9-40 Active band-pass-filter frequency limitations.

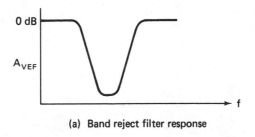

(a) Band reject filter response

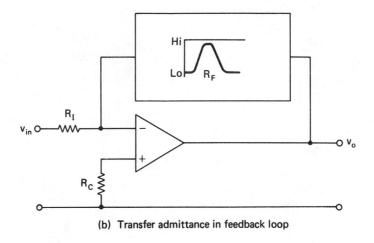

(b) Transfer admittance in feedback loop

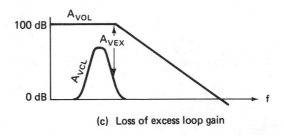

(c) Loss of excess loop gain

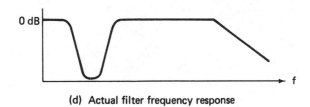

(d) Actual filter frequency response

Fig. 9-41 Active band-reject-filter frequency limitations.

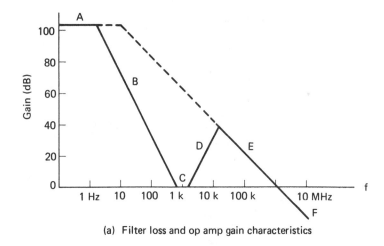

(a) Filter loss and op amp gain characteristics

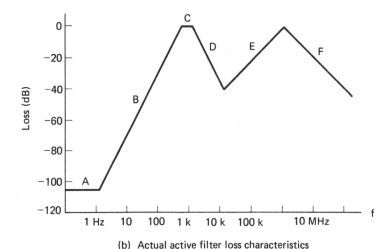

(b) Actual active filter loss characteristics

Fig. 9-42 Actual two-pole active Butterworth bandwidth filter response.

EXAMPLE 9-31

A two-pole Butterworth band-pass filter has a resonant frequency of 1 kHz and a bandwidth of 200 Hz. A μA 741 op amp is used for the active element in this filter; it has a gain–bandwidth product of 3 MHz. Determine if the filter can maintain a loss characteristic on the high-pass side of 60 dB.

Solution

Because the filter is a two-pole Butterworth, the skirts have a loss slope of 40 dB/decade; this slope begins on the high side at the cutoff frequency, 1.1 kHz. Figure 9-42 illustrates the loss characteristic, inverted, on the same plot as the gain-bandwidth characteristic for the 741 op amp. From the figure it can be seen that the filter is able to reject low-frequency signals to 105 dB but will only reject high-frequency signals to 40 dB.

Active Filters

The author hopes that the reader is provided with enough information concerning active filters that further material may be read and, by relating to topics in this chapter, a greater understanding of this extensive subject may be obtained.

REFERENCES

1. **Benjamin C. Kuo,** *Automatic Control Systems* (Englewood Cliffs, N.J.: Prentice-Hall, Inc., 1962)
2. **Frederick E. Terman,** *Electronic and Radio Engineering, Second Edition* (New York: McGraw-Hill Book Company, 1955)
3. **Franklin F. Kuo,** *Network Analysis and Synthesis, Second Edition* (New York: John Wiley & Sons, Inc., 1966)
4. **Mohammed S. Ghausi,** *Principles and Design of Linear Active Circuits* (New York: McGraw-Hill Inc., 1965)
5. **Erich Christian** and **Egon Eisenmann,** *Filter Design Tables and Graphs* (New York: John Wiley & Sons, Inc., 1966)
6. **Gobind Daryanani,** *Principles of Active Network Synthesis and Design* (New York: John Wiley & Sons, Inc., 1976)
7. **Adel S. Sedra** and **Peter O. Brackett,** *Filter Theory and Design: Active and Passive* (Portland, Oreg.: Matrix Publishers, Inc., 1978)

PROBLEMS

1. A high-fidelity amplifier has a frequency-response plot as illustrated in Fig. P9-1. Draw this plot on an *s*-plane in three dimensions.

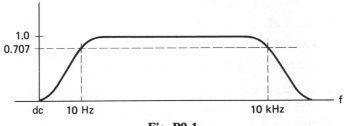

Fig. P9-1

2. A point on the *s*-plane represents a pendulum with a damper.

   a. Show the position of the point on the *s*-plane that represents the undamped pendulum.

   b. Show the relative shift of the point on the *s*-plane as the damper is applied.

3. Draw poles at the two points in three dimensions for Problem 2.

4. Show a mirror and window on a *s*-plane in three dimensions for Problem 2.

   a. How many poles are represented on the diagram?

   b. Identify the axis of the *s*-plane that separates the pole pairs.

5. A pole close to the damping axis represents a(an):

   a. Low-frequency oscillation.
   b. Undamped oscillation.
   c. High-frequency oscillation.
   d. Steady oscillation.
   e. Damped frequency oscillation.

6. A pole far from the damping axis represents a(an):

   a. High-frequency oscillation.
   b. Damped frequency oscillation.
   c. Steady oscillation.
   d. Undamped oscillation.
   e. Low-frequency oscillation.

7. A pole close to the frequency axis represents a(an):

   a. Steady oscillation.
   b. Low-frequency oscillation.
   c. High-frequency oscillation.
   d. Damped frequency oscillation.
   e. Undamped oscillation.

8. A pole far from the frequency axis represents a(an):

   a. Low-frequency oscillation.
   b. Undamped oscillation.
   c. High-frequency oscillation.
   d. Steady oscillation.
   e. Damped frequency oscillation.

9. A pole on the frequency axis represents a(an):

   a. Undamped oscillation.
   b. High-frequency oscillation.
   c. Damped frequency oscillation.
   d. Steady oscillation.
   e. Low-frequency oscillation.

10. $V_o/V_{in} = s + 4$.

   a. Is this a pole or a zero?
   b. Plot the point on the s-plane.

11. $V_o/V_{in} = 1/(s + 5)$.

   a. Is this a pole or a zero?
   b. Plot the point on the s-plane.

12. A transfer function is represented by the equation $T(s) = s/(s + 3.3)$.

    a. How many points are on the s-plane?

    b. Is there a pole or a zero at the origin?

    c. Plot the points on the s-plane using the correct symbol for each.

13. A transfer function is represented by the equation $1/(s^2 + 4s + 53)$.

    a. How many points are represented?

    b. Are they poles or zeros?

    c. Solve for the equation roots ($s = -\sigma \pm j\omega$).

    d. Plot the points on the s-plane.

14. For a transfer function in the form $T(s) = s^2/(s^2 + 8s + 50)$:

    a. How many points are represented?

    b. Place the points in their $s = -\sigma \pm j\omega$ format.

    c. Plot the points on the s-plane using the appropriate symbol for each.

15. The negative reciprocal of the transfer admittance is given by (see **Rule 6**, Section 6.3) $-1/y_{21(s)} = s^2/(s^2 + 8.5s + 56)$.

    a. How many points are represented?

    b. Are they poles or zeros? How many of each?

    c. What is the form of the denominator equation? Solve for the roots of the equation.

    d. Plot σ_z, σ_p, and ω_p on the s-plane.

16. The roots of a transfer function equation are as follows:

$$\sigma_z = -2.8$$
$$\sigma_p = -1.3$$
$$\omega_p = \pm j3.7$$

Reconstruct the transfer-function equation.

17. The denominator of a transfer function is $s^2 + 10^4 s + 10^8$; the numerator is $s + 100$.

    a. Write the transfer function.

    b. Find the values of σ_z, ω_o, and ζ.

    c. Find the value of s_1 and s_2 for the denominator.

18. Find the actual value of the frequency on the $j\omega$-axis ($s = j\omega$). (*Hint*: $\sigma = 0$ on the $j\omega$-axis.)

19. A resonant circuit has $\zeta = 0.02$ and $\omega_o = 5000$.

    a. Find the roots of s_1 and s_2.

    b. Reconstruct the characteristic equation.

    c. Plot s_1 and s_2 on the s-plane.

20. A series RLC circuit that resonates at 3185 Hz has $C = 0.22 \ \mu F$ and $L = 200$ mH.

    a. If $Q = 100$, calculate R.

    b. If $Q = 10$, calculate R.

c. If $R = 150$ Ω, calculate Q.

d. Find ζ for parts (a), (b), and (c).

21. If the resonant circuit for Problem 20 is initially energized with 3 mJ:

    a. Calculate the initial voltage across C.

    b. Calculate the voltage across the capacitor during the second and third cycles for $Q = 30$.

    c. Calculate the voltage across the capacitor during the second and third cycles for $Q = 20$.

    d. Plot the decrease in peak values for parts (b) and (c) on the same curves.

    e. Comment on the relative decrease in successive peaks for the two values of Q. Make a generalization of the relationship between the decline in ac capacitor voltages versus time with changes in Q.

22. Using the data from Problem 20:

    a. Calculate the values of the roots s_1 and s_2 to the characteristic equation for $Q = 15$.

    b. Calculate s_1 and s_2 for $Q = 20$.

    c. Plot s_1 and s_2 on an s-plane for parts (a) and (b).

    d. Which component of the root exhibits the greatest change on Q curves?

    e. Relate the shift of the poles on the s-plane in part (c) with your plots in part (e) of Problem 21.

23. Determine the bandwidth of the resonant circuit of Problem 20.

    a. For $Q = 15$ and $Q = 150$.

    b. Calculate the roots of the characteristic equation (s_1 and s_2) in terms of bandwidth for both Q values in part (a).

    c. Plot the east-face response for both Q values in part (a) opposite the pole positions in part (c) of Problem 22.

    d. Relate the shift of the poles in the s-plane to the change in shift of the east-face response.

    e. As the bandwidth becomes broader, what happens to the frequency of the peak on the east-face response?

24. A one-pole low-pass filter has a transfer function of $V_o/V_{in} = 100/(s + 100)$.

    a. If $R = 15$ kΩ, find C.

    b. Find the filter cutoff frequency.

    c. Find the location of the pole on the s-plane.

    d. Sketch the east-face response and the s-plane and show the positions of the cutoff frequency.

25. Draw the active filter version of the low-pass filter in Problem 24. Find the value of the current-compensation bias resistor.

26. A one-pole low-pass filter has $R = 100$ kΩ and $C = 0.1$ μF.

    a. Write the transfer-function equation for the filter.

    b. Let $s = j\omega$ in the transfer function, and find the value of V_o/V_{in} at $\omega = 55$, 2150, 30,000, and 300,000 rad/s.

326   *Active Filters*

c. Determine the relationship between the gain (loss) at 30,000 and 300,000 rad/s. Find the decrease of gain in dB as the frequency increases. Do you recognize the relationship?

d. Based on your results in part (c), what is the gain in dB at 150 kHz?

e. What effect does the zero of transmission have on the frequency response?

27. A two-pole low-pass filter is shown in Fig. P9-27.

   a. Determine the transfer function of the filter.

   b. What is the distinctive feature of a low-pass transfer function?

   c. Calculate ω_o, ζ, σ_p, $j\omega_p$, and jf_p.

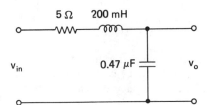

Fig. P9-27

28. A Sallen-Key low-pass filter is shown in Fig. P9-28.

   a. Find R_o, R_c, L, and C.

   b. Find ω_o, f_{op}, σ_p, and ζ.

   c. Write the transfer-function equations with numerical coefficients for all terms.

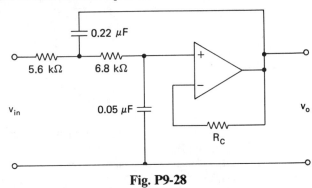

Fig. P9-28

29. For Problem 28, plot the location of the poles on the s-plane.

30. Calculate values for L_n and C_n for the three-pole normalized Butterworth low-pass filter using Table 9-1.

31. Calculate the L_n and C_n values for the four-pole normalized Butterworth low-pass filter.

32. Calculate the L_n and C_n values for the six-pole normalized Butterworth low-pass filter.

33. Calculate the angles between pole locations for a 10-pole Butterworth filter. Plot the poles on the unit circle (see Fig. 9-11).

34. Reconstruct the normalized low-pass Butterworth polynomials for the 10-pole filter of Problem 33. Use the relationships $-\sigma = 1 \cos \theta$ and $\omega = 1 \sin \theta$, where θ is the angle between the $-\sigma$ axis and a radius line to the pole. [*Hint*: Use Eq. (9-3).]

35. The circuit shown in Fig. P9-35 must have a cutoff frequency of 520 Hz. The capacitive reactance of C_3 should be roughly 30 kΩ. Calculate the values of C_3, R_3, and R_c. Select C_3 from Appendix B2 (film). Select R_3 from the 1% table of Appendix B1, and choose R_c from the 5% table.

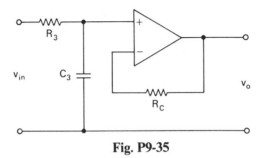

Fig. P9-35

36. Calculate the values of all components for the two-pole low-pass Butterworth filter shown in Fig. P9-36. The cutoff frequency is at 500 Hz. Let the capacitive reactance of C_2 fall between 12 and 25 kΩ. Select appropriate capacitors from Appendix B2 and resistors from the 1% table of Appendix B1.

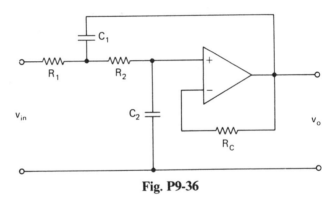

Fig. P9-36

37. A three-pole active low-pass Butterworth filter has the circuit shown in Fig. P9-37. The cutoff frequency is at 200 Hz; C_2 and C_3 should have a capacitive reactance within the range 18 to 25 kΩ. Calculate the component values, and select appropriate R and C values from Appendices B1 and B2.

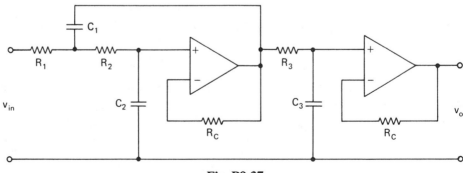

Fig. P9-37

Active Filters

38. Select component values for a six-pole low-pass Butterworth filter with a cutoff frequency of 680 Hz. The capacitive reactance of C_2 should fall between 15 and 30 kΩ.

    a. Draw the circuit and label the components.

    b. Select an appropriate C_2 value from Appendix B2 (film).

    c. Calculate the size of C_1, and select values from Appendix B2.

    d. Calculate values for R_1 and R_2. Select values from the 1% table of Appendix B1.

39. Determine L_n and C_n values for the four-pole normalized low-pass Chebyshev polynomials with 0.5 dB ripple.

40. For Problem 39, determine the loss required for the circuit to simulate a passive filter circuit. If $R_B = 12$ kΩ, calculate R_A and select a value from Appendix B1 (1%).

41. Determine the normalized Chebyshev polynomials for a five-pole filter with 0.25 dB passband ripple using the Butterworth polynomials and the five steps described in the chapter.

42. Determine the normalized Chebyshev polynomials for the six-pole filter with 1.0 dB passband ripple using the Butterworth polynomials and the five steps described in the chapter.

43. A three-pole low-pass filter must have the steepest possible cutoff slope ($-$dB/decade). Which of the first five filter approximations would best represent this?

44. A low-pass filter must have the fastest possible rise in response to an input step function. Choose a filter type to meet this specification without regard to any other characteristics.

45. Which low-pass filter type will yield the steepest possible cutoff without any ripple?

46. What is the effect on frequency response for the low-pass filter as the poles move:

    a. Away from the frequency axis?

    b. Closer to the frequency axis?

47. What is the effect on step response as the poles move:

    a. Away from the frequency axis?

    b. Toward the frequency axis?

48. Repeat Problem 37 for the three-pole Chebyshev filter with 0.5 dB ripple.

49. Repeat Problem 37 for the three-pole optimal filter.

50. Repeat Problem 37 for the three-pole parabolic filter.

51. Calculate the voltage-divider ratio for the four -pole Chebychev low-pass filter with 0.5 dB ripple that will allow for adjustment of the gain which will result in a simulation of the passive version.

52. Why does a Cauer low-pass filter have a sharper cutoff response than the Chebyshev filter version?

53. Determine the value of ω_o for the numerator and the product of L_{nz} and C_{nz} for the normalized Cauer polynomials having three poles with $\Omega_s = 1.5$.

54. Verify that the numerator multiplier for the two-pole normalized Cauer polynomial with $\Omega_s = 3.0$ is correct. (*Hint*: Observe the value of A_{max}.)

55. What is the effect on the Cauer filter skirt steepness as Ω_s becomes larger?

56. A one-pole high-pass active filter has a cutoff frequency of 340 Hz.

    a. Calculate the RC time constant.

    b. Write the transfer function of the filter.

    c. Show the circuit of the filter.

    d. Construct the s-plane diagram. Show the location of any poles and zeros.

    e. Plot the east-face frequency response, and indicate the cutoff frequency.

57. A one-pole high-pass Chebyshev filter with 0.25 dB passband ripple has a cutoff frequency of 500 Hz. A 0.033 μF capacitor is used in the circuit. Calculate the value of R_3 and R_c. Select a 1% resistor from Appendix B1 for R_3, and a 5% resistor for R_c.

58. Write the normalized high-pass transfer function for a one-pole Chebyshev filter with 1.0 dB passband ripple.

59. An RLC high-pass passive filter has $R_o = 5\ \Omega$, $L = 0.2$ H, and $C = 0.2$ F. Write the transfer-function equation.

60. What is the distinctive feature of a high-pass transfer function?

61. One of two sections of a four-pole high-pass active filter circuit is shown in Fig. P9-61; it has a parabolic response shape. The resonant frequency (f_o) is 560 Hz; C_1 and C_2 are both 0.033 μF.

    a. Calculate the values of R_1 and R_2 for each section.

    b. Calculate the values of R_o, L, and C for each section.

    c. Calculate the values of ω_o, σ_P, and ζ for each section.

    d. Plot the poles and zeros on an s-plane.

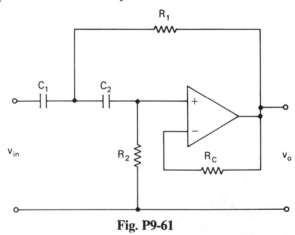

Fig. P9-61

62. The three-pole normalized low-pass transfer function has a denominator of $(s = 1)(s^2 + 1.582s + 2.507)$.

    a. Calculate the coefficients for the normalized high-pass transfer function. Write the transfer function in proper high-pass form.

    b. Using the high-pass transfer function obtained in part (a), plot the poles and zeros on an s-plane to scale.

    c. Draw the east-face response. Remember that f_c is not in the same location as for the high-pass Butterworth.

d. To find the response at cutoff, let $S = j1$ and calculate the response using the transfer function from part (a).

63. Determine the pole and zero locations for the normalized four-pole high-pass parabolic filter functions. Plot the poles and zeros on an s-plane.

64. For the high-pass filter function in Problem 63, $f_0 = 320$ Hz.

    a. Show the four-pole active circuit.

    b. Let $C_1 = C_2 = 0.022$ μF and calculate the balance of the component values.

    c. Select appropriate values from Appendix B1 (1%).

65. What are the distinctive features of the band-pass transfer function?

66. What is the difference between a low-pass response function and a band-pass response function?

67. An active band-pass filter is shown in Fig. P9-67.

    a. If $Q = 2$, find the ratio of R_f to R_I.

    b. If $f_0 = 140$ Hz, find the bandwidth.

    c. Find the high and low cutoff frequencies (f_{ch} and f_{cl}).

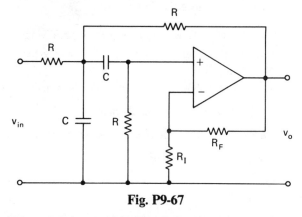

Fig. P9-67

68. Determine the circuit values for a three-pole parabolic and band-pass filter. The center frequency (f_0) is 460 Hz, and the bandwidth is 80 Hz. Let $C = 0.022$ μF, where each filter section has the circuit shown in Fig. P9-68.

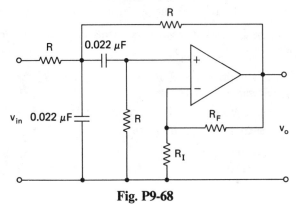

Fig. P9-68

69. Repeat Problem 68 for the three-pole optimal filter. What happens to the high and low pole positions when $\omega_2 < 1$? Does this prevent you from calculating component values for the filter?

70. A five-pole band-pass filter with a center frequency of 720 Hz has a -3 dB bandwidth of 100 Hz and a -20 dB bandwidth of 240 Hz. Find the value of the shape factor.

71. For the term "biquadratic" equation:

   a. Define it in its most general sense; give examples.
   b. Define it as it applies to biquadratic filters.
   c. What is the defining factor between the definitions in parts (a) and (b)?

72. For a non-minimum-phase transfer function, define how it differs from a minimum-phase function.

73. Determine the component values for a twin-tee notch filter with a notch frequency of 620 Hz and a gain off resonance of 5. Let $C = 0.047\ \mu\text{F}$.

74. A band-reject filter with a parabolic response is formed by combining the band-pass filter of Problem 68 with a notch-depth amplifier and a subtractor circuit (see Fig. 9-27). The rejection at resonance should be 67 dB.

   a. Calculate the gain change required in the input divider and voltage follower.
   b. Draw the entire circuit. Show the location and value of any parts required in the circuit.
   c. which component should be altered if the notch depth is to be changed to 80 dB?

75. How does a state-variable filter get its name?

76. What is biquad? How can a number of state variables in the circuit be determined?

77. What is the Q-enhancement effect?

78. Which filter function can be represented in the KHN and the A-M biquads but not in the T-T biquad?

79. What effects in the A-M biquad circuit are reduced by using a quad monolithic op amp?

80. Determine the value of ω_{zn}, ω_{on}, Q, M, ω_{mn}, σ_{pn}, and ω_{pn} for the normalized low-pass Cauer polynomial with two poles having $A_{max} = 0.5$ dB and $\Omega_S = 2.0$; check Q. Construct a diagram showing the relative positions of the normalized frequencies.

81. Using the normalized low-pass constants determined in Problem 80, let $C = 0.01\ \mu\text{F}$ and:

   a. Calculate the values of R, R_1, R_2, R_3, and R_4 for a low-pass notch filter with a peak amplitude at 420 Hz.
   b. Select appropriate resistors from Appendix B1 (1%).
   c. Calculate f_o, f_z, f_p, and f_{j1} for the filter.

82. Determine the function ω_{zn}, ω_{on}, Q, M, ω_{mn}, σ_{pn}, ω_{pn}, and ω_{Pn} for the normalized high-pass notch filter; check Q. Use the two-pole Cauer polynomial with $A_{max} = 0.5$ dB and $\Omega_s = 2.0$.

332

83. Using the normalized high-pass constants from Problem 82:

   a. Determine the resistor values for a high-pass notch filter with a peak amplitude at 300 Hz where $C = 0.047 \, \mu F$.

   b. Select values from Appendix B1 (1%).

   c. Calculate f_o, f_z, f_p, f_P, and f_{j1}.

84. When a low-pass notch, a band-pass, and a high-pass notch filter are used to form a sharp Cauer band-pass filter, what are the relative positions of the low- and high-pass notch filters?

85. A single-pole all-pass equalizer must create a phase shift of $+90°$ at 720 Hz. If a 0.022 μF capacitor is used:

   a. Draw the circuit.

   b. Calculate the size of the resistor to be used.

86. A single-pole all-pass equalizer for use in quadrature phase detector must have a phase shift of $-30°$ at 327 Hz. The capacitor's reactance should be approximately 22 kΩ.

   a. Draw the circuit.

   b. Select an appropriate film capacitor from Appendix B2.

   c. Calculate the resistor size and select a 1% resistor from Appendix B1.

87. A two-pole all-pass equalizer has $RC = 1.5$ ms and $Q = 5$.

   a. Calculate the phase at 0 Hz (dc), 100 Hz, 200 Hz, 500 Hz, and 1000 Hz.

   b. Plot the phase shift opposite the pole-zero diagram.

   c. Determine where the minimum and maximum phase change occurred.

88. A three-pole Chebychev low-pass filter has a cutoff frequency of 300 Hz. How much rejection, in dB, can be expected if a μA 741 op amp is used? A minimum open loop gain of 50,000 is presumed to be available. The gain–bandwidth product is 1.5 MHz.

89. A three-pole active high-pass Butterworth filter has a cutoff frequency of 3000 Hz.

   a. What is the effective closed loop gain at 1000 Hz?

   b. If the op amp has a gain–bandwidth product of 10^5 Hz, how much excess loop gain is available for control at 1000 Hz?

   c. Using Eq. (5-8) where $(R_F + R_I)/R_I = A_{ref}$, calculate the percent error in the filter's amplitude response at 500 Hz.

   d. Repeat part (c) at 2 kHz.

90. A four-pole Butterworth band-pass filter has a resonant frequency of 3 kHz and a bandwidth of 240 Hz. A μA 741 op amp with a gain–bandwidth product of 1.5 MHz is used in all four sections of the filter. Determine the greatest loss characteristic that can be maintained on the high side.

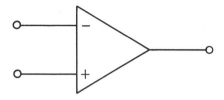

10 **Oscillators**

10.1   Introduction

The study of oscillators is extensive. Oscillator circuits have been developed that will provide various output wave shapes. Circuits are available to operate in various ranges, from fractions of a hertz to many gigahertz. As you have probably surmised, oscillators that use operational amplifiers as the active element are limited in frequency to below 1 MHz; often, these circuits are limited to the audio range or below. This chapter is devoted to circuits that use op amps as the active element. Thus this chapter is devoted to oscillators that provide various output waveshapes at audio frequencies and below.

10.2   Creating an Oscillator

In Section 5.2.4, the topic of op amp oscillation was first mentioned and methods of preventing its occurrence were presented. Oscillation was defined as occurring when the loop phase was allowed to become $0°$ while an excess loop gain existed. This definition of an oscillator is still valid, but now we are going to present methods of causing that situation to occur under controlled conditions. The components in the feedback loop have a phase shift, which permits the loop phase to become $0°$ while an excess loop gain at a particular frequency exists; this will be frequency of oscillation. The question is often asked: What makes the oscillator begin oscillating; why does it not just sit there? The answer is—noise—from the op amp and circuit components (see Chapter 5). This noise has a broad spectrum of frequencies, including the frequency of oscillation. When the oscillator circuit is first turned on, these frequencies pass from the op amp output through the network back to the op amp input. The frequency of oscillation passes at 0° and is amplified by the op amp. Each time this frequency goes around the loop, it is amplified again and the oscillation at the frequency where the loop phase is 0° builds up to full amplitude. This chapter follows the active-filter chapter because some of the techniques used in active filters are also used in oscillators; an active band-pass filter will oscillate at its resonate frequency if the Q is sufficiently high (greater than 20).

10.3   Bounding the Amplitude

If one were to place an oscillator on the s-plane, the pole–zero diagram would be that of a one-pole pair band-pass filter (see Fig. 9-22) with the poles slightly to the right of the $j\omega$-axis. The poles should be on the $j\omega$-axis, so that the oscillation remains at a constant amplitude (see Fig. 9-2a). Most active sine-wave oscillators have output amplitudes that would rise to saturation where the waveform clips on the top and bottom peaks. This clipping is the way the oscillator has of moving the poles back from the right half of the s-plane to being on the $j\omega$-axis. This is an undesirable situation, as the user usually does not want a clipped waveform. Thus methods must be used that will move the poles back to being on the $j\omega$-axis without causing the output waveform to clip. This is called "bounding" the output amplitude and is accomplished in many ways, each unique to a particular oscillator circuit. All amplitude bounding has one thing in common; it causes the op amp circuit to lose sufficient excess loop gain for oscillation at large amplitudes, before the op amp clips due to saturation. It is a method of creating a "soft" clipping and always distorts the output waveform to some extent. Thus a sine-wave oscillator that is bounded has a waveform where the sine-wave peaks are slightly flattened. The amount of flattening depends on the degree of softness of the output bounding.

10.4   Sine-Wave Oscillators

A sine-wave oscillator is one in which the loop phase is caused to become 0° at a particular frequency. A network of passive components is connected between the output and input of the op amp which provides 180° of phase shift at only one frequency; the op amp provides the other 180° between its inverting input and the output.

10.4.1   Twin-Tee Oscillator

The *twin-tee* network is a sharp band-reject or notch filter. As mentioned in Section 9.5, if this notch network is placed in the feedback loop of an amplifier, the inverse of its transfer function will be created by **Rule 6** and the circuit will become a band-pass filter with a high Q. This circuit will oscillate at the *twin-tee* notch frequency given by

$$f_0 = \frac{1}{2\pi RC} \tag{10-1}$$

where the output sine wave will be clipped, without bounding. The bounding network, also shown in Fig. 10-1, begins limiting the output voltage at an amplitude

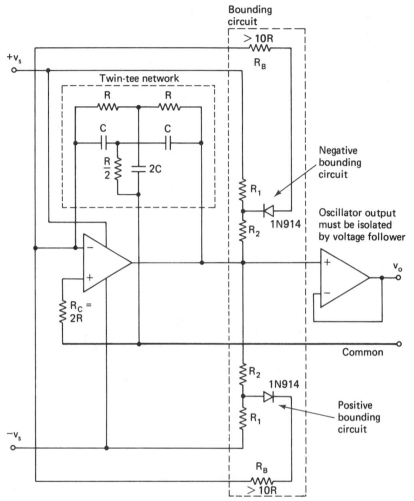

Fig. 10-1 Twin-tee sine-wave oscillator with bounding.

(v_o) given by

$$R_2 = R_1\left(\frac{v_o - 0.7\,\text{V}}{V_S + 0.7\,\text{V}}\right) \tag{10-2}$$

where the supply voltage to the op amp is $\pm V_S$ and the 0.7 V is the drop across each diode. It does this by reducing the twin-tee network Q to a level where the circuit will not oscillate. The actual value of the bounding resistor (R_B) is greater than $10R$, but must be adjusted empirically to limit oscillation without causing excessive distortion. This is the least desirable of the sine-wave oscillators, as it is difficult to adjust the network for a particular frequency.

An example will help to illustrate the method of component selection for this circuit.

EXAMPLE 10-1

A *twin-tee* oscillator has a frequency of 120 Hz. The op amps have supply voltages of ± 10 V. The output amplitude should begin bounding at ± 7 V. Determine all component values.

Solution

Let $X_c = 20$ kΩ,

$$C = \frac{1}{2\pi(120)(20 \text{ k}\Omega)} = 0.06 \ \mu\text{F}$$

Use the 0.05 μF capacitor from Appendix B2, as the 0.1 μF is also a standard value and is twice its size. Then $C = 0.05 \ \mu$F and $2C = 0.1 \ \mu$F.

$$R = \frac{1}{2\pi(120)(0.05 \ \mu\text{F})} = 26.5 \text{ k}\Omega$$

Use 26.7 kΩ from Appendix B1; use 13.3 kΩ for $R/2$. Let $R_I = 27$ kΩ (5%); R_2 is then found using Eq. (10-2) as

$$R_2 = (27 \text{ k}\Omega)\left(\frac{7 \text{ V} - 0.7 \text{ V}}{10 \text{ V} + 0.7 \text{ V}}\right) = 15.9 \text{ k}\Omega$$

Use 16 kΩ. Begin by letting $R_B = 270$ kΩ and adjusting for the best sine wave.

10.4.2 Wien Bridge Oscillator

This circuit is probably the most popular, as it requires the fewest components of any sine-wave oscillator. Figure 10-2a shows the Wien bridge circuit in the form that gave it the name "bridge" and Figure 10-2b shows the actual circuit; they are the same circuit. The frequency of oscillation is given by

$$f_0 = \frac{1}{2\pi\sqrt{R_1 C_1 R_2 C_2}} \tag{10-3}$$

where the change in any circuit component value will cause a change in frequency; thus if $R_1 = R_2 = R$ and $C_1 = C_2 = C$, the frequency of oscillation is given by

$$f_0 = \frac{1}{2\pi RC} \tag{10-4}$$

The phase shift through the network from v_0 to the noninverting op amp input is $0°$ at resonance, while the gain resistors make up for the loss through the network to provide the necessary excess loop gain for oscillation. The loss through the network is $R_2/(R_1 + 2R_2)$; therefore, the noninverting gain must be 3 for oscillation. Thus

$$\frac{R_F}{R_I} + 1 = \frac{R_1}{R_2} + 2$$

for oscillation, or

$$\frac{R_F}{R_I} = \frac{R_1}{R_2} + 1 \tag{10-5}$$

when $R_1 = R_2$, then the noninverting gain from **Rule 3** is

$$\frac{R_F + R_I}{R_I} = 3 \tag{10-6}$$

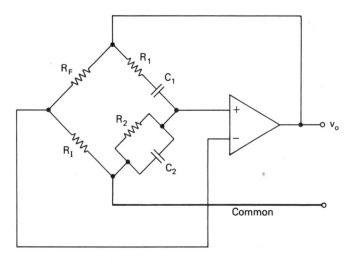

(a) Circuit in bridge form

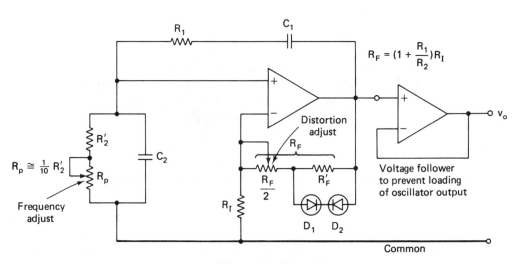

$$R_F = (1 + \frac{R_1}{R_2})R_I$$

Voltage follower
to prevent loading
of oscillator output

(b) Actual oscillator circuit

Fig. 10-2 Wien bridge oscillator.

One would think that if a gain of 3 is sufficient for oscillation, a larger gain would be better to assure oscillation. This is true, but at the expense of increased sine-wave distortion; the gain must be just 3 for a pure sine wave.

The bounding circuit is composed of R'_F, D_1, and D_2; the zener-diode avalanche voltage (V_Z) should be

$$V_Z = \left(\frac{R'_F}{R_F + R'_F + R_I} \right) V_O - 0.7 \text{ V} \qquad (10\text{-}7)$$

The value of R_2 should be 5% less than the calculated value or two 1% values (in

Appendix B1) below the calculated value. This permits the potentiometer (R_2') to have a $\pm2\%$ range for adjusting the frequency.

An example should bring the use of these relationships into perspective.

EXAMPLE 10-2

Select the component values for a Wien bridge oscillator that operates at a frequency of 211 Hz. The op amp supply voltages are ±15 V and v_o (peak) should be 10 V.

Solution

X_C for the capacitor should be about 20 kΩ. Thus

$$C = \frac{1}{2\pi(211)(20 \text{ k}\Omega)} = 0.038 \ \mu\text{F}$$

Use 0.05 μF from Appendix B2 (film). Then $C_1 = C_2 = C = 0.05 \ \mu$F. Using Eq. (10-3), we obtain

$$R_1 R_2 = \frac{1}{4\pi^2(211)^2(0.05 \ \mu F)^2} = 2.276 \times 10^8 (\Omega)^2$$

Let $R_1 = 10.0$ kΩ, then $R_2 = 22.76$ kΩ; select a 22.1 kΩ and a 1 kΩ pot for R_2. From Eq. (10-5), $(R_1/R_2) + 1 = 1.439$, thus $R_F = (1.439)R_I$; let $R_I = 10$ kΩ. Then $R_F = 14.3$ kΩ, use a 7.15 kΩ 1% and a 10 kΩ pot for R_F. From Eq. (10-7), $V_Z = [7.15/(14.3 + 10)]$ 10 V $-$ 0.7 V = 2.24 V; use three forward-biased diodes for V_Z.

10.4.3 Phase-Shift Oscillator

The third type of sine-wave oscillator for use with op amps is the phase-shift oscillator. Two circuits exist for this type of oscillator. The first type, illustrated in Fig. 10-3a, uses a three-section, two-port, phase-shifting network as the input to an inverting-type amplifier (see Section 6.3). If the R's and C's in the two-port network are each respectively equal, the frequency of oscillation is given by

$$f_0 = \frac{1}{2\pi\sqrt{3} \ RC} \tag{10-8}$$

One resistor can be reduced slightly in value and a series potentiometer added to provide a frequency adjust. The phase shift through the network is 180°, 60° for each of the three sections. The signal loss through the network is $\frac{1}{12}$, which must be made up through gain in the inverting amplifier; thus, from **Rule 6**

$$R_F = 12R \tag{10-9}$$

for oscillation to occur. Bounding should be applied to this circuit by splitting the feedback resistor into two equal parts totaling more than $12R$ and connecting two zener diodes (see Fig. 10-3a) each having a voltage of

$$V_Z = \frac{v_{o(\text{peak})}}{2} \tag{10-10}$$

An example will show the method of finding component values.

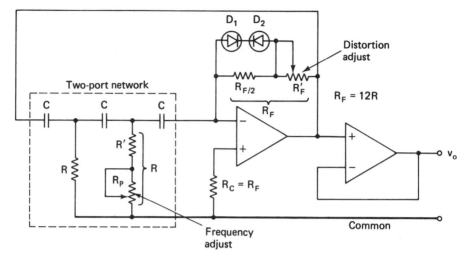

(a) Two-port phase shift network on amplifier

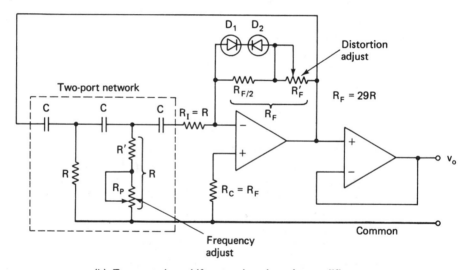

(b) Two-port phase shift network on inverting amplifier

Fig. 10-3 Phase-shift oscillator.

EXAMPLE 10-3

Determine the component values for a phase-shift oscillator circuit of Fig. 10-3a, which operates at 30 Hz and has power-supply voltages of ±15 V on the op amp. The output voltage should be 10 V peak.

Solution

$X_C = 2$ kΩ, as we want R to be small because of the large R_F which we must use

$$C = \frac{1}{2\pi(30)(2 \text{ k}\Omega)} = 2.6 \ \mu\text{F}$$

Use the 2.0 μF capacitor from Appendix B2 (film). Then from Eq. (10-8),

$$R = \frac{1}{2\pi\sqrt{3}\,(30)(2\ \mu F)} = 1.53\ k\Omega$$

Use the 1.54 kΩ (1%) resistor from Appendix B1. Let $R' = 1$ kΩ (1%) and use a 1 kΩ potentiometer for R_p. $R_F = (12)(1.53\ k\Omega) = 18.36\ k\Omega$; let $R_F/2 = 9.1$ kΩ (5%) and use a 10 kΩ potentiometer for R'_F. The zener-diode voltage for the bounding is given by Eq. (10-10) as $V_D = 10\ V/2 = 5$ V. Use 5 V TRW zener diodes from Appendix B4. R_C, the offset current compensation resistor, is then 18 kΩ (5%).

The second type of phase-shift oscillator uses a three-section, two-port, phase-shifting network on the input of an inverting amplifier, as shown in Fig. 10-3b.

Again the phase shift through the network is 180°, but the signal loss through the network is now $\frac{1}{29}$. If all R's and C's, respectively, are equal, and if $R_I = R$, then

$$f_0 = \frac{1}{2\pi\sqrt{6}\,RC} \tag{10-11}$$

and from **Rule 2**

$$R_F = 29\ R \tag{10-12}$$

for oscillation to just occur. The same method of bounding this amplifier should be used for the previous circuit. This circuit, Fig. 10-3b, uses one more resistor than the one in Fig. 10-3a. The frequency can be adjusted by replacing one of the shunt resistors in the network with a slightly smaller resistor and potentiometer.

An example will show the use of Eqs. (10-11) and (10-12).

EXAMPLE 10-4

Select the component values for an oscillator that operates at 60 Hz. Let $C = 1$ μF, $V_S = \pm 15$ V, and $v_o = 10$ V peak.

Solution

From Eq. (10-11),

$$R\frac{1}{2\pi\sqrt{6}\,(1\ \mu F)(60\ Hz)} = 1.083\ k\Omega$$

Use the 1.07 kΩ resistor from Appendix B1. Let $R' = 806$ Ω (1%) and use a 1 kΩ potentiometer for R_p. From Eq. (10-12), $R_F = 29(1.08\ k\Omega) = 31.4\ k\Omega$; use a 16 k$\Omega$ (5%) for $R_F/2$ and use a 20 kΩ potentiometer for R'_F. The two zener diodes have a 5 V breakdown voltage.

Now that we have the ability to develop an oscillator, it is time to demonstrate the usefulness of the single-pole all-pass equalizer (Section 9.4.4).

EXAMPLE 10-5

Using the oscillator in Example 10-4 and two single-pole all-pass equalizers, construct a three-phase signal generator at 60 Hz.

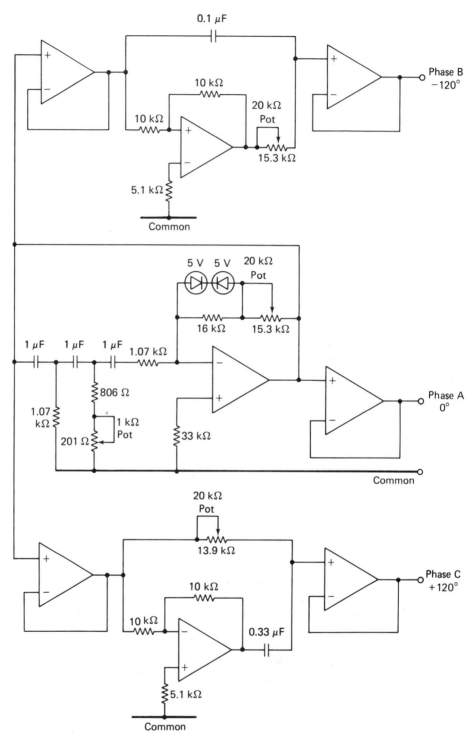

Fig. 10-4 Three-phase 60 Hz generator (Example 10-5).

Solution

A three-phase generator has phases as follows: $0°$, $+120°$, and $-120°$. The oscillator output will be the $0°$ reference phase (A), phase B will be $-120°$, and phase C will be $+120°$.

Using Eq. (9-11b) for phase B, $(2\pi \cdot 60\ \text{Hz})RC =$

$$\tan\left(\frac{-120°}{2} + 90°\right) = 0.577$$

$$RC = \frac{0.577}{377} = 1.53 \times 10^{-3}\ \text{s}$$

Let $C = 0.1\ \mu\text{F}$; then

$$R = \frac{0.00153}{0.1\ \text{F}} = 15.3\ \text{k}\Omega$$

Use a 20 kΩ potentiometer adjusted for 15.3 kΩ.

Phase C components are calculated using Eq. (9-119a), where

$$(2\pi \cdot 60\ \text{Hz})RC = \tan\left(\frac{+120°}{2}\right) = 1.732$$

$$RC = \frac{1.732}{377} = 4.59 \times 10^{-3}\ \text{s}$$

Let $C = 0.33\ \mu\text{F}$; then

$$R = \frac{0.00459}{0.33\ \mu\text{F}} = 13.92\ \text{k}\Omega$$

Use a 20 kΩ potentiometer adjusted for 13.9 kΩ. The total circuit is illustrated in Fig. 10-4.

10.5 Relaxation Oscillator

A relaxation oscillator is one that operates by the alternate charging and discharging of a capacitor. The capacitor may be either charged and discharged to zero volts or first charged to one polarity and then charged to the other polarity. The frequency of a relaxation oscillator is determined by the time to charge the capacitor before it begins to discharge. Since in all relaxation oscillators the capacitor charges through a resistor in the circuit, the time of oscillation is determined by both the resistor and capacitor values. A relaxation oscillator requires an active threshold circuit which will stay in one state while the capacitor charges, and only changes states after the threshold has been passed. The capacitor voltage then operates between the threshold voltage and zero volts or between two threshold voltages in the case of a bipolar oscillator. Figure 10-5 illustrates a relaxation oscillator composed of a Schmitt trigger circuit with the series RC network connected between the circuit output and ground. The capacitor alternately charges to the plus and minus threshold voltages, where the Schmitt trigger changes its output state. The frequency of oscillation is given by

$$f_0 = \frac{1}{2RC \ln\left[(2R_I/R_F) + 1\right]} \qquad (10\text{-}13)$$

The capacitor may be either a film or a polarized type. The frequency is dependent on the stability of the resistors and capacitor for its stability. Thus if a polarized capacitor is used, one should expect a low level of stability, but lower frequencies

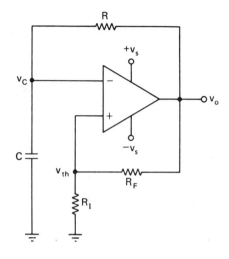

(a) Circuit diagram

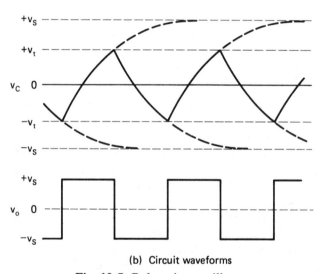

(b) Circuit waveforms

Fig. 10-5 Relaxation oscillator.

can be achieved. The relaxation oscillator is commonly called an astable multivibrator, as it does not have a stable state.

EXAMPLE 10-6

A relaxation oscillator must have a resonant frequency of 100 Hz. $R_F = R_I = 10$ kΩ.

a. Calculate the RC time constant necessary for the oscillator.
b. If $C = 1$ μF, calculate the value of R.

Solution

a. Rearranging Eq. (10-13), we obtain

$$RC = \frac{1}{2f_0 \ln[(2R_I/R_F) + 1]} = \frac{1}{2(100)\ln(3)} = \frac{1}{219.7} \ 0.00455$$

b. $R = RC/C = 0.00455/1 \ \mu F = 4.55 \ k\Omega$. Use the $4.53 \ k\Omega$ from Appendix B1.

The one-shot is actually a monostable multivibrator, as it has one stable state and one astable state. The circuit remains in the stable state until an external trigger pulse causes the state to change. It then remains in the astable state for the length of

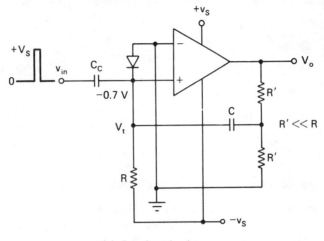

(a) One shot circuit

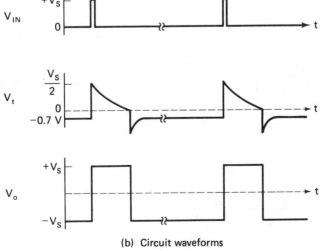

(b) Circuit waveforms

Fig. 10-6 One-shot (monostable multivibrator).

time an RC circuit requires to charge to a threshold voltage; the circuit then returns to the stable state awaiting another trigger pulse. The circuit illustrated in Fig. 10-6a operates in this manner, as can be seen from the waveforms in Fig. 10-6b. The length of time the one-shot remains in its timing (unstable) state is given by

$$t = 0.7RC \qquad (10\text{-}14)$$

An example will help to illustrate the use of Eq. (10-14) in selecting components for the one-shot.

EXAMPLE 10-7

A signal processing circuit requires a 100 ms delay after a comparator changes state from LO to HI.

    a. Determine the RC time constant.
    b. If $C = 1\ \mu\text{F}$, determine the value of all other circuit components.

Solution

    a. Using Eq. (10-14), $RC = 100\ \text{ms}/0.7 = 143\ \text{ms}$.
    b. $R = 143\ \text{ms}/1\ \mu\text{F} = 143\ \text{k}\Omega$. Use a 150 k$\Omega$, 5% resistor. Let $R' = 10\ \text{k}\Omega$.

10.6 Triangular, Square-Wave Oscillator

Students who have been exposed to laboratory work connected with the courses prerequisite to the material in this text will have seen or used a sine, square, or triangular waveform function generator. These instruments have a frequency range from fractions of a hertz to 1 MHz. Readers who have studied the material in Chapters 2 through 7 have all the necessary knowledge to develop one of these function generator circuits.

The basic oscillator is an integrator with a square wave on its input. The integrator output is a triangular waveform. The trick involved with one of these generators is to develop the input square wave for the integrator. Figure 10-7a illustrates the circuit for the square and triangular portion of the function generator. A careful analysis of the figure reveals that it is composed of three separate states:

    1. Integrator.
    2. Schmitt trigger.
    3. Inverter.

These three stages form the necessary frequency control and positive feedback necessary for oscillation. The circuit operates as follows:

    1. The positive portion of the square-wave cycle is applied to the integrator input.

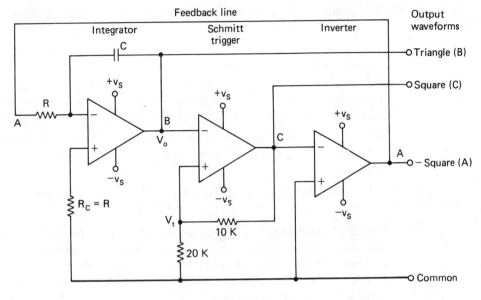

Feedback line

Integrator   Schmitt   Inverter   Output
                 trigger   waveforms

(a) Function generator circuit

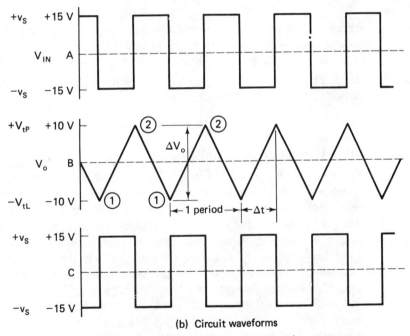

(b) Circuit waveforms

Fig. 10-7 Triangular, square-wave function generator.

2. The integrator output slews, in a straight line, downward in voltage at a rate determined by the RC time constant of the integrator.

3. At a particular negative voltage the Schmitt trigger changes state. This voltage represents the peak excursion of the integrator.

4. The output square wave from the Schmitt trigger is out of phase with the square wave necessary for the integrator input. An inverter (zero-crossing detector) reverses this phase and it becomes acceptable as the integrator input.

5. When the Schmitt trigger switches states, the polarity of voltage of the integrator is changed also. The integrator now begins slewing in the opposite direction.

Figure 10-7b illustrates the various waveforms connected with the generator in Fig. 10-7a. A series of orderly steps are necessary for developing one of these circuits:

1. The input square-wave voltage to the integrator is established by the output saturation voltages of the inverter op amp. The output voltage of the integrator stops slewing when the Schmitt trigger switches state. These two conditions permit the determination of the RC time constant of the integrator as follows. Equation (6-25) defines the RC time constant of the integrator given one input dc voltage and the corresponding output slew rate; rearranged, it is

$$RC = -\frac{\Delta t}{\Delta v_o}(V_{IN}) \qquad (6\text{-}18)$$

where V_{IN} is one saturation voltage from the inverter, Δv_o is the output voltage from $-v_o$ (peak) to $+v_o$(peak), and Δt is the time between peaks or $\frac{1}{2}$ (period) of the waveform, as illustrated in Fig. 10-7b.

2. Δv_o is first determined by defining the resistor values in the Schmitt trigger circuit using Eq. (7-20), which, when rearranged, becomes

$$\frac{R_F}{R_I} + 1 = \frac{V_S}{V_{tH}} = \frac{-V_S}{V_{tL}} \qquad (7\text{-}20)$$

where V_S is the op amp saturation voltage.

3. The resistor (R) in the integrator is usually a potentiometer, to permit a variation in the oscillation frequency. R has a low- and a high-value limit, dependent on the op amp parameters. The low limit is dependent on the amount of current that can be delivered by the preceding op amp stage; it usually has a low limit of 1 kΩ. The high limit is dependent on the op amp input bias current I_B (see Section 3.3.1). When the input current to the integrating resistor (R) becomes the same order of magnitude as I_B, the amount of current through the integrating capacitor (C) is no longer equal to that in R and the output waveform is no longer linearly related to V_{IN}. For instance the input bias current to a μA 741 op amp is specified as being 80 nA maximum. The minimum integrating current to the integrator should be 0.8 μA. The maximum value for R is then given by

$$R_{max} \le \frac{V_{IN}}{10(I_B)} \qquad (10\text{-}15)$$

FG 501
1 MHz FUNCTION GENERATOR

0.001 Hz to 1 MHz
Five Waveforms
Vcf and Gated Burst
Hold Mode

The FG 501 produces low-distortion sine, square, triangle, pulse, and ramp waveforms from 0.001 Hz to 1 MHz. An external vcf input permits control of the output frequency from an external voltage source. Frequency sweep up to 1000:1 ratio may be accomplished by applying a voltage ramp to the vcf input. A hold control allows the operation of the generator to be halted instantaneously at any point in its cycle. Release of the hold will then allow the operation to continue normally. A gate input is provided to allow "burst" or single cycle operation, with the phase of the generator output at the start of the burst controllable over a ±90° range. Output signal voltage is adjustable to 10 V p-p into a 50-ohm load, with dc offset also adjustable up to ±3.75 V into 50 Ω.

FG 501 Function Generator . . . .

Fig. 10-8 Tektronix FG 501 function generator. (Courtesy of Tektronix Inc., Beaverton, Oreg.)

4. the capacitor (C) is usually switched in powers of 10 to provide for "range switching" the oscillator frequency. The op amp slew rate usually limits the highest frequency of the oscillator, but C should not be less than 100 times the typical stray capacitance (20 pF). There is no limit on the maximum size of C, except that it must be a low-leakage "film" capacitor.

A function generator capable of sine, square, or triangular output waveform can be constructed from this basic square and triangle generator by adding a sine shaper (see Section 7.3.2). A generator of this type is manufactured by Tektronix, Inc., as their model number FG 501 illustrated in Fig. 10-8. It covers the frequency range of 0.001 Hz to 1 MHz in the nine power-of-10 ranges determined by an integration capacitor selection. The frequency dial, integrator resistor (R), is calibrated from 0.1 to 10 or a range of 100:1. A selector switch permits sine, square, and triangular waveforms as well as some others. The output voltage can be varied in ac amplitude and dc offset by independent controls. This particular generator has additional features not pertinent to this chapter. An integrated circuit version is made by Exar Linear (XR-2206).

Using the μA 741 op amp and a power-supply voltage of ± 15 V, the maximum range of the integration resistor (R) is 1 kΩ to 18 MΩ, using Eq. (10-14, or 18,000:1. Thus the 100:1 dial range of the FG 501 is very conservative and is probably chosen more for the linear region of a single-turn potentiometer than the maximum possible resistance change.

An example will help to illustrate the method of determining the component values for a generator similar to the FG 501.

Triangular, Square-Wave Oscillator　　　　　　　　　**349**

EXAMPLE 10-8

A CA 3130 COS/MOS op amp is used as the active element in a square and triangular waveform generator. The integrating potentiometer is 100 kΩ and is linear from 1 to 100 kΩ. The slew rate of a CA 3130 is 10 V/µs. The supply voltages are ±7 V and the Schmitt trigger should switch at ±5 V. The frequency dial is calibrated from 0.3 to 30; the generator must operate from 0.3 Hz to some maximum frequency.

 a. Determine the maximum frequency of operation.

 b. Determine the integrating capacitor values and the number of ranges (powers of 10).

 c. Determine the ratio of resistor values for the Schmitt trigger.

Solution

 a. The slew rate is 10 V/µs. Using Eq. (5-16), period $(T) = 2[+5 - (-5)]/(10 \text{ V}/\mu s) = 2$ µs; thus the minimum period is 2 µs and the maximum frequency is 500 kHZ (use 300 kHz as the upper frequency limit).

 b. $V_{\text{IN}} = \pm 7$ V; $v_o = \pm 5$ V.

$$\Delta t_{\text{min}} = \frac{3.33 \ \mu s}{2} = 1.66 \ \mu s$$

$f_{\text{min}} = 0.3$ Hz; maximum period $= 1/f_{\text{min}} = 3.33$ s. $\Delta t_{\text{max}} = 3.33/2 = 1.66$ s. Using Eq. (6-18), we obtain

$$RC_{\text{min}} = -\frac{1.66 \ \mu s}{+10 \text{ V}}(-7 \text{ V}) = 1.16 \ \mu s$$

$$C_{\text{min}} = \frac{1.16 \ \mu s}{1 \text{ k}\Omega} = 0.0116 \ \mu F$$

$$RC_{\text{max}} = \frac{1.66 \text{ s}}{+10 \text{ V}}(-7 \text{ V}) = 1.16 \text{ s}$$

$$C_{\text{max}} = \frac{1.16 \text{ s}}{100 \text{ k}\Omega} = 11.6 \ \mu F$$

 c. Using Eq. (7-20) for the Schmitt trigger, we obtain

$$\frac{R_F}{R_I} = \frac{+7 \text{ V}}{+5 \text{ V}} - 1 = 0.4$$

Let $R_F = 12$ kΩ and $R_I = 30$ kΩ.

10.7 Voltage-Controlled Oscillators

A voltage-controlled oscillator (VCO) is one in which the input voltage is varied and the frequency of oscillation changes accordingly. Voltage-controlled oscillators that employ op amps as the active element fall into a specific category; they use an integrator, with a reset switch on the integration capacitor, which has a varying voltage on its input.

The voltage-controlled oscillator falls into the general category of relaxation oscillators, which operate by allowing a capacitor to charge and then resetting the capacitor when it reaches a preset voltage (gained a preset charge). This type of circuit falls into the more general category of astable multivibrators. An astable multivibrator is a circuit employing a digital flip-flop, cross-connected with capacitors in such a manner that it would always change state after the capacitor discharged (or charged). These circuits have largely been replaced by monolithic chips which oscillate at a specific frequency depending upon the values of an external capacitor and resistor connected to its leads. But, the method has not changed; the device operates by discharging (or charging) a capacitor for a specific length of time.

The ideal situation for a voltage-controlled oscillator is to have the frequency change linearly as the voltage changes; this can be accomplished within limits. Figure 10-9a illustrates the circuit of a voltage-controlled oscillator. The circuit is composed of three parts:

1. Integrator.
2. Schmitt trigger.
3. Capacitor reset switch (transistor).

The circuit operates as follows:

1. A dc voltage is applied to the integrator input, which causes its capacitor to charge in a linear fashion.
2. When the capacitor has charged to a preset voltage level, which is the Schmitt trigger threshold voltage, the Schmitt trigger changes to the triggered state.
3. During the time the trigger is in the triggered state, the transistor switch is saturated and discharging the capacitor.
4. The trigger circuit is biased in such a manner that when the capacitor is fully discharged, it changes state back to the stable (untriggered) state.

Both the charge cycle and the discharge cycle take time to complete. If the discharge time were zero, the VCO frequency would be linear with input dc voltage, as the integrator capacitor charge time, to a specific voltage, is linear with a dc voltage. The discharge time is usually constant regardless of the charge time. Thus the frequency becomes more nonlinear as the discharge time becomes a larger proportion of the charge time, at higher frequencies. Many monolithic devices employ "current-controlled oscillators" in their system. One must remember that the input resistor on an integrator serves only to convert the input voltage to a current which then passes through the integrating capacitor. If the resistor could be eliminated and the current still passes through the capacitor, the integrator operation would remain unchanged; it then becomes a current-controlled oscillator (CCO). Thus the VCO and CCO are, for all practical purposes, identical.

The waveform period is the sum of the integrating capacitor charge time and discharge time, as illustrated in Fig. 10-9b. If the input voltage to the integrator is positive, the Schmitt trigger must be biased negatively such that its high-level threshold (V_{tH}) is at zero volts and the low-level threshold (V_{tL}) is between zero and $-V_S$. If the integrator input voltage is negative, then V_{tL} is at zero volts and V_{tH} is

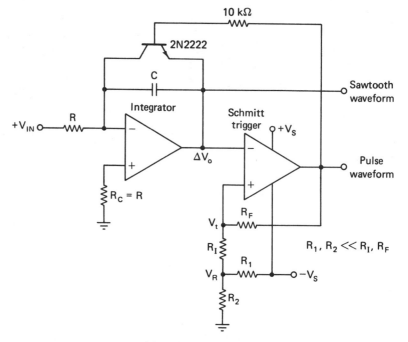

(a) VCO circuit diagram

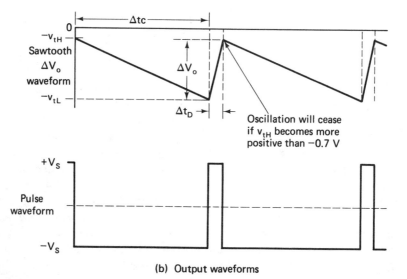

(b) Output waveforms

Fig. 10-9 Voltage-controlled oscillator.

between zero and $+V_S$. The circuit that will be analyzed has a positive input voltage to the integrator. The charge time (t_C) is given by Eq. (6-18), rearranged here as

$$\Delta t_C = -RC\left(\frac{\Delta v_o}{V_{IN}}\right) \tag{6-18}$$

The capacitor discharge time (t_D) is usually limited by the current capability of the switch transistor in saturation and given by Eq. (4-1), rearranged here as

$$\Delta t_D = C\left(\frac{\Delta v_C}{I_C}\right) \tag{4-1}$$

where the quantity $\Delta v_C/I_C$ is representative of the saturation resistance of the switch transistor; a 2N2222 is able to carry about 1 A of current at a very fast switch rate. The frequency of VCO is then given by

$$f = \frac{1}{\Delta t_C + \Delta t_D} \tag{10-16}$$

while the dynamic range of a VCO is the range of frequencies over which the oscillator can operate while maintaining a linear relationship between the frequency and input voltage. The upper limit on the dynamic range occurs as the charge time approaches the discharge time; an arbitrary upper limit is established at either the point where the charge time is 10 times the discharge time or the op amp is slew rate limited.

For the discharge time,

$$f_H = \frac{1}{11(t_D)} \tag{10-17}$$

The low-frequency limit is determined, as in the square–triangle oscillator, by the point where the input current has reduced to 10 (input-bias current), or

$$f_L = \frac{10}{C}\left(\frac{I_B}{\Delta v_0}\right) \tag{10-18}$$

The dynamic range (DR), in dB, of the frequency range of the VCO is then given by

$$DR = 20\log_{10}\left(\frac{f_H}{f_L}\right) \tag{10-19}$$

An example will help to illustrate the use of these equations.

EXAMPLE 10-9

A VCO, using a μA 741 op amp, must operate at a high frequency of 10 kHz with an output voltage of 10 V p-p. It uses a 2N2222, which will discharge the integrating capacitor at 1 A. The input voltage is 10 V when the output frequency is 10 kHz; $V_s = \pm 15$V.

a. Determine the size of the integrator capacitor and resistor.
b. Determine the resistor values for the Schmitt trigger.
c. Determine the dynamic range of the VCO and the minimum input voltage to the integrator.

Voltage-Controlled Oscillators   **353**

Solution

a. From Eq. (10-17),

$$t_D = \frac{1}{11(10 \text{ kHz})} = 9.1 \ \mu s$$

Then, from Eq. (4-1),

$$C = \frac{(9.1 \ \mu s)(1 \text{ A})}{10 \text{ V}} = 0.91 \ \mu F$$

Equation (10-16) yields the charge time as

$$\Delta t_C = \frac{1}{f} - \Delta t_D = 100 \ \mu s - 9.1 \ \mu s = 91 \ \mu s$$

and Eq. (6-18) gives R as

$$R = -\left(\frac{\Delta t_C}{C}\right)\left(\frac{V_{IN}}{\Delta v_o}\right) = -\left(\frac{91 \ \mu s}{0.91 \ \mu F}\right)\left(\frac{10 \text{ V}}{-10 \text{ V}}\right) = 100 \ \Omega$$

(*Note:* This stage and the one preceding this stage must be able to drive $I = 10 \text{ V}/100$ $\Omega = 100 \text{ mA.}$)

b. Use the method outlined in Example 7-11. From Eq. (7-15), let $R_I = 10 \text{ k}\Omega$.

$$|0.7 \text{ V} - (-10.7 \text{ V})| = 2\left(\frac{10 \text{ k}\Omega}{R_F + 10 \text{ k}\Omega}\right)|15 \text{ V}|$$

$$R_F = 2\left(\frac{10 \text{ k}\Omega}{9.3 \text{ V}}\right)(15 \text{ V}) - 10 \text{ k}\Omega \simeq 20 \text{ k}\Omega$$

From Eq. (7-16),

$$V_R = \left(\frac{-0.7 - 10.7}{2}\right)\left(\frac{20 \text{ k}\Omega + 10 \text{ k}\Omega}{20 \text{ k}\Omega}\right) = -7.50 \text{ V}$$

Thus, form a resistive voltage divider between -15 V and common, using resistor values 10 times lower in value than R_R and R_I, which provides -7.50 V at the division point.

$$\frac{R_2}{R_1 + R_2} = \frac{-7.50 \text{ V}}{-15 \text{ V}} = 0.50$$

$$\frac{R_1}{R_2} + 1 = \frac{1}{0.5} = 2.0 \qquad R_1 = R_2$$

Let $R_2 = 1 \text{ k}\Omega$; then $R_1 = 1.00 \text{ k}\Omega$.

c. From Eq. (10-18),

$$f_L = \left(\frac{10}{0.91 \ \mu F}\right)\left(\frac{80 \text{ nA}}{10 \text{ V}}\right) = 0.088 \text{ Hz}$$

Then from Eq. (10-19),

$$DR = 20 \log_{10}\left(\frac{10 \text{ kHz}}{0.088 \text{ Hz}}\right) = 101.1 \text{ dB}$$

Equation (6-18) rearranged yields the input voltage as

$$t_C(\text{at } 0.088 \text{ Hz}) = 11.36 \text{ s}$$

$$V_{IN} = -RC\left(\frac{\Delta v_0}{\Delta t_C}\right)$$

$$= -(100 \ \Omega)(0.91 \ \mu F)\left(\frac{-10 \text{ V}}{11.36 \text{ s}}\right) = 80 \ \mu V$$

PROBLEMS

1. Is it possible for an amplifier and an oscillator to have the same circuit?

2. An active circuit has a reactive network in the negative feedback loop.

    a. What phase shift through the network would cause oscillation?

    b. The network has a loss of 45 dB at the oscillation frequency. What is the closed loop gain (give a ratio) in the amplifier?

3. An amplifier in a sine-wave oscillator circuit must maintain a gain of at least 7 for oscillation. An technician has set the gain to 10 to guarantee oscillation. What do you notice about the circuit operation?

4. The oscillator of Problem 3 uses an ideal op amp with infinite supply voltages.

    a. Where are the poles located for the oscillator?

    b. Where are the poles located when the supply voltages are set for ± 12 V?

5. Define the term "bounding."

6. A twin-tee oscillator operates at a resonate frequency of 340 Hz. The twin-tee network uses a 0.033 μF and a 0.022 μF capacitor. The supply voltages are at ± 10 V.

    a. Calculate the resistor values for the twin-tee network.

    b. $R_1 = 47$ kΩ, and the sine wave should have a peak value of 7 V. Find R_2 and the approximate value of R_B.

    c. Draw the circuit.

7. A Wien bridge oscillator has $R_1 = 1.5$ kΩ, $R_2 = 4.7$ kΩ, $C_1 = 0.022$ μF, and $C_2 = 0.033$ μF. Find the frequency of oscillation and the ratio of R_f to R_I.

8. A Wien bridge oscillator must oscillate at 750 Hz.

    a. The capacitors both have the same value, and their capacitive reactance is approximately 15 kΩ. If $R_1 = R_2 = R_I$, calculate the value of all components; select 1% values.

    b. What component (give value) must be added to allow for fine tuning the frequency?

    c. Calculate the value of the zener diode in the bounding circuit if the output amplitude is to be 10 V peak.

9. A phase-shift oscillator using a two-port network on the summing junction (Fig. 10-3a) must oscillate at 120 Hz with an amplitude of 5 V peak. If a 0.47 μF capacitor is used, calculate the value of the components in the circuit.

10. A phase-shift oscillator using an inverting amplifier (Fig. 10-3b) must oscillate at 47 Hz. Calculate the value of the other circuit components used if $C = 0.05$ μF. The peak amplitude must be at 15 V.

11. A two-phase oscillator circuit drives a 60 Hz servomotor. The motor requires 0° and $+90°$ for operation. The reactance of the oscillator capacitor should be roughly 7 kΩ. The reactance of the phase-shift capacitor should be 20 kΩ. The output amplitude of the oscillator and the phase shifter should be 8 V peak.

    a. Draw the circuit you would see.

    b. Calculate all component values.

12. A relaxation oscillator must produce a 15 Hz square wave. A 4 μF filter capacitor is used, and $R_I = R_f = 25$ kΩ. Calculate the value of R.

13. A one-shot must produce a delay of 75 ms. If $C = 2$ μF, calculate the values of all other components.

14. A variable-frequency triangular-square waveform generator uses a 250 kΩ potentiometer which is linear from 4k to 200 kΩ. The supply voltages are ± 15 V, and the Schmitt trigger switches at ± 7 V. The frequency dial is calibrated from 0.1 to 5. The op amp used is a MC 1436G with a 2 V/μs slew rate and a 40 nA input bias current.

   a. Determine the maximum frequency of operation.

   b. Determine the value of the integrating capacitor and the number of ranges (powers of 10).

   c. Determine the ratio of resistor values for the Schmitt trigger.

   d. Draw the circuit diagram.

15. The FET sine shaper from Problem 7-13 is connected to the integrator output from Problem 14.

   a. Draw a circuit that will permit waveform selection with equal peak outputs for all waveforms.

   b. Draw an output circuit that will provide for amplitude and offset adjustment.

16. A VCO using an LM 208 op amp must operate at a high frequency of 20 kHz with an output voltage of 10 V p-p. It uses a 2N2222 that discharges the integrating capacitor at 500 mA. The input voltage is 8 V when the output frequency is 20 kHz. The supply voltages are at ± 15 V.

   a. Determine the sizes of the resistor and integrating capacitor to be used.

   b. Calculate the resistor values for the Schmitt trigger.

   c. What is the dynamic range and minimum input voltage for the VCO?

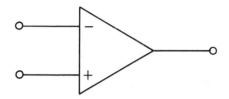

11 Linear Devices

11.1 Introduction

Since the introduction of the first edition of this text, many changes have occurred in the linear amplifier area. One of the most notable of these is the introduction and continuing development of high-frequency op amps. While it was prudent to discuss only special op amp variances in the first edition, the emphasis placed on high-frequency op amps seems to be dominating the new development now. Thus a section devoted to high-frequency op amps will be added. The phase-locked loop still maintains a strong place in the industry; it will be retained.

11.2 Operational Transconductance Amplifier

The operational transconductance amplifier (OTA) is a unique device employing a differential amplifier with a variable current constant current tail (W) and three current mirrors,* X, Y, and Z (variable/constant current sources), as illustrated in Fig. 11-1a. Two types of current mirrors are used: two-transistor and three-transistor mirrors. The two-transistor current mirror illustrated in Fig. 11-1b has an output current that is related to the input current by the relationship

$$I_o = I_{in}\left(\frac{h_{fe}}{h_{fe} + 2}\right) \qquad (11\text{-}1)$$

The three-transistor current mirror (Fig. 11-1c) has the relationship

$$I_o = I_{in}\left(\frac{h_{fe}^2 + 2h_{fe}}{h_{fe}^2 + 2h_{fe} + 2}\right) \qquad (11\text{-}2)$$

*RCA Application Note: ICAN-6667, Appendix 1.

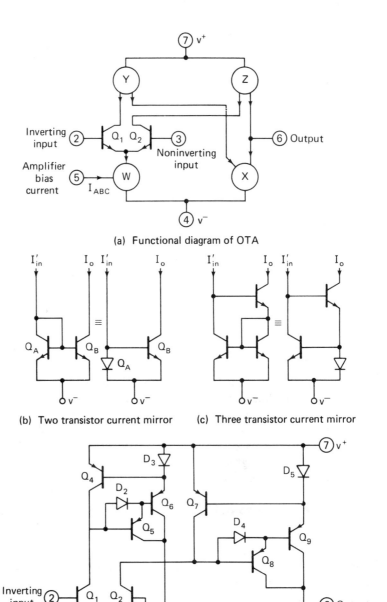

(a) Functional diagram of OTA

(b) Two transistor current mirror   (c) Three transistor current mirror

(d) Schematic diagram of OTA

Fig. 11-1 Operational transconductance amplifier (OTA). (Courtesy of RCA Solid State, Somerville, N.J.)

In Eq. (11-1), I_0 is within 2% of I_{in} when h_{fe} is 100; for Eq. (11-2), I_o is within 0.02% of I_{in} when h_{fe} is 100. Thus both current mirrors are quite accurate with reasonable h_{fe}'s. The two-transistor current mirror was first introduced as the constant-current tail for a differential amplifier in Section 1.5.4; it has an output impedance greater than 100 MΩ. The output impedance of a three-transistor current mirror is greater than 1000 MΩ (see Section 1.5.4).

EXAMPLE 11-1

The transistors in a linear monolithic circuit using current mirrors all have $h_{fe} = 80$.

    a. Find the output current for the two-transistor current mirror, if $I_{in} = 100 \ \mu A$.

    b. Find the output current for the three-transistor current mirror if the input current is 1 mA.

Solution

a. $I_o = 100 \ \mu A \left(\dfrac{80}{80 + 2} \right) = 97.56 \ \mu A$

b. $I_o = 1000 \ \mu A \left[\dfrac{80^2 + 2(80)}{80^2 + 2(80) + 2} \right] = 1000 \ \mu A \ (0.9997) = 999.7 \ \mu A$

Figure 11-1d illustrates the schematic of the CA 3080 OTA device. A current, driven into the base of the "current tail" transistor, is labeled "amplifier bias current (I_{ABC}) and causes the open loop gain to vary in proportion to this bias current. The device is constructed to deliver an output current which is dependent, through the transconductance (g_m), on both the differential input voltage (v_{id}) and the amount of bias current (I_{ABC}). An output voltage is obtained by connecting a load resistor, through which the output current passes.

The input bias terminal (pin 5) has an "active" diode (see Fig. 1-8d) connected to the negative supply voltage ($-V_S$). The bias current is usually obtained by passing current through a resistor connected between the input bias terminal and the positive supply voltage ($+V_S$). Thus the value of the bias resistor (R_{ABC}) is

$$R_{ABC} = \frac{+V_S - (-V_S) - 0.71 \ V}{I_{ABC}} \Omega \tag{11-3}$$

EXAMPLE 11-2

An OTA connected between ±7 V power supplies. Find the value of the bias resistor to yield a 200 μA bias current.

Solution

$$R_{ABC} = \frac{+7 \ V - (-7 \ V) - 0.71 \ V}{200 \ \mu A}$$

$$= \frac{13.29 \ V}{200 \ \mu A} = 66.4 \ k\Omega$$

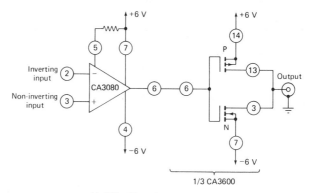

(a) OTA with one inverter output driver

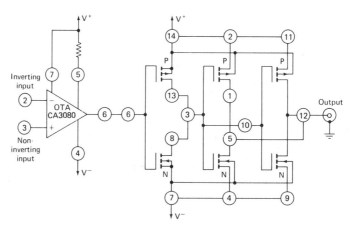

(b) OTA with three inverter output drivers

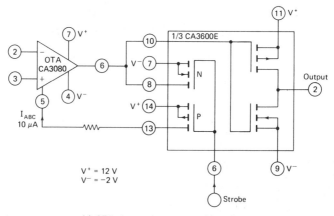

(c) OTA micropower comparator with strobe

Fig. 11-2 Applications of OTA as op amp and comparator. (Courtesy of RCA Solid State, Somerville, N.J.)

Since the OTA has such a large open loop output impedance, it is best used in a closed loop mode with an output current driver state with a high input impedance. One such device is the COS/MOS CA 3600 three-inverter array (transistor array). With one of these used as the output stage for the OTA, the open loop gain becomes 130 dB; this circuit is illustrated in Fig. 11-2a. Each stage of the CA 3600 can deliver 6 mA to a load resistor. A connection that uses all three sections of the CA 3600 is illustrated in Fig. 11-2b. The circuit is comprised of the OTA driving one inverter, which in turn is driving the other two inverters in parallel. This connection will deliver 12 mA to a load resistor and has an open loop gain of 160 dB.

The transconductance of the CA 3080 is given by

$$g_m = \frac{\Delta i_0}{\Delta V_{id}} = (19.2) I_{ABC} \quad S \text{ (Siemens)} \tag{11-4}$$

where the range of I_{ABC} is 0 to 1 mA. The output impedance of the OTA is

$$R_{out} = \frac{7.5 \times 10^3}{I_{ABC}} \Omega \tag{11-5}$$

where, again, the range of I_{ABC} is 0 to 1 mA. The output can source or sink 650 μA into the load resistor. This permits open loop gains of up to 100,000 (100 dB) with a large load resistor.

The slew rate in an OTA is a function of the amplifier bias current (I_{ABC}) and varies according to the equation

$$SR = (65 \times 10^3) I_{ABC} \quad V/\mu s \tag{11-6}$$

where I_{ABC}, again, varies from 0 to 1 mA.

EXAMPLE 11-3

An OTA has a bias current of 200 μA.

    a. Calculate the transconductance.
    b. Calculate the output impedance.
    c. Calculate the slew rate.

Solution

    a. $g_m = (19.2)(200 \ \mu A) = 3840 \ \mu S$
    b. $R_{out} = \dfrac{7.5 \times 10^3}{200 \ \mu A} = 37.5 \ M\Omega$
    c. $SR = (65 \times 10^3)(200 \ \mu A) = 13 \ V/\mu s$

The gain–bandwidth is specified as 2 MHz; if f_{GB} is calculated using Eq. (5-13a) with a slew rate of 65 V/μs ($I_{ABC} = 1$ mA), the gain–bandwidth is 10 MHz. Thus it is not quite certain where the actual gain–bandwidth lies, but the slew rate should be a more accurate indicator than this published specification in this case. The input bias terminal can be used as a "strobe" to turn the OTA on and off. With zero current into the bias terminal, the output impedance of the OTA is a virtual open circuit with an output impedance of 1000 MΩ. An example of this is shown in Fig.

11-2c, where the OTA and two stages of CA 3600 array are used as a "micropower comparator." The open loop gain of the OTA is given by

$$A_{\text{VOL}} = g_m R_L \tag{11-7}$$

and if R_L is the high input impedance to an inverter 3N138 FET (5 MΩ) source follower, the open loop gain is 100,000 (100 dB). If a CA 3600 inverter with a gain of 32 and an input impedance of 5 MΩ, the open loop gain is 3,200,000 (130 dB).

EXAMPLE 11-4

A COS/MOS amplifier having an input impedance of 6 MΩ and a gain of 100 is used as the load for the OTA in Example 11-3. Calculate the circuit gain.

Solution

$$A_{\text{VOL}} = g_m R_L = (3840 \ \mu\text{S})(6 \ \text{M}\Omega) = 23,000$$

$$(23,000)(100) = 2.3 \times 10^6 \quad \text{or} \quad 127 \ \text{dB gain}$$

This large gain, coupled with a 13 V/μs slew rate, yields an excellent op amp.

11.3   Current-Differencing (Norton) Amplifier

The current-differencing amplifier, also called a Norton amplifier, operates by sensing the difference between the two input currents. The device has a part number of LM 1900/2900/3900. This amplifier is intended to be operated from a single power supply; the negative supply terminal of the amplifier is grounded. This means that the dc output voltage is *always* above ground potential and the device must usually be ac-coupled at both the input and output. Since the op amp is a dc-coupled device, the current-differencing amplifier presents some new problems and circuit limitations.

The input stage to the Norton amplifier is a two-transistor current mirror, where the current into the inverting input (see Fig. 11-3a) is equal to the bias current into a noninverting input. The output stage is also a current source, so it may assume any potential between its two saturation limits. The dc bias of the Norton amplifier is established by connecting a resistor from a positive voltage source to noninverting (bias) input. This current passes through a forward-biased diode and the input voltage is +0.7 V. The voltage on the inverting input is also 0.7 V, as the Q_3 base–emitter is forward-biased. The output voltage should be half of the supply voltage for maximum voltage swing, and the current into the inverting input equals that into the noninverting (bias) input. Making the value of R_F one-half the bias resistor value causes a voltage drop of one-half the supply voltage to be across it, and since the input side is nearly zero volts, the output side is at $+V_S/2$. Thus as a general rule,

$$R_B = 2R_F \tag{11-8}$$

Once the circuit has been biased in this manner, it operates like an op amp and the circuits have a similarity. The output stage has two current amplifiers, which have different slew rates. A positive output swing has a slew rate of 0.5 V/μs; a negative

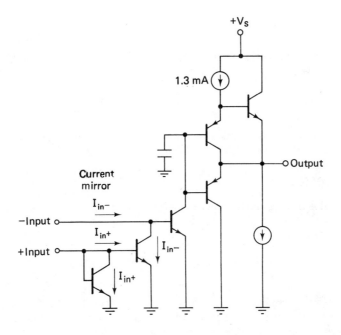

(a) Norton amplifier schematic

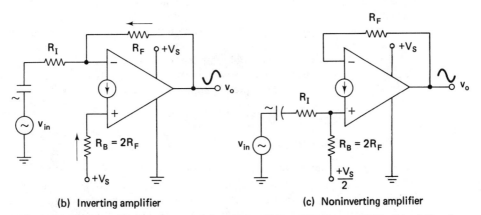

(b) Inverting amplifier

(c) Noninverting amplifier

Fig. 11-3 Current-differencing (Norton) amplifier. (Courtesy of National Semiconductor Corporation, Santa Clara, Calif.)

output swing has a 20 V/μs slew rate. Naturally, the slowest rate limits the speed of the amplifier in ac applications.

The inverting amplifier is shown in Fig. 11-3b. A signal voltage at the input causes an input current to flow through R_I. Since the bias current through R_B into a noninverting input is constant, the current through R_F must reduce as much as the input current increases; this causes a signal inversion to occur on the device output. The inverting gain is

$$A_{\text{VCL}} = -\frac{R_F}{R_I} \qquad \text{(inverting gain)} \qquad (11\text{-}9)$$

where the value of the bias resistor does not affect the gain because of the low impedance of the forward-biased diode at the amplifier input.

EXAMPLE 11-5

A current-differencing amplifier must have an inverting gain of -7 and an input impedance of 10 kΩ. Calculate the value of the feedback and bias resistors.

Solution
$$R_F = (7)(10 \text{ k}\Omega) = 70 \text{ k}\Omega$$
$$R_B = 2R_F = 140 \text{ k}\Omega$$

The noninverting amplifier operates in a similar manner. The input signal is applied through R_I to the noninverting input together with the bias current, as illustrated in Fig. 11-3c. The signal current adds to or subtracts from the bias current and the current through R_F must follow it, as the current mirror demands current equality. The resulting noninverting gain is

$$A_{\text{VCL}} = +\frac{R_F}{R_I} \qquad \text{(noninverting gain)} \qquad (11\text{-}10)$$

where the value of the bias resistor does not affect the gain equation because of the low input impedance across the forward-biased diode. Notice that for both the inverting and noninverting amplifiers, the input signal is ac-coupled. The output, although not shown in the Fig. 11-3 b and c, is also ac-coupled. These circuits may be used as dc amplifiers, but operation is not as precise as with a standard op amp with dual supplies. Care must be taken to offset the input signal dc average to be compatible with the bias current. These devices lend themselves to high-pass and band-pass filters, as these circuits are inherently ac-coupled. The low-pass and band-reject (or notch) filters present some special problems because of the required dc offset.

11.4   Low-Voltage Op Amps

The most recent development in operational amplifiers at the time this text was being written are those that operate at low voltages. They have most of the same characteristics as those of standard op amps but will operate with a power-supply voltage across the device of as low as 1.1 V. This means that they will operate on a single dry-cell battery. The battery drain is 0.5 mA maximum. This opens up a whole new field of applications for operational amplifiers in battery-operated equipment, heart pacers, hearing aids, and so on.

11.4.1 Op Amp and Voltage Reference

The first of these devices to be announced was the National Semiconductor LM 10. This device was developed by Bob Widlar as a consultant to National Semiconductor (see the Preface). The LM 10 has two separate devices within a single package, as illustrated in Fig. 11-4a. It has an op amp with a 200 mV reference source on the noninverting input and an op amp with all inputs available. Through the use of input and feedback resistors, the output of the reference amplifier can be made any voltage within the power-supply rail limits. This is like having two op amps within

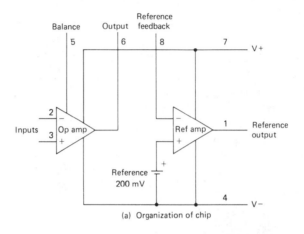

(a) Organization of chip

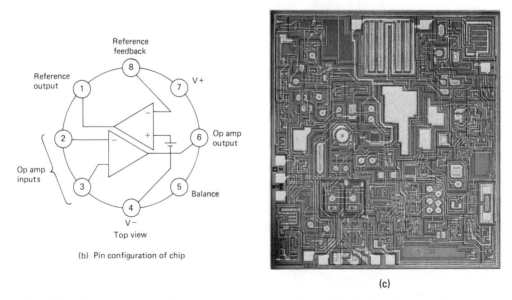

(b) Pin configuration of chip

(c)

Fig. 11-4 LM 10 op amp and voltage reference. (Part (c) Photo-micrograph. Courtesy of National Semiconductor Corporation, Santa Clara, Calif.)

one package. The application note which introduces the op amp states:

> A new approach to op-amp design and application has been taken here. First, the amplifier has been equipped to function in a floating mode, independent of fixed supplies. This, however, in no way restricts conventional operation. Second, it has been combined with a voltage reference, since these two functions are often interlocked in equipment design. Third, the minimum operating voltage has been reduced to nearly one volt. It will be seen that these features open broad new areas of application.

The pin configuration is illustrated in Fig. 11-4b; here it can be seen that the power-supply leads, the inputs, and output of the op amp are in the conventional positions for a LM 101 or LM 108 op amp (TO-5 can). Fig. 11-4c is a photomicrograph of the integrated-circuit chip where the layout of the individual components can be seen. The die size is 97 × 105 mils, which is large for an op amp. But with the low-voltage design, which increases the yield; the 3-inch wafers; and modern processing, the selling price is reasonable.

The design of the internal circuitry is unconventional. The op amp portion uses a differential-amplifier, but the constant current tail is not a current mirror. Instead, the device has a bias bus which runs throughout and provides base bias for constant-current-state transistors all along the signal path; thus the output stage can swing to within 50 mV of the supplies with a 50 μA load or to within 0.4 V with a 20 mA load.

EXAMPLE 11-6

The circuit shown in Fig. 11-5 is a D-cell with a 1.5 V terminal voltage and 3 A-h capacity. The input signal is a 1 kHz sine wave with a 0.74 peak voltage

    a. Will the output of the amplifier saturate at the peaks?

    b. What is the expected lifetime of the circuit in years?

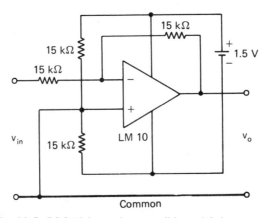

Fig. 11-5 LM 10 inverting amplifier with battery supply.

Solution

a. The current through the feedback resistor at the peak input voltage is 49.3 μA. The output voltage will go to ± 0.745 V before saturation occurs. No, it will not saturate.

b. The two 15 kΩ centering resistors draw 50 μA; the effective feedback circuit current is $(0.707)(50 \ \mu A)$. The total effective circuit current is 85 μA.

$$\text{hours} = \frac{3 \ A\text{-h}}{85 \ \mu A} = 35,300 \text{ h} \qquad \text{years} = \frac{35,300 \text{ h}}{8760 \text{ h/year}} = 4 \text{ years}$$

The input common-mode voltage swing extends to the supply voltages. The op amp is internally compensated for a constant gain–bandwidth product of 200 kHz. The slew rate is specified as 0.15 V/μs. These two specifications are not as good as conventional op amps, but the balance of the specifications are as good as the LM 108.

A voltage reference is built in and can be used as though it were external to the regulator (see Section 4.5.2). A good example of its use is illustrated in the circuit for a high-voltage regulator with a bootstrapped control amplifier. The circuit as it appears in the National Semiconductor Application Note TP-14 is shown in Fig. 11-6a. A redrawn version is shown in Fig. 11-6b. There it can be seen that the reference amplifier is connected as a voltage follower with a 0.2 V input voltage; the output voltage of the reference amplifier will therefore be 0.2 V, by *Rule 1*. Since the 0.2 V internal reference is connected to the output voltage, the reference-amplifier output voltage will be 0.2 V above the bootstrapped regulator output voltage. The divider ratio is 500 kΩ/502 kΩ; thus the input voltage to the noninverting input to the op amp portion of the chip is $(v_o + 0.2 \text{ V})(500/502) = v_o$, since the inverting input is connected to the bootstrapped regulator output and *Rule 1* still applies. Thus the output voltage must satisfy the relationship, which is true only when $v_o = 50$ V; or $(50 + 0.2)(500/502) = (50.2)(500/502) = 50$ V. The input voltage to the bootstrapped regulator must be sufficiently large to permit the darlington pass transistor stage to operate in the linear region; thus the input voltage must be 3 V greater than the output voltage, or 53 V. The divider ratio must be such that the input voltage to the divider is always 0.2 V greater than the voltage at the division point; thus

$$R_1 = \left(\frac{0.2 \text{ V}}{v_o} \right) R_2 \tag{11-11}$$

EXAMPLE 11-7

Calculate the ratio of R_2 to R_1 for a bootstrapped power supply, using the LM 10, with an input voltage of $+18$ V and a regulated output voltage of $+12$ V.

Solution

$$\frac{R_2}{R_1} = \frac{12 \text{ V}}{0.2 \text{ V}} = 60$$

Although somewhat complicated, this circuit is an excellent illustration of the versatility of the LM 10.

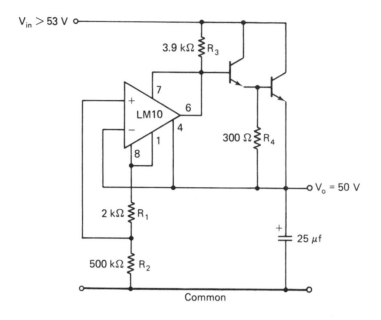

(a) Regulator circuit (application note TP-14)

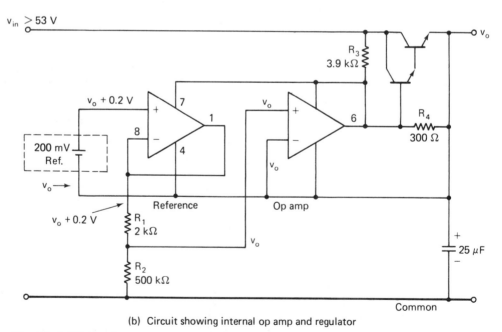

(b) Circuit showing internal op amp and regulator

Fig. 11-6 High-voltage regulator with bootstrapped control amplifier. (Courtesy of National Semiconductor Corporation, Santa Clara, Calif.)

11.4.2 Low-Power MAXCMOS Op Amp

Intersil Corporation developed a series of CMOS op amps which they call MAXCMOS OP AMPS. These devices will operate on power-supply voltages that range from $\pm \frac{1}{2}$ to ± 8 V. The device has many of the good characteristics of CMOS op amps (see Section 2.7.6) plus some additional ones. The input common-mode voltage range is limited to ± 0.1 V, with a ± 0.5 V supply voltage. The output will swing to within 10 mV of the power-supply rails with a 100 kΩ load resistor, and it appears that the output swing is ± 0.3 V peak with a 10 kΩ load resistance, or will swing to within 200 mV of the power-supply rails with a ± 0.5 V supply. The input bias currents are 1 pA. It can be programmed to one of several supply currents by the attachment of a jumper to a pin on the chip. These are:

1. 10 μA.
2. 100 μA.
3. 1 mA.

The constant current slew rate and the gain–bandwidth product vary with the programmed supply current. The slew rate varies from 0.016 V/μs to 0.16 V/μs to 1.6 V/μs as the supply current increases. The gain–bandwidth varies from 44 kHz, through 480 kHz, to 1.4 MHz as the supply current increases. The typical open loop gain at dc is 100 dB, the typical CMRR is 90 dB, and the typical PSRR is 88 dB, which correlate closely with a standard op amp. One version of the MAXCMOS of amp (7632) has no frequency compensation; thus to remain stable (not oscillate) the closed loop gain (A_{VCL}) must be greater than 20 at $I_Q = 1$ mA, greater than 10 at $I_Q = 100$ μA, and greater than 5 at $I_Q = 10$ μA to prevent sufficient excess loop gain for oscillation to occur. The quad devices have internal compensation and have fixed programming currents (I_Q). They are numbered 7641 ($I_Q = 1$mA) and 7642 ($I_Q = 10$ μA).

Close examination of the specifications reveal that the device is expected to operate with very large feedback resistors. The larger signal voltage gain (A_{VOL}) drops to 60 dB (1000) when the supply voltage is ± 0.5 V and the load (feedback) resistor is as low as 100 kΩ; the available output voltage is only 100 mV peak at this supply voltage. At larger supply voltages, the performance is better; A_{VOL} is 68 dB (2512) with supply voltages of ± 5 V and a load (feedback) resistor of 10 kΩ.

The peak output voltage at the larger supply voltage is 1 V below the rail voltage or ± 4 V for ± 5 V supply voltages. Thus the low-voltage advantages gained through the use of this device are gained at the cost of low output voltage and low open loop gain. It appears that this device is expected to be used for dc control applications or below audio-frequency ac, as the open loop gain at audio frequencies is virtually unusable. A good example of the use of a MAXCMOS op amp is shown in Fig. 11-7; notice that the circuit is designed to operate at 10 Hz.

11.5 Stabilized Operational Amplifiers

A stabilized operational amplifier is one where an ac amplifier, together with an electrically operated switching system called a chopper, is used to correct for any dc

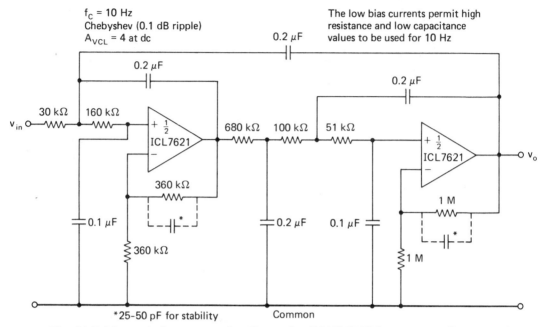

Fig. 11-7 Five-pole low-pass active filter using MAXCMOS op amps. (Courtesy of Intersil, Inc., Cupertino, Calif.)

offsets or offset drift characteristics. These are generally small systems composed of several components. As a generalization, they contain a dc amplifier around which an ac amplifier is placed. Chopping switches compare the present input voltage against one just previously obtained and stored in a capacitor; the ac amplifier corrects for any difference between the two. Figure 11-8a illustrates the simplified block diagram, and Fig. 11-8b illustrates the waveforms. These systems are inherently stable and reduce low-frequency changes, such as offset drifts, to an imperceptible amount. High-frequency changes are permitted to occur; thus the system operates as normal op amp.

11.5.1 Chopper-Stabilized Op Amp

The Texas Instrument TL089 is a linear integrated monolithic chopper-stabilized operational amplifier. In Fig. 11-8a, A1 is the dc amplifier and A2 is the chopper-stabilized ac amplifier. The chopper can be seen as two single-pole—double-throw switches operated by the chopper drive (oscillator) circuit. The actual device is connected using three capacitors as illustrated in Fig. 11-8c. The two 0.1 μF (C_F and C_{SH}) are of the low-leakage type (Appendix B2, film) and the 500 μF (C_T) is a polarized capacitor (tantalum).

The input offset voltage for the TL089 is 150 μV maximum, an order of magnitude better than typical monolithic op amps. The offset drift is 0.25μ V/° C, which is nearly two orders of magnitude better than the μA 741 and one order better than the LM 208. For applications where the offset drift is critical, this device

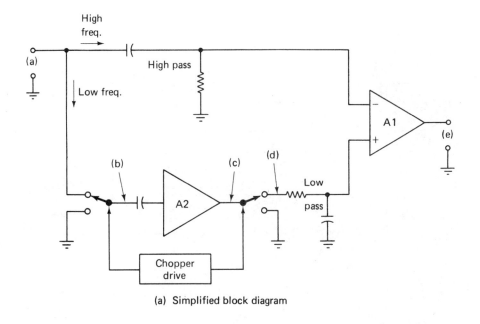

(a) Simplified block diagram

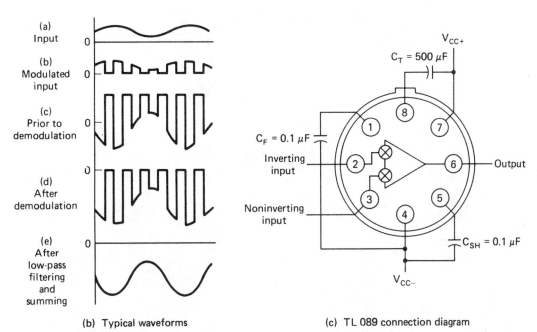

(b) Typical waveforms

(c) TL 089 connection diagram

Fig. 11-8 Chopped-stabilized operational amplifier. (Courtesy of Texas Instruments Inc., Dallas, TX.)

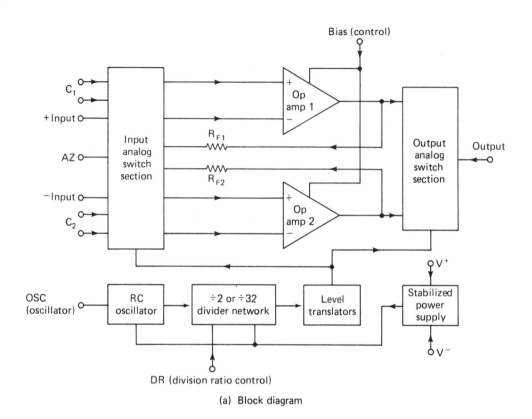

(a) Block diagram

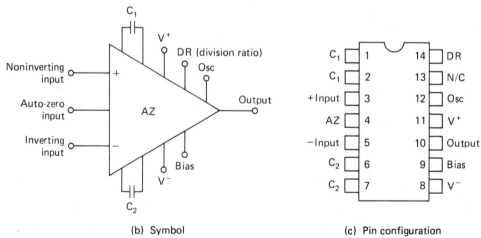

(b) Symbol

(c) Pin configuration

Fig. 11-9 Commutating auto-zero op amp. (Courtesy of Intersil Inc., Cupertino, Calif.)

provides a significant advantage. The Intersil ICL 7650 is another example of a chopper-stabilized op amp.

11.5.2 Commutating Auto-Zero Op Amp

Intersil, Inc., has also developed a new op amp system with the same advantages as a chopper-stabilized op amp. It is called a commutating auto-zero (CAZ) op amp; it is actually a small system, as it contains two op amps, an oscillator, a digital counter, and some electronic switches (CMOS transmission gates), as illustrated in Fig. 11-9a. The device symbol and pin labels are shown in Fig. 11-9b; Fig. 11-9c shows the pin configuration.

The main advantage of this device is extremely low input offset voltage, low long-term drift phenomena, and low-temperature effects. Its disadvantage is the large low-leakage (film) capacitors required for the auto-zeroing effects of the device; these should be 1 μF film capacitors (see Appendix B2). The basic amplifier has one more input than does a regular op amp—the AZ, or auto-zero input. The voltage at the AZ input is that voltage to which each of the internal op amps must be auto-zeroed; it is usually zero volts (common).

The CAZ amplifier concept, like the chopper-stabilized amplifier, offers a number of advantages as compared to a bipolar or FET input op amp:

1. Input offset voltages are between 1000 and 10,000 times less.
2. Long-term drift is dramatically reduced.
3. Temperature drift effects are reduced by 100 times or more.
4. Supply-voltage sensitivity (PSRR) is reduced.

A good example of the use of this kind of device is shown in Fig. 11-10 for the circuit of dual-slope A/D converter. This circuit uses the 19-bit Intersil ICL 7109 dual-slope ADC chip and the ICL 7600/7601 CAZ chip. The full 19 bits (288, 524 counts) is not usable, as this would require that the CAZ, with a gain of 100, divide the input signal into 0.1 μV steps. According to the application note, counts of 4096 (12 bits) are reasonable. This means that the lowest seven significant outputs of the digital register are ignored; the least significant bit is then bit 8.

11.6 High-Frequency Op Amps

Many manufacturers now produce op amps having gain bandwidths well above 50 MHz and having slew rates of several hundred volts per microsecond. These use the standard configuration of the op amp, but employ internal high-frequency methods to obtain the large bandwidths.

11.6.1 Cascode Amplifier

The cascode amplifier is a connection that eliminates the Miller effect causing high-frequency rolloff to occur. A cascode amplifier is a common emitter–common base joining of two transistors or FETS. The circuit is shown in Fig. 11-11. Stage 1

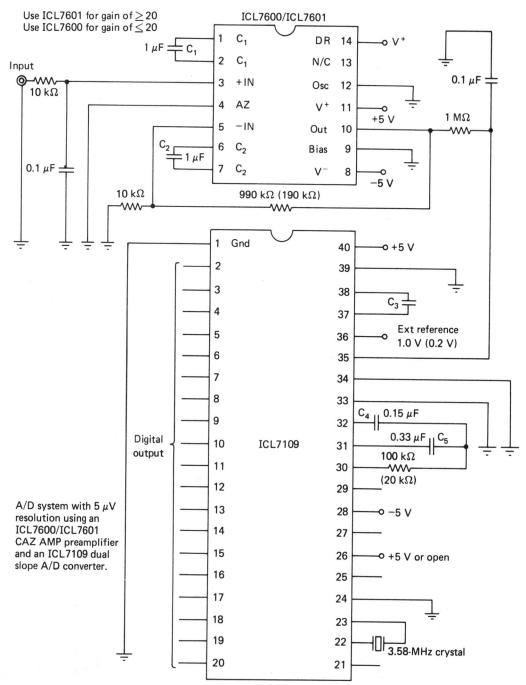

Fig. 11-10 Dual-slope A/D converter using CAZ. (Courtesy of Intersil, Inc., Cupertino, Calif.)

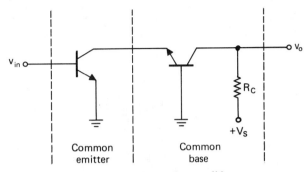

Fig. 11-11 Cascode amplifier.

is a common-emitter (or common-drain) amplifier. This amplifier normally has a voltage gain of

$$A_V = -h_{fe}\frac{R_L}{h_{ie}} \tag{11-12}$$

but the second stage has an input impedance of $[h_{ie}/(h_{fe} + 1)]$, which is R_L of the first stage; thus the first-stage gain is

$$A_V = -h_{fe}\frac{R_L}{h_{ie}} = -h_{fe}\frac{\left(\dfrac{h_{ie}}{h_{fe}+1}\right)}{h_{ie}} \approx -1 \tag{11-13}$$

and the second-stage gain is

$$A_V = +\left(h_{fe} + 1\right)\frac{R_C}{h_{ie}} \approx +h_{fe}\frac{R_C}{h_{ie}} \tag{11-14}$$

Thus the total amplifier gain is

$$A_V = -h_{fe}\frac{R_C}{h_{ie}} \tag{11-15}$$

which is the same gain as a common emitter amplifier but has *no Miller effect* because of the unity voltage gain in the first stage. Thus the Miller multiplication of the feedback capacitance, which usually takes place, is eliminated. These amplifiers can operate to several hundred megahertz and sometimes beyond 1 GHz. An example of this type of op amp can be seen in the CA 3040 video amplifier illustrated in Fig. 11-12.

EXAMPLE 11-8

Calculate the gain of the input cascode amplifier with $+6$ V connected to pin 2 and -6 V connected to pins 5 and 11. The collector current in Q_5 and Q_6 is 1 mA and $h_{fe} = 100$.

Solution

The signal first passes through two emitter followers (Q_1 and Q_2) each having a gain of unity. It then passes through the first stages of the cascode amplifiers (Q_3 and Q_4), which also have gains of unity. The second stages of the cascode amplifiers each conduct 1 mA, so

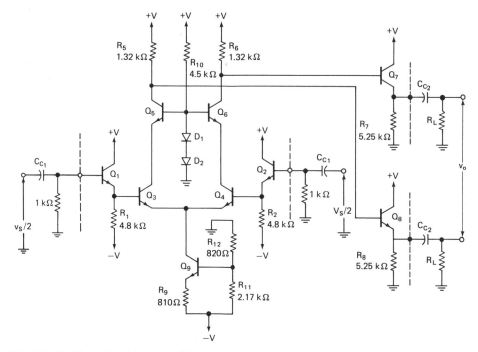

Fig. 11-12 Cascade video amplifier integrated circuit: CA 3040. (Courtesy of RCA Corporation, Somerville, N.J.)

$h_{ie} = 2.6$ kΩ. The output emitter followers have a very high input impedance due to the emitter resistors, so no appreciable loss of gain occurs due to loading. The gain of the cascode second stages is found, using Eq. (11-14), to be 50.

11.6.2 Standard Op Amps Employing Process Techniques

The Harris Corporation produces a series of op amps that employ a special technique for obtaining high frequencies. This is a process they call "dielectric isolation." The process is best described in their analog data book as*

1. Almost all op amp designs require at least one PNP transistor in the signal path. Typical junction isolated (J.I.) op amps must use a lateral PNP which inherently has very low frequency response, limiting typical compensated bandwidth to 1 MHz. The dielectric isolation (D.I.) process makes it practical to build a vertical PNP with much higher bandwidth, making possible compensated op amp bandwidths of 12 MHz or higher (Figs. 11-13, 14). Also, transistor collector to substrate capacitance is 2/3 less using D.I., further enhancing high-frequency performance.

2. Other devices such as optimally specified MOS or JFET transistors may be fabricated on the same chip. Isolated diffused and thin film resistors are also practical.

*Courtesy Harris Corporation.

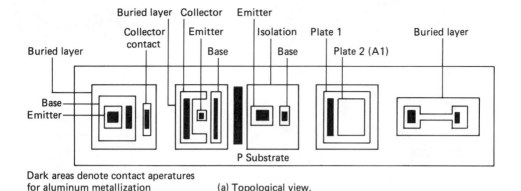

Buried layer Collector Emitter

Collector | Emitter | Isolation Plate 1 Buried layer
contact | Base | Base Plate 2 (A1)

Buried layer

Base
Emitter

P Substrate

Dark areas denote contact aperatures
for aluminum metallization   (a) Topological view.

| NPN | Lateral PNP | Substrate PNP | MOS | Diffused |
| transistor | transistor | transistor | capacitor | resistor |

Collector   Collector

Epitaxial   Emitter   Emitter Emitter   Plate 2(Al)   Resistor
layer   SiO2   Base   Base   Base   Plate 1

P+ N+ P N+ P+ P P N⁺ P+   P N⁺ P+   N⁺   P+   P   P+

Isolation diffusion   N⁺ buried layer
P substrate

(b) Cross-sectional view.

Fig. 11-13 Structures of various components formed in the junction-isolation process. (Courtesy of Harris Semiconductor, Melbourne, Fla.)

3. The isolation removes the possibility of parasitic SCRs which might create latchup under certain sequences of power and signal application.

4. Leakage currents to the substrate under high-temperature conditions are greatly reduced. While the Harris circuits were not specifically designed for operating temperatures greater than $+125°$, many have shown superior performance.

11.6.3   Special Op Amps Employing Current Feedback

The Comlinear Corporation has invented and developed a technique of current feedback that extends frequency responses and slew rates. This technique is described in their Application Note 300-1 and is presented here.

If Eq. (2-24) is rewritten where $G = 1/\beta = (R_F/R_I + 1)$ and $A_{(s)} = N_{(s)}/D_{(s)}$, where $N_{(s)} = A_{\text{VOL}}$ and $D_{(s)}$ represents the poles of the open loop transfer function, the result is as follows:

$$\frac{V_o}{V_{\text{in}}} = \frac{A}{1 + A\left(\dfrac{1}{G}\right)} = G\frac{N_{(s)}}{N_{(s)} + GD_{(s)}} \qquad (11\text{-}16)$$

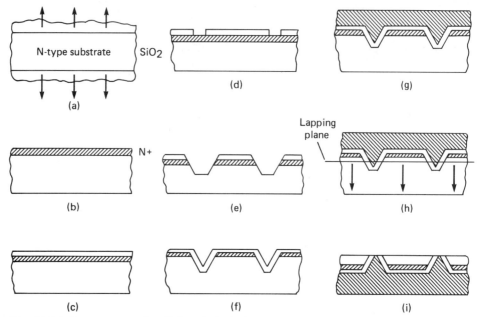

Fig. 11-14 Process steps for dielectric isolation: (a) surface preparation; (b) N-buried layer diffusion; (c) masking oxide; (d) isolation pattern; (e) silicon etch; (f) dielectric oxide; (g) polycrystalline deposition; (h) backlap and polish; and (i) finished slice. (Courtesy of Harris Semiconductor, Melbourne, Fla.)

Comlinear has made the following observations:

> Thus it can be seen that G not only scales the magnitude of the gain (as desired) but it also multiplies the effect of $D_{(s)}$ on the closed loop response. The locations of the closed loop poles are now a function G; thus, if an application requires a large G, the poles will be at a lower frequency than for a low value of G. This is the chief failure of the conventional high-speed op amp designs.

Figure 11-16 shows the closed loop frequency response of a conventional op amp for various gains. Notice that increasing the gain decreases the bandwidth.

Compensation

Conventional high speed op amps are unstable at most gain settings so compensation must be used. Dominant pole compensation creates a very low frequency pole that limits bandwidth so severely that the problematic high frequency response poles are no longer dominate. Unfortunately, compensation does not allow complete control of the pole and zero locations, thus simultaneously optimizing bandwidth, gain flatness, and settling time [*Author's note*: function of gain margin] is difficult if not impossible.

External Compensation

Compensation is usually connected external to the amplifier so the designer can tailor the response to the application. Unfortunately, conventional op amps are very sensitive to stray reactances in the PC board layout, temperature, loading, and a variety of other

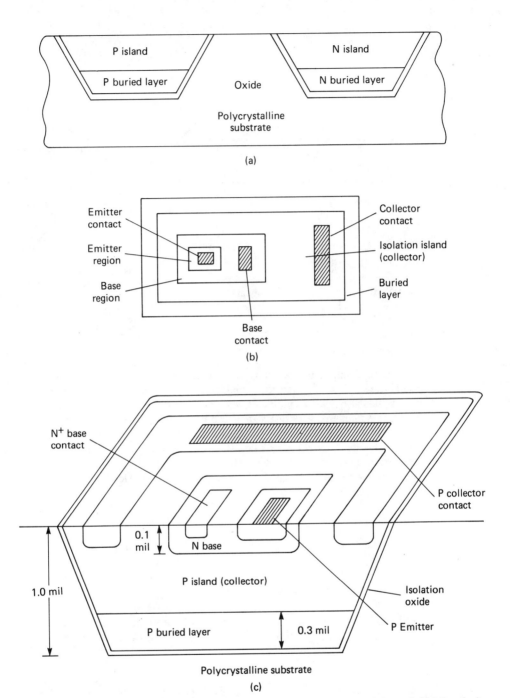

Fig. 11-15 High-frequency process: (a) cross-sectional view of P and N islands for *PNP* and *NPN* transistors; (b) topological view showing relative placement of transistor regions; (c) cross-sectional view of high-frequency *PNP* device formation in the D.I. process. (Courtesy of Harris Semiconductor, Melbourne, Fla.)

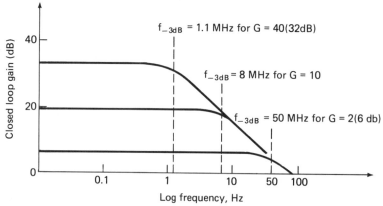

Fig. 11-16 Illustrating reduction in bandwidth as gain increases. (Courtesy of Comlinear Corporation, Fort Collins, Colo.)

factors which make reliable compensation difficult. In addition, production can become expensive as each op amp compensation network must be individually "tweaked" to the desired response.

Slew Rate Limiting

In order to obtain high open loop gain, several internal gain stages must be used. As a result, transit times through the amplifier are large. In addition to reducing the phase margin, this also leads to problems when large or fast rise time signals are present at the input. These large or very fast signals can cause an internal gain stage to saturate or be cut-off before feedback can propagate back to the input to reduce the error signal. To prevent this behavior the slew rates of the internal stages are simply limited [*Author's note*: using constant current slew rate] so nonlinear behavior cannot occur before the signal propagates through the amplifier. Consequently the slew rate and signal bandwidth of the op amp are reduced severely.

The Comlinear Innovation in Op Amps

Most of the problems with conventional op amps discussed up to now are either directly or indirectly caused by the limitation of having G, the gain setting, affect the frequency response. If G could be removed from the denominator of the expression, performance and ease of use could be extended dramatically.

Figure 11-17 illustrates the Comlinear op amp and shows an unusual (and patented) circuit configuration. The circuit description is as follows:

Input Buffer

The input buffer is a unity-gain voltage amplifier that is connected across the inputs of the op amp. In operation, the buffer forces V_2 to equal V_1 *independent of any feedback through* R_2. This causes the inverting input to have a very low input impedance. When feedback around the loop is applied, the impedance of this node is reduced further and V_2 becomes a "virtual ground" with respect to V_1. This low impedance allows current to flow easily into or out of the inverting input.

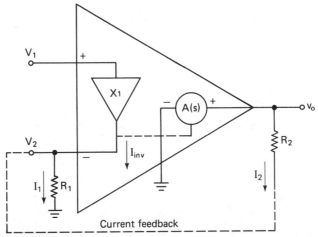

Fig. 11-17 The drawing of the Comlinear op amp shows an unusual (and patented) circuit configuration. The main parts are described in the text. (Courtesy of Comlinear Corporation, Fort Collins, Colo.)

Transimpedance Amplifier

This amplifier is a transresistance amplifier as described in Section 2.6.3 and is the gain block inside Comlinear op amps. In operation, the transimpedance amplifier *senses* the inverting input current (literally the current into or out of the inverting input) and transforms this current into the output voltage. The transfer function of this transimpedance amplifier is $A_{(s)}$; the units are ohms.

Feedback Mechanism

Feedback in the form of a current is applied through R_2 to the inverting input.

Mathematics of the Comlinear Op Amp

$$I_{inv} = I_1 - I_2$$

$$I_{inv} = \frac{V_2}{R_1} - \frac{V_0 - V_2}{R_2}$$

Then, since $V_0 = I_{inv} A_{(s)}$ and $V_2 = V_1$ (because of the buffer)

$$\frac{V_0}{A_{(s)}} = V_1 \left(\frac{1}{R_1} + \frac{1}{R_2} \right) - \frac{V_0}{R_2}$$

Then, rearranging

$$\frac{V_0}{V_1} = \frac{\dfrac{R_1 + R_2}{R_1 R_2}}{\dfrac{1}{R_2} + \dfrac{1}{A_{(s)}}} = \frac{1 + \dfrac{R_2}{R_1}}{1 + \dfrac{R_2}{A_{(s)}}}$$

Again, letting $1 + \dfrac{R_2}{R_1} = G$,

$$\frac{V_0}{V_1} = \frac{G}{1 + \dfrac{R_2}{A_{(s)}}}$$

Letting $A_{(s)} = \dfrac{N_{(s)}}{D_{(s)}}$

$$\frac{V_0}{V_1} = G\frac{N_{(s)}}{N_{(s)} + R_2 D_{(s)}}$$

A Comparison: Conventional vs. Comlinear

The closed loop gain equations are now in the same form and can be compared directly.

Conventional

$$\boxed{\frac{V_0}{V_1} = G\frac{N_{(s)}}{N_{(s)} + (G) D_{(s)}}} \qquad (11\text{-}17)$$

Comlinear

$$\boxed{\frac{V_0}{V_1} = G\frac{N_{(s)}}{N_{(s)} + (R_2) D_{(s)}}} \qquad (11\text{-}18)$$

R_2 has replaced G in the frequency-dependent portion of the transfer function. Since R_2 can be held constant, whereas G cannot, the pole locations and hence the performance can be held constant. ($G = 1 + R_2/R_1$, so the gain setting can be varied by changing R_1.)

Results

The results of this new approach to op amp design are dramatic:

Elimination of the Gain–Bandwidth Product

The frequency response of the Comlinear op amp changes very little when the gain is increased by a factor of 10, from a gain of 4 to a gain of 40. Other specifications are similarly unaffected by gain changes. A similar change in gain with a conventional op amp could reduce the bandwidth by a factor of 10, even under ideal single pole conditions. As shown in Fig. 11-18, the -3 dB bandwidth remains constant over a wide range of gains. Small second order effects explain the slight deviation in performance from that predicted by the equations.

Excellent Performance

The high-speed performance of Comlinear op amps is dramatically better, usually by an order of magnitude, than conventional op amps. A typical rise time for a 5 V output step can be as low as 1.6 ns, for example; compare this to a conventional op amp where rise times of 30 to over 100 ns are common. Other outstanding specifications include settling times which can be as low as 10 ns for a 0.02% tolerance and slew rates that range from 3000 V/μs to over 7000 V/μs, depending on which model is chosen. (The slew rate was intentionally kept very fast so it would not limit the response of the amplifiers under large signal conditions. Bandwidth, not slew rate, controls amplifier response and keeps it linear.)

Linear Phase

Although phase linearity is a rarely mentioned specification, it is very important for signal fidelity. Comlinear op amps have excellent phase linearity, usually the deviation from linear phase is less than two degrees from dc to over 50% of the bandwidth. Conventional

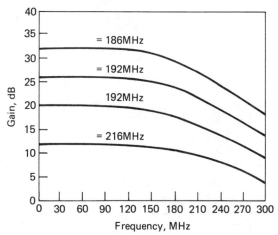

Fig. 11-18 Gain vs. frequency for Comlinear op amp. (Courtesy of Comlinear Corporation, Fort Collins, Colo.)

op amps however, must use techniques like ac feedforward which increase bandwidth but degrade phase linearity; thus, signal fidelity is sacrificed.

Predictable Performance

Since most of the specifications are virtually independent of gain setting, the performance of Comlinear op amps remains consistent even with varying circuit configurations. (As design requirements change, adjustments in gain can usually be achieved by a change in just one resistor, with a conventional op amp a change in gain could require a redesign of the entire circuit.) In addition, Comlinear is able to provide complete specifications for different gain settings.

Internal Compensation

With R_2 fixed and the poles of the frequency response consequently fixed, internal compensation is provided to simultaneously optimize bandwidth, settling time, linearity, and distortion. Since the compensation is internal, Comlinear op amps can save time and money in production with no compensation networks to "tweak." (Although Comlinear uses an internal high precision resistor for R_2, on many models an external resistor of another value can be used. Since this would change the pole locations, the internal compensation is made accessible to the user through one package pin. One external capacitor is then used to re-optimize the performance. Thus, both consistency and flexibility can be achieved.)

Standard Usage

Determining the gain for Comlinear op amps is the same as that for conventional op amps, where $G = 1 + R_2/R_1$. The same ease of use applies equally well to both inverting and differential configurations.

The Fallacy of the Gain–Bandwidth Product

The gain–bandwidth product has for years been a key specification for op amps; in fact, the concept of the gain–bandwidth product is often a major topic in basic op amp tutorials. Unfortunately, this often touted specification means little to the engineer who must work with very high speed op amps. For most high speed op amps the gain–bandwidth product is actually very misleading.

High-Frequency Op Amps   **383**

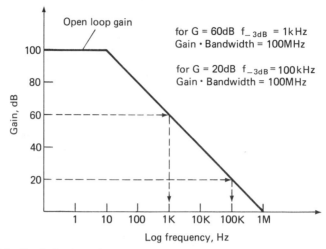

Fig. 11-19 Idealized single-pole op amp. (Courtesy of Comlinear Corporation, Fort Collins, Colo.)

The basis for the gain–bandwidth product is the assumption that the open loop gain rolls off due to a single pole. When the assumption is valid, as is the case with some low frequency op amps, the gain–bandwidth product concept is also valid. This is shown below in Fig. 11-19. Having frequency performance depend on gain is troublesome, but at least with a single pole rolloff the bandwidth is easily determined and stability is assured. High-speed op amps, however, have several poles before unity gain crossover is reached. Figure 11-20 shows how multiple pole rolloff severely degrades high-frequency performance. Clearly, a conventional op-amp with a gain–bandwidth product of 1GHz will not yield a bandwidth of 500 MHz at a gain of 2. This is why most manufacturers specify their gain–bandwidth products at very high gains, typically 1000.

A much more useful (and accurate) way to show frequency performance is to actually show the performance for various gains. This allows the engineer to fully characterize the

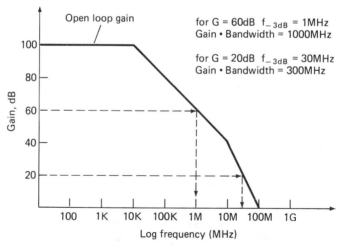

Fig. 11-20 Idealized double-pole op amp. (Courtesy of Comlinear Corporation, Fort Collins, Colo.)

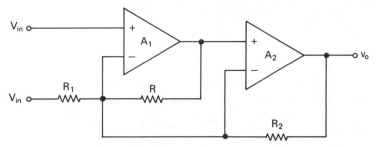

Fig. 11-21 Response-gain independent (RGI) amplifier.

amplifier without having to actually test the device. Although the performance of Comlinear amplifiers varies little with changes in gain, Comlinear provides complete specifications for at least three representative gain settings.

11.6.4 Response-Gain Independent Operational Amplifier

The author has a patent application pending on a variation of the Comlinear op amp. In this scheme, two voltage followers are employed as illustrated in Fig. 11-21. The first acts as a unity-gain stage, similar to the first stage of the Comlinear amplifier. This drives the second follower with the two inverting inputs joining as the amplifier summing junction. An input resistor, connected to the summing junction, acts as the circuit inverting input. The circuit performs in an identical fashion to the Comlinear amplifier described in the preceding section. The equations of the noninverting RGI amplifier are derived in Appendix A22 and the results are duplicated below:

$$\frac{V_o}{V_{ni}} = \left(\frac{R_2 + R_1}{R_1} \right) \left(\frac{R}{R_2} \right) \left[\frac{\dfrac{N_2}{D_2}}{1 + \left(\dfrac{N_2}{D_2} \right) \left(\dfrac{R}{R_2} \right)} \right] \qquad (11\text{-}19)$$

Now

$$A_{OL} = \frac{N_2}{D_2} \qquad \beta = \frac{R}{R_2} \qquad G = \frac{R_2 + R_1}{R_1}$$

The important effect of these terms is that the -3 dB frequency rolloff is dependent upon (R/R_2) and not G. Since R and R_2 can be controlled, the gain (G) is independent of the bandwidth.

EXAMPLE 11-9

An RGI amplifier has its dominate pole at 1.2 MHz and the second and third poles are at 50 MHz and 250 MHz, respectively. The amplifier has a gain of 100, $R = R_2 = 50$ kΩ, and $A_2 = N_2/D_2 = 3.72 \times 10^{24}/(S + 2\pi \cdot 1.2 \text{ MHz})(S + 2\pi \cdot 50 \text{ MHz})(S + 2\pi \cdot 250 \text{ MHz})$. Find the value of R_1 and the bandwidth at a gain of 10.

$$N_2 = (10^5)(2\pi \cdot 1.2 \text{ MHz})(2\pi \cdot 50 \text{ MHz})(2\pi \cdot 250 \text{ MHz})$$

$$D_2 = (S + 2\pi \cdot 1.2 \text{ MHz})(S + 2\pi \cdot 50 \text{ MHz})(S + 2\pi \cdot 250 \text{ MHz})$$

The dominant pole is at 1.2 MHz and the frequency rolloff is at -20 dB/decade until 50 MHz, which is 1.6 decades above 1.2 MHz. Thus a gain of 10, the bandwidth is 10 times the bandwidth at a gain of 100, which is 10(1.2 MHz) or 12 MHz. R_1 is found from (50 k$\Omega/R_1 + 1) = 100$, $R_1 = 50$ k$\Omega/99 = 505$ Ω.

The value of an amplifier of this type is that it is stable out to a gain of $f_{0°}$ $\cong \sqrt{50 \text{ MHz} \times 250 \text{ MHz}} = 111.8$ MHz, where the closed loop gain is nearly unity; thus it is stable (see Fig. 11-22).

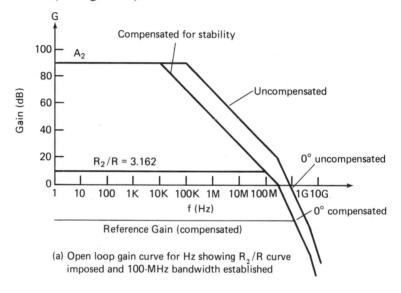

(a) Open loop gain curve for Hz showing R_2/R curve imposed and 100-MHz bandwidth established

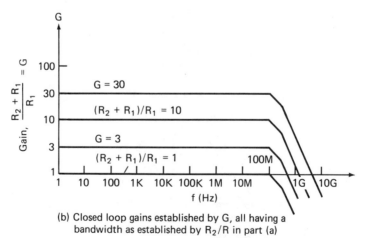

(b) Closed loop gains established by G, all having a bandwidth as established by R_2/R in part (a)

Fig. 11-22 Gain and response characteristics for RGI amplifier.

The inverting form of the RGI amplifier (see Appendix A23) is similar to the noninverting form, the difference being that $G = (-R_2/R_1)$.

$$\frac{V_o}{V_{ni}} = \left(-\frac{R_2}{R_1}\right)\left(\frac{R}{R_2}\right)\left[\frac{\dfrac{N_2}{D_2}}{1 + \left(\dfrac{N_2}{D_2}\right)\left(\dfrac{R}{R_2}\right)}\right] \qquad (11\text{-}20)$$

Now

$$A_{OL} = \frac{N_2}{D_2} \qquad \beta = \frac{R}{R_2} \qquad G = -\frac{R_2}{R_1}$$

Again, the -3 dB frequency rolloff is dependent on (R/R_2) and not G. As the technology progresses, the high-frequency op amp is going to become ever more available. Those on the market in the late 1980s cost over $100 each, and some cost over $200 each. This will change as the demand becomes greater and more are produced. Others will enter the field, if only as second sources, and that will increase competition and drive down the price. By the early 1990s, op amps with a 1 GHz bandwidth will be commonplace; the price will be under $100 and may even be under $10. At that point, we will see a whole new generation of high-frequency applications appear that are now thought impossible. You, the reader, will experience this advance in the analog state of the art.

11.7 Phase-Locked Loops

A phase-locked loop (PLL) is actually a nonlinear control system in which the analysis is beyond the scope of this book, as nonlinear control systems constitute a complex, highly mathematical subject. The phase-locked loop has two regions of operation: the nonlinear region, where it is not in lock, and the linear region, where the loop is locked. By limiting this discussion to the region where the loop is locked, the topic can be simplified considerably. It is also the region in which most of our interest lies anyway.

Every control system has three essential elements:

1. Differencing circuit.
2. Amplifier, with frequency rolloff.
3. Feedback circuit.

The first attempt at presenting the operational amplifier as a control system was done in Section 2.6, where, in Fig. 2-9, the three elements of a control system were clearly defined. It now becomes evident that up to now we have been discussing linear control systems without using that definition. The task of transferring our present knowledge to another type of linear control system, possessing the same three essential elements, is no longer formidable. A phase-locked loop, IN LOCK, is a linear control system possessing the same three elements shown in Fig. 11-23a; the circuits within the three areas are different (i.e., they are not the same circuits as for

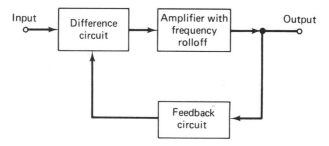

(a) Closed loop control system

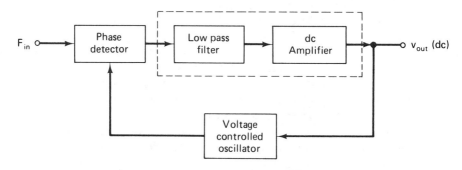

(b) Phase locked loop system

Fig. 11-23 Linear control systems.

an operational amplifier circuit). Figure 11-23b illustrates a basic phase-locked-loop circuit. The three essential elements now become four, as the amplifier and frequency rolloff, element 2, are broken in two separate parts:

1. Differencing circuit—wide-frequency-band phase detector.
2. Frequency rolloff—low-pass filter.
3. Amplifier—wide-frequency-band dc amplifier.
4. Feedback network—voltage-controlled oscillator (VCO).

The elements of PLL are, at first, strange. The input to the PLL circuit is a frequency, not a dc voltage, and the circuit operates in the following manner:

1. The incoming frequency is one input to the phase detector (differencing element).
2. The output from the VCO, also a frequency, is the second input to the phase detector.
3. When the PLL is not in lock, the output from the phase detector is a difference frequency or "beat frequency" with a dc offset that is amplified by the amplifier and bandlimited by the low-pass filter. The dc offset passes through the filter and high beat frequencies are filtered out.

4. The resulting dc signal voltage causes the VCO to begin varying its frequency in the direction of the incoming frequency. When the two frequencies, incoming and VCO, are the same, the beat frequency becomes 0 Hz(dc) and the VCO now locks to the incoming signal frequency. It remains there as long as the dc voltage, into the VCO, corrects for any deviations in either incoming or VCO frequencies. The PLL is now locked and becomes a linear control system.

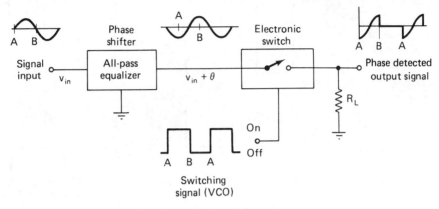

(a) Block diagram of phase detector

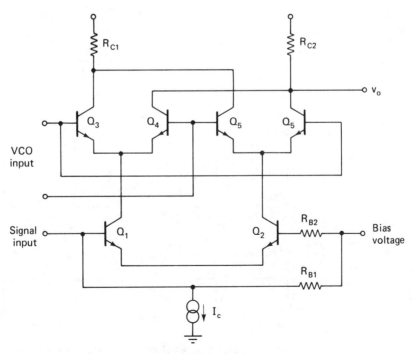

(b) Integrated phase detector

Fig. 11-24 Phase-detector circuit. (Courtesy of Signetics Corporation, Sunnyvale, Calif.)

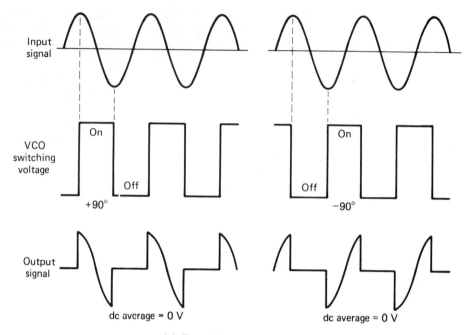

Input
signal

VCO
switching
voltage

On

Off

+90°

On

Off

−90°

Output
signal

dc average = 0 V

dc average = 0 V

(a) Phase detector at quadrature phase

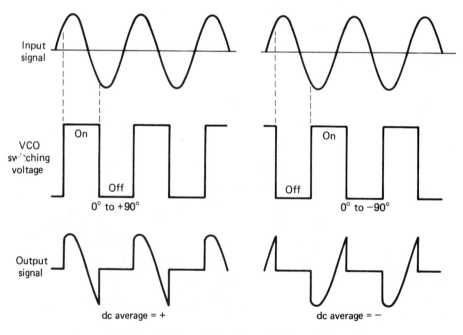

Input
signal

VCO
switching
voltage

On

Off

0° to +90°

On

Off

0° to −90°

Output
signal

dc average = +

dc average = −

(b) Phase detector with leading or lagging switching voltage

Fig. 11-25 Phase-detector waveforms.

Just as an op amp circuit can be changed by taking the output from different locations, the PLL system has several output locations. The output from the amplifier or low-pass filter and the input to the VCO is a dc voltage which is proportional to the amount of shift the VCO must have to remain in lock. If the input frequency shifts, the dc voltage will shift to correct the VCO for the frequency change. Thus one method of applying a signal to the incoming frequency is to slightly alter its frequency; this is called frequency modulation (FM) if the frequency is shifting in proportion to a modulation frequency or voltage. It is frequency shift keying (FSK) if the frequency is shifting a discrete amount in response to a digital logic level change. The actual integrated phase detector is shown in Fig. 11-24b. The output can be taken from the VCO output in which case it is a frequency which is 90° out of phase with the incoming signal frequency. The reason for the quadrature (90°) phase relationship can be seen by observing the dc average of Fig. 11-25a. A phase detector can be thought of as a circuit where the incoming signal frequency is switched on and off by the VCO frequency. When the VCO (switching) frequency is in the quadrature phase to the incoming signal frequency, the dc average is zero. As the phase shifts away from the quadrature relationship, the dc average changes with it, as illustrated in Fig. 11-25b.

EXAMPLE 11-10

Determine the polarity of the average (dc) output voltage in a phase detector for the following situations:

a. The switching signal lags the sine wave by 30°.
b. The switching signal leads the sine wave by 60°.
c. The switching signal leads the sine wave by 120°.
d. The switching signal lags the sine wave by 150°.

Solution

a. +
b. +
c. −
d. −

Thus one sure way to observe whether the PLL is locked is to observe the two inputs to the phase detector with a dual-trace oscilloscope and look for a quadrature-phase relationship.

Just as the VCO in Section 10.6 can operate over different frequency ranges through the selection of different RC time constants, the VCO in a PLL can be changed in frequency by selecting an RC time constant. The VCO in a phase-locked loop has a "free-running" or "center" frequency (f_0), which is the frequency of the VCO when not locked to the incoming signal frequency. As the incoming frequency approaches the free-running frequency of the VCO, the different (beat) frequency from the phase detector begins pulling the frequency of the VCO toward

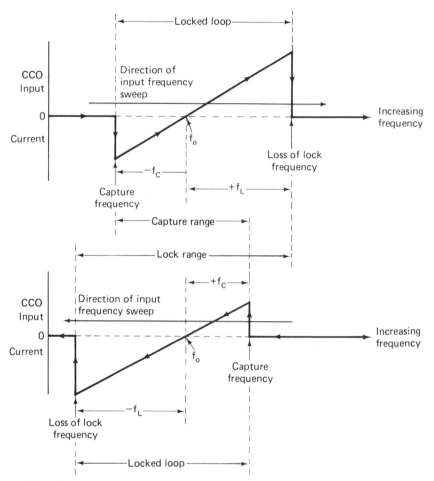

Fig. 11-26 Capture and lock range. (Courtesy of Signetics Corporation, Sunnyvale, Calif.)

a locked condition. The range over which this pulling action occurs is called the "capture" or "lock-in" range. The loop becomes locked when the VCO has been pulled to the exact frequency and quadrature phase of the incoming signal. The "lock," "tracking," or "hold-in" range is the range of incoming signal frequencies over which the loop will remain locked. The capture range is always smaller than the lock range, as it is more difficult to attain lock than to retain lock. Figure 11-26 shows the relationship between the capture range and the lock range. It can be seen from Fig. 11-26 that as an incoming signal frequency is increasing past the free-running frequency, the loop will capture the signal at a frequency close to the free-running frequency and then lose lock at a frequency on the other side of the free-running frequency farther away. Where it loses lock is at the end of the lock range, and this is always a broader range than the capture range. As a signal frequency is moving down in relation to the free-running frequency, the loop will capture inside the lock range frequency and then lose lock beyond the lower capture frequency. Both the capture range and the lock range are determined by the cutoff frequency of the low-pass filter, as that determines the bandwidth of the PLL.

The PLLs will be presented in this discussion: the 565 and 567. The 567 PLL will be presented first, as it is the most basic in operation. Figure 11-27 shows the block diagram of the 567 PLL. There are six individual sections to this PLL. The first four form the actual phase-locked-loop circuit:

1. PLL phase detector.
2. Low-pass filter (R_2 and C_2).
3. Dc amplifier.
4. Current-controlled oscillator (CCO) (see Section 10.6).

The other two sections,

5. Quadrature-phase detector.
6. Amplifier (comparator).

are inserted to detect the locked condition. The loop phase detector yields zero volts, in lock, where the incoming signal and CCO output are in quadrature phase. The "quadrature" phase detector yields zero volts when the incoming signal and CCO output are exactly in or out of phase and yields a positive or negative voltage out when they are in quadrature phase. Thus the comparator output (pin 8) is LO when the PLL is locked and HI when the PLL is not in lock. These two stages are *not* part of the PLL: they serve only to give an output indication of lock.

The free-running or center frequency (f_0) is first determined by selecting the RC time constant of the CCO using

$$f_0 = \frac{0.97}{R_1 C_1} \quad \text{Hz} \tag{11-21}$$

The value of C_2 is determined through the use of two distinct equations.

1. Where the input signal amplitude is less than 200 mV rms, the value of C_2 is found from

$$C_2 = \frac{(1145)(V_{in} \text{ rms})}{[(\%BW)(f_0)]^2} \quad \text{F} \tag{11-22a}$$

which represents the sloped lines on the curve of Fig. 11-28a. The bandwidth in this region is proportional to the input signal amplitude; this is an undesirable situation, as a predictable bandwidth is usually required.

2. The bandwidth becomes constant after the input signal amplitude exceeds 200 mV rms. The desired bandwidth, in percent of f_0, is first chosen. An appropriate constant from the vertical-line portion of Fig. 11-28a, corresponding

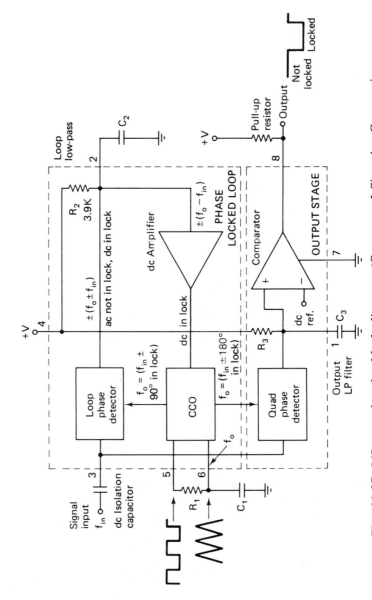

Fig. 11-27 567 tone decoder block diagram. (Courtesy of Signetics Corporation, Sunnyvale, Calif.)

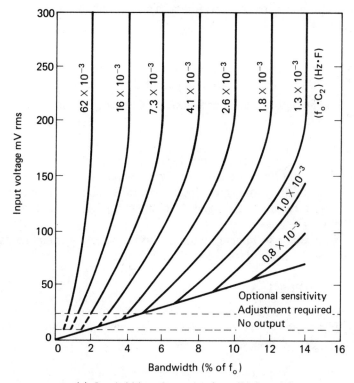

(a) Bandwidth vs. input signal amplitude and C_2

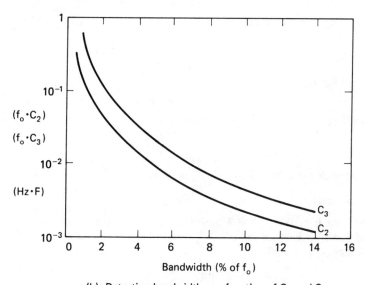

(b) Detection bandwidth as a function of C_2 and C_3

Fig. 11-28 Bandwidth and V_{in} curves for selecting C_2 and C_3. (Courtesy of Signetics Corporation, Sunnyvale, Calif.)

to the desired bandwidth, is then selected and inserted into

$$C_2 = \frac{\text{constant}}{f_0} \quad \text{F} \qquad \qquad (11\text{-}22\text{b})$$

which yields the value of C_2. This is most likely the case for the use of the PLL, as the user wants a bandwidth, independent of input signal amplitude.

3. The output stage bandwidth is established by selecting the value of C_3 given by

$$C_3 = 2C_2 \quad \text{F} \qquad \qquad (11\text{-}23)$$

This filter prevents the logic level, out of the comparator driven by the quadrature phase detector, from chattering near the edges of lock. It assures that only two discrete logic levels will be on the output lead. The bandwidth as a function of C_2 and C_3 is shown in Fig. 11-28b.

An example will illustrate the method of selecting components for the case where the input signal voltage is less than 200 mV.

EXAMPLE 11-11

Determine the circuit component values for a 567 tone decoder with a free-running frequency of 1000 Hz and a bandwidth of 100 Hz. The input signal voltage is 100 mV rms.

Solution

Let $R_1 = 4.3 \text{ k}\Omega$; then using Eq. (11-21), we obtain

$$C_1 = \frac{0.97}{(4.3 \text{ k}\Omega)(1000 \text{ Hz})} = 0.23 \ \mu\text{F}$$

Using Eq. (11-13a) yields

$$C_2 = \frac{(1145)(100 \text{ mV})}{[(10\%)(1000)]^2} = 1.14 \ \mu\text{F}$$

From Eq. (11-14),

$$C_3 = 2(1.14 \ \mu\text{F}) = 2.3 \ \mu\text{F}$$

Use a pull-up resistor on pin 8 of 20 $\text{k}\Omega$.

EXAMPLE 11-12

Determine the values of C_2 and C_3 using the curve in Fig. 11-28a.

Solution

Find the intersection of 10% bandwidth and the 100-mV rms input signal. This intersection falls halfway between the two f_0C_2 curves, 1.3×10^{-3} and 1.8×10^{-3}. The geometric mean of these two values is

$$\text{mean constant} = \sqrt{(1.3)(1.8)} \times 10^{-3} = 1.53 \times 10^{-3}$$

Then using Eq. (11-22b), we obtain

$$C_2 = \frac{1.53 \times 10^{-3}}{1000} = 1.53 \ \mu\text{F}$$

C_3 is found, again using Eq. (11-23), as

$$C_3 = 2(1.53 \ \mu F) = 3 \ \mu F$$

The first question that arises from this example is: Which value is correct? The most likely answer is the one using the curves, as the equation is an approximation to the curves.

The next example illustrates the use of the curves for finding component values when the input signal is greater than 200 mV rms.

EXAMPLE 11-13

Determine the component values for a tone decoder using the 567 PLL when the input signal is 1 V rms, f_0 is 100 kHz, and the bandwidth is 5% of f_0.

Solution

Let $R_1 = 5.1 \ k\Omega$; then using Eq. (11-21), we obtain

$$C1 = \frac{0.97}{(5.1 \ k\Omega)(100 \ kHz)} = 0.0019 \ \mu F$$

The curves in Fig. 11-28a are vertical for input voltages above 200 mV rms. Thus 5% falls halfway between the two constants, 7.3×10^{-3} and 16×10^{-3}. Determine the geometric mean of the two constants using

$$\text{mean constant} = \sqrt{(7.3)(16)} \times 10^{-3} = 10.8 \times 10^{-3}$$

Then using Eq. (11-22b), we obtain

$$C_2 = \frac{10.8 \times 10^{-3}}{100 \ kHz} = 0.11 \ \mu F$$

Then using Eq. (11-23) yields

$$C_3 = 2(0.108 \ \mu F) = 0.22 \ \mu F$$

Use a 20 kΩ pull-up resistor on pin 8.

The 567 PLL is called a tone decoder, as its primary purpose is to select a specific tone (frequency) from a band of frequencies with a particular bandwidth (deviation from the specific tone). When the tone is detected and the PLL has locked onto it, the output logic level is LO; when lock is lost, the logic level is HI. The 565 PLL, illustrated in Fig. 11-29a, is more versatile. The lead from the VCO to the loop phase detector is broken and both ends of the lead are brought out on pins. If pins 4 (VCO output) and 5 (loop phase detector input) are connected together, the loop is closed. By inserting a digital counter between the VCO output and the loop phase detector as shown in Fig. 11-29a, the circuit becomes a frequency multiplier. This complicates the component selection, as the VCO free-running frequency must be at the multiplied frequency while the bandwidth is selected for the input or reference frequency (unmultiplied). This circuit is regularly used as a frequency synthesizer for citizen-band radios by providing the counter with selectable division ratios. The voltage levels are not compatible with TTL logic (see Section 8.2) inputs; thus a transistor-level shifter is used at pin 4 to drive the counter as illustrated in Fig. 11-29b.

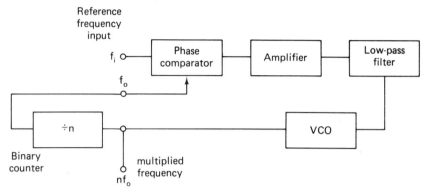

(a) Block diagram showing insertion of digital counter (divider)

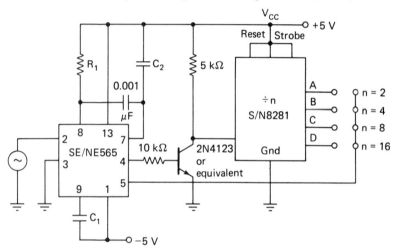

(b) Circuit showing insertion of level shifter and counter

Fig. 11-29 565 PLL frequency synthesizer or multiplier and counter. (Courtesy of Signetics Corporation, Sunnyvale, Calif.)

The phase-locked loop is a versatile device used mainly in communications circuits. The tone decoder is actually a very sharp band-pass filter. If the output of the tone decoder were used to drive an electronic switch on the input signal, only frequencies within the lock range of the PLL would be passed. The filter would have an instantaneous cutoff at the band edges; it would be an "ideal" band-pass filter. An interesting use of this concept is the Touch-Tone* decoder shown in Fig. 11-30. The depression of a button on a Touch-Tone telephone keyboard creates two audio tones: each key selects a "row" tone and a "column" tone in a matrix configuration. The decoder must have circuits that will detect these tones, individually where two exist simultaneously. The outputs of all the decoder circuits, four row and three column tones, are fed to a logic decoder which yields a logic HI when two tones are detected for the particular number of that output.

*Registered trademark, Bell Telephone Laboratories.

398

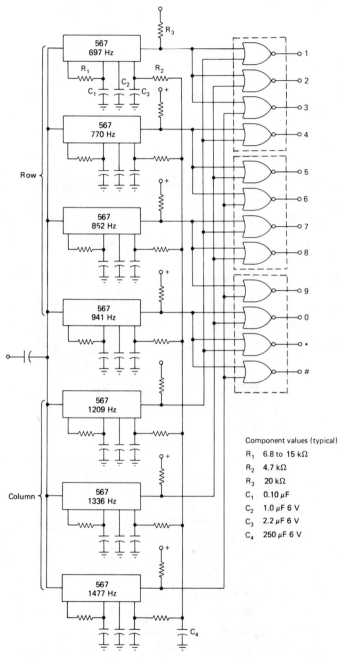

Fig. 11-30 PLL Touch-Tone decoder. (Courtesy of Signetics Corporation, Sunnyvale, Calif.)

A frequency-shift-keying (FSK) decoder can be implemented with one tone decoder. The input signal is two tones, close to each other in frequency. An example is the "Space" and "Mark" frequencies of a teletype signal; they are 1070 Hz and 1270 Hz, respectively. The tone decoder's free-running frequency is centered at 1170 Hz and the bandwidth is made sufficiently large to encompass both tones. The tone decoder will then lock onto either frequency, but in so doing the VCO input voltage shifts to move the VCO frequency to maintain lock. This shifting voltage is the FSK decoder output voltage. It is then passed through a comparator to "square it up" and used as the 0 and 1 outputs.

PROBLEMS

1. Find the output current for the two-transistor current mirror with $h_{fe} = 150$ and $I_{in} = 2.0$ mA.

2. Find the output-to-input current ratio for the three-transistor current mirror if $h_{fe} = 150$.

3. Find the value of R_{ABC} for an OTA with supply voltages of ± 15 V and a bias current of 200 μA.

4. What is the transconductance of the OTA in Problem 3?

5. What is the output impedance of the OTA in Problem 3?

6. Calculate the slew rate of the OTA in Problem 3.

7. The OTA of Problem 3 is connected to an output stage having an internal impedance of 3 MΩ. Calculate the voltage gain of the two stages.

8. The Norton amplifier shown in Fig. 11-3b has $R_I = 50$ kΩ and an inverting gain of -5.

   a. Calculate R_F.

   b. Calculate R_B.

9. Find R_I and R_B for a noninverting Norton amplifier with $R_f = 100$ kΩ and a gain of 10.

10. Can a current-differencing amplifier be used as a low-pass or band-reject filter having a frequency response down to dc with 0 V offset? Explain.

11. If a low-voltage op amp continuously draws 1 mA from a single D-size battery rated at 2 A-h, how many weeks will the battery last?

12. An LM 10 low-voltage op amp must be connected as an inverting amplifier with a gain of -2.5 and an offset of $+3.5$ V on all output signals.

    a. Draw the circuit diagram.

    b. What are the values of the other components in the circuit if $R_f = 200$ kΩ?

13. The circuit shown in Fig. 11-5 uses a D-cell with a 1.5 V terminal voltage and a 4 A-h capacity. The input signal is an 800 Hz sine wave with a peak voltage of 0.30 V.

    a. Will the output of the amplifier saturate at the peaks of the signal voltage?

    b. What is highest frequency that the circuit will pass at 0.3 V peak?

    c. What should be the expected lifetime in years of the circuit?

14. An ICL 7611 BMTY is battery-operated and connected similar to Problem 13. It will be used to amplify a sine-wave voltage, and it operates with a load current of 75 μA.

a. Determine the largest peak value of the input sine wave for a battery voltage of 2 V.

b. Find the maximum gain attainable in the circuit if a 1 kΩ input resistor is used.

15. Compare the gain–bandwidth, slew rate, and offset drift characteristics of TI's TL 089 chopper-stabilized op amp and Fairchild's μA 741 op amp.

16. How does the Intersil ICL 7600 commutating auto-zero op amp differ from the TI 089 chopper-stabilized op amp in number of required external components?

17. A cascode amplifier has $R_L = 10$ kΩ, $\beta = 80$, and $h_{ie} = 3$ kΩ.

a. Study your circuits book and explain why no Miller effect occurs.

b. Calculate the cascode amplifier voltage gain.

c. If $c_{be} = 10$ pF, $c_{bc} = 3$ pF, and $c_{ce} = 8$ pF, $R_b = 500$ kΩ, calculate the high-frequency rolloff point for a *common-emitter* amplifier. (*Note:* You must first calculate the common-emitter amplifier gain.)

d. Repeat part (c) for a *cascode* amplifier and compare the two high-frequency rolloff points.

18. An RGI amplifier has its dominate pole at 2 MHz and the second and third poles are at 30 MHz and 500 MHz, respectively. The amplifier has a gain of 50, $R = R_2 = 60$ kΩ, and $A_2 = N_2/D_2 = 7.4415 \times 10^{28}/(s + 2\pi \cdot 2 \text{ MHz})(S + 2\pi \cdot 30 \text{ MHz})(S + 2\pi \cdot 500$ MHz).

a. Explain the constituent parts of the coefficient part of A_2 (i.e., the 7.4415×10^{28}).

b. Find the value of R_1 for a closed loop gain of 5.

c. Find the bandwidth for part (b).

19. What common aspects do an inverting op amp amplifier and a phase-locked loop have?

20. State the functions of each of the four elements in a phase-locked loop.

21. What is the phase relationship between the input signal and the VCO output signal when the PLL is in lock?

22. Determine the polarity of the average (dc) output voltage in a phase detector for the following situations:

a. The switching signal lags the sine wave by 35°.

b. The switching signal leads the sine wave by 70°.

c. The switching signal leads the sine wave by 105°.

d. The switching signal lags the sine wave by 140°.

23. Determine the circuit components for a tone decoder using a 567 with a free-running frequency of 1500 Hz and a bandwidth of 100 Hz. The input signal voltage is 50 mV rms.

24. Determine the values of C_1 and C_2 using the curves in Fig. 11-28 for the tone decoder in Problem 23.

25. Determine the component values for a tone decoder using the 567 PLL when the input signal is 2 V rms. f_0 is 200 kHz and the bandwidth is 10.5% of f_0.

Closed Loop Input Impedance of Noninverting Amplifier:

Derivation of Equation (2-49)

$$v_{Id} = v_B - v_A$$
$$= -(v_o - v_{in}) \qquad \text{in Fig. 2-17a} \qquad (2\text{-}1)$$

Thus

$$v_o - v_{in} = v_{Id}$$

i_{in} (flowing into the v_{in} lead) splits two ways at the $+$ input:

1. Part of it flows up through the R_{in} resistor connected to the $(+)$ terminal.
2. Part of it flows through R_d to v_o via the $(-)$ terminal.

Thus

$$i_{in} = \frac{v_{in}}{R_{in}} + \frac{v_{id}}{R_d} = \frac{v_{in}}{R_{in}} + \frac{v_{ojA}}{R_d} \quad \text{as } v_{Id} = \frac{v_o}{A} \qquad (2\text{-}3b)$$

Divide both sides by v_{in}:

$$\frac{i_{in}}{v_{in}} = \frac{1}{v_{in}}\left(\frac{v_{in}}{R_{in}} + \frac{v_{o/A}}{R_d} \right) = \frac{1}{R_{in}} + \frac{v_o/v_{in}}{A R_d} = \frac{A R_d + (v_o/v_{in})R_{in}}{(A R_d)(R_{in})}$$

where $A = A_{VOL}$ and $v_o/v_{in} = A_{VCL}$. Also, $A_{VEX} = A_{VOL}/A_{VCL}$. Thus divide top and bottom of i_{in}/v_{in} by (v_o/v_{in}) to yield

$$\frac{i_{in}}{v_{in}} = \frac{(A_{VEX})R_d + R_{in}}{(A_{VEX})R_d R_{in}} \qquad Z_{in} = \frac{(A_{VEX}R_d)(R_{in})}{A_{VEX}R_d + R_{in}}$$

which can be recognized as the parallel combination of $A_{VEX}R_d = R_{dF}$ and R_{in}.

Appendix A2

Closed Loop Output Impedance:

Derivation of Equation (2-50)

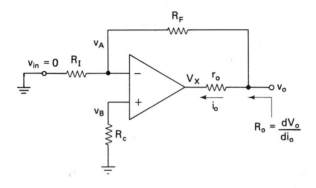

$$v_{id} = v_B - v_A \qquad v_B = 0 \qquad v_{id} = -v_A$$

$$v_x = Av_{id} = -Av_A$$

Since $v_{id} = 0$, $v_A = v_o\left(\dfrac{R_I}{R_F + R_I}\right)$, and $dv_A = dv_o\left(\dfrac{R_I}{R_F + R_I}\right)$,

$$dv_x = -Adv_A$$

$$di_o = \frac{dv_o - dv_x}{r_o} \qquad r_o di_o = dv_o - dv_x$$

$$r_o di_o = dv_o - (-Adv_A) = dv_o + A\left[dv_o\left(\frac{R_I}{R_F + R_I}\right)\right]$$

$$r_o di_o = dv_o\left(1 + \frac{AR_I}{R_F + R_I}\right)$$

$$R_o = \frac{dv_o}{di_o} = \frac{r_o}{1 + \dfrac{AR_I}{R_F + R_I}} = \frac{r_o}{1 + \dfrac{A}{\left(\dfrac{R_F}{R_I} + 1\right)}} = \frac{r_o}{1 + A_{\text{VEX}}}$$

This derivation is applicable for both inverting and noninverting amplifiers.

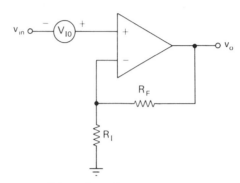

For the *noninverting* amplifier, the offset voltage is immediately apparent from the following example:

$$A_V = \frac{R_F + R_I}{R_I}$$

$$v_{out} = (v_{in} + V_{IO})\left(\frac{R_F + R_I}{R_I}\right)$$

$$v_{out} = \left(\frac{R_F + R_I}{R_I}\right)v_{in} + \left(\frac{R_F + R_I}{R_I}\right)V_{IO}$$

Thus it becomes apparent that an offset voltage is multiplied by the voltage follower gain at the same time that the input voltage is multiplied by that gain.

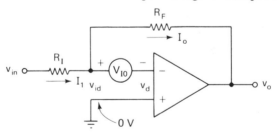

For the *inverting* amplifier, it is not quite as apparent, but the following derivation will serve to verify the equation and show that it is the same as for the noninverting case.

$$v_{Id} = v_d + V_{IO}$$

$$v_d = \frac{v_o}{-A} = -\frac{v_o}{A}$$

$$v_{Id} = -\frac{v_o}{A} + V_{IO}$$

$$I_1 = I_0 = \frac{v_{in} - v_{Id}}{R_I} = \frac{v_{Id} - v_o}{R_F} \qquad \text{Eq. (2.7a) where } v_{Id} = v_A$$

$$v_{in}R_F - v_{Id}R_F = v_{Id}R_I - v_oR_I$$

$$v_{in}R_F - \left(-\frac{v_o}{A} + V_{IO}\right)R_F = \left(-\frac{v_o}{A} + V_{IO}\right)R_I - v_oR_I$$

$$Av_{in}R_F + v_oR_F - AV_{IO}R_F = -v_oR_I + AV_{IO}R_I - Av_oR_I$$

$$v_o(-R_I - R_F - AR_I) = v_{in}(AR_F) - V_{IO}(AR_F + AR_I)$$

$$v_o = \frac{AR_F}{[R_I(1 + A) + R_F]}(v_{in}) - \frac{-A(R_F + R_I)}{[R_I(1 + A) + R_F]}(V_{IO}) \quad \text{when } A \to \infty$$

$$v_o = \left(-\frac{R_F}{R_I}\right)v_{in} + \left(\frac{R_F + R_I}{R_I}\right)V_{IO}$$

Thus it is shown that even though the gain equation from input to output changes for the inverting and noninverting amplifier configurations, the gain equations for both types are identical for the offset voltage generator.

Appendix A4

Output Offset Due to Bias Current:

Derivation of Equation (3-5)

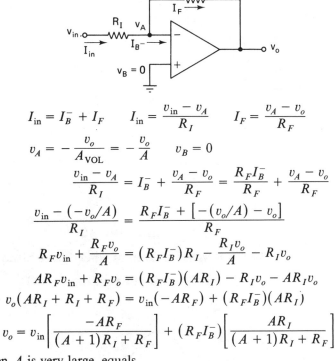

$$I_{in} = I_B^- + I_F \qquad I_{in} = \frac{v_{in} - v_A}{R_I} \qquad I_F = \frac{v_A - v_o}{R_F}$$

$$v_A = -\frac{v_o}{A_{\text{VOL}}} = -\frac{v_o}{A} \qquad v_B = 0$$

$$\frac{v_{in} - v_A}{R_I} = I_B^- + \frac{v_A - v_o}{R_F} = \frac{R_F I_B^-}{R_F} + \frac{v_A - v_o}{R_F}$$

$$\frac{v_{in} - (-v_o/A)}{R_I} = \frac{R_F I_B^- + [-(v_o/A) - v_o]}{R_F}$$

$$R_F v_{in} + \frac{R_F v_o}{A} = (R_F I_B^-)R_I - \frac{R_I v_o}{A} - R_I v_o$$

$$AR_F v_{in} + R_F v_o = (R_F I_B^-)(AR_I) - R_I v_o - AR_I v_o$$

$$v_o(AR_I + R_I + R_F) = v_{in}(-AR_F) + (R_F I_B^-)(AR_I)$$

$$v_o = v_{in}\left[\frac{-AR_F}{(A + 1)R_I + R_F}\right] + (R_F I_B^-)\left[\frac{AR_I}{(A + 1)R_I + R_F}\right]$$

which, when A is very large, equals

$$v_o = v_{in}\left(-\frac{R_F}{R_I}\right) + R_F I_B^-$$

Appendix A5

Derivation of Transfer Function for Low-Pass *RC* Filter Section:

Derivation of Equation (5-1)

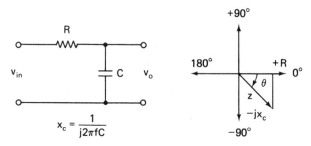

$$x_c = \frac{1}{j2\pi fC}$$

By the voltage-divider rule,

$$v_o = \frac{1/j2\pi fC}{R + 1/j2\pi fC}v_{in} \qquad \frac{v_o}{v_{in}} = \frac{1/j2\pi fC}{R + 1/j2\pi fC}$$

Divide the top and bottom by *R*:

$$\frac{v_o}{v_{in}} = \frac{1/j2\pi fRC}{1 + 1/j2\pi fRC}$$

Multiply the top and bottom by *jf*:

$$\frac{v_o}{v_{in}} = \frac{1/2\pi RC}{1/2\pi RC + jf}$$

$$f \ll \frac{1}{2\pi RC} \qquad \frac{v_o}{v_{in}} = \frac{1/2\pi RC}{1/2\pi RC} = 1\underline{/0^\circ}$$

$$f = \frac{1}{2\pi RC} \qquad \frac{v_o}{v_{in}} = \frac{f}{f + jf} = \frac{1}{1 + j1} = \frac{1}{\sqrt{2}\underline{/45^\circ}} = 0.707\underline{/-45^\circ}$$

$$f \gg \frac{1}{2\pi RC} \qquad \frac{v_o}{v_{in}} = \frac{1/2\pi RC}{jf} = \frac{1/2\pi RC}{f\underline{/+90^\circ}} = \left(\frac{1}{2\pi RC}\right)\left(\frac{1}{f}\right)\underline{/-90^\circ}$$

Appendix A6

Maximum Slew Rate of Sine-Wave Output:

Derivation of Equation (5-17)

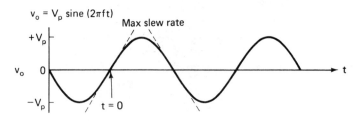

Take the derivation of v_o with respect to time (t) and evaluate the derivative at the zero crossing $(2\pi ft = 0)$.

$$\frac{d}{dt}\left[V_p\sin(2\pi ft)\right] = 2\pi f V_p\cos(2\pi ft)$$

which, evaluated at $t = 0$, is

$$\text{slope (slew rate) at zero crossing} = 2\pi f V_p \quad \text{V/s}$$

The units must be volts/microsecond, so multiply by 10^{-6} and change the units to volts/μs; thus the slew rate at the zero crossing becomes

$$\max\sin(\text{SR}) = 2\pi(10^{-6})(V_p)(f_{\text{Hz}}) \quad \text{V/}\mu\text{s}$$

When the op amp output is slew-rate-limited because the op amp slew rate is slower than the sine-wave zero-crossing slew rate, the resulting waveform is as follows:

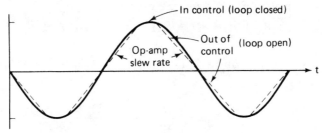

Distorted sine wave due to slew rate limiting

Appendix A7

Input Impedance to Difference Amplifier:

Section 6.2.1

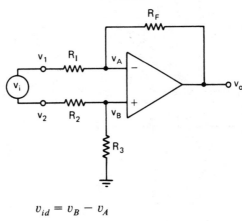

$$v_{id} = v_B - v_A$$

$$v_{id} = -\frac{v_O}{A} \quad \text{if } v_B = 0 \text{ V}$$

Deductive reasoning

1. Since $v_A \simeq v_B$ by **Rule 1**, the potential difference between v_A and v_B is nearly zero.

2. R_I and R_2 are then hypothetically connected together at the summing junction (v_A) and the noninverting input (v_B).

3. The loop impedance between v_1 and v_2 is $R_I + R_2$.

Mathematical derivation

1. $\Delta v_O = -\Delta i (R_F + R_I + R_2 + R_3) + \Delta v_i$

2. $\Delta v_{Id} = -\Delta i (R_I + R_2) + \Delta v_i = -\dfrac{\Delta v_O}{A}$

3. $\Delta v_O = A \Delta i (R_I + R_2) - A \Delta v_i$
   Equating Eqs. (1) and (3), we get

4. $-\Delta i (R_F + R_I + R_2 + R_3) + \Delta v_i = A \Delta i (R_I + R_2) - A \Delta v_i$
   or

$$\Delta v_i (A + 1) = \Delta i (A + 1)(R_I + R_2) + \Delta i (R_F + R_3)$$

Then

$$R_{\text{diff}} = \frac{\Delta v_i}{\Delta i} = \frac{(A + 1)(R_I + R_2)}{A + 1} + \frac{R_F + R_3}{A + 1}$$

and

$$R_{\text{diff}} = R_I + R_2 + \frac{R_F + R_3}{A + 1}$$

When A is large,

$$R_{\text{diff}} \simeq R_I + R_2$$

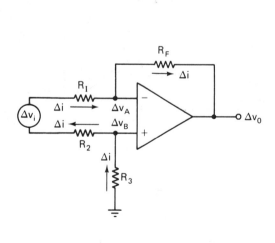

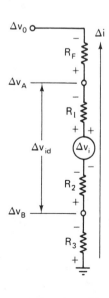

Appendix A8

Two Op Amp Differential-Mode Amplifier:

Derivation of Equation (6-1)

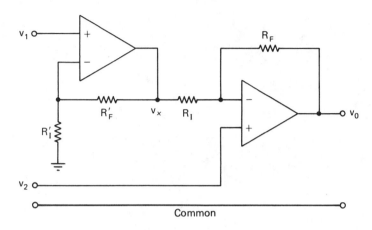

1. $v_x = \left(\dfrac{R'_F + R'_I}{R'_I}\right)v_1$ by **Rule 3**

2. $v_O = \left(\dfrac{R_F + R_I}{R_I}\right)v_2 + \left(-\dfrac{R_F}{R_I}\right)v_x$ by **Rules 2, 3, and 5**

Inserting Eq. (1) into (2), we get

3. $v_O = \left(\dfrac{R_F}{R_I} + 1\right)v_2 + \left(-\dfrac{R_F}{R_I}\right)\left(\dfrac{R'_F + R'_I}{R'_I}\right)v_1$

   or

$$v_O = \left(\dfrac{R_F}{R_I} + 1\right)v_2 - \left(\dfrac{R_F}{R_I} + \dfrac{R'_F R_F}{R_I R'_I}\right)v_1$$

If $R'_F = R_I$ and $R'_I = R_F$, then

$$v_O = \left(\dfrac{R_F}{R_I} + 1\right)(v_2 - v_1) \tag{6-1}$$

Appendix A9

Variable-Gain Differential-Mode Amplifier:

Derivation of Equation (6-2)

Using **Rules 2, 3, and 5**, we obtain

$$v_A = \left(\dfrac{R_A + R_B}{R_B}\right)v'_1 + \left(-\dfrac{R_A}{R_B}\right)v'_2 = v'_1 + \left(\dfrac{R_A}{R_B}\right)(v'_1 - v'_2)$$

$$v_C = \left(\dfrac{R_C + R_B}{R_B}\right)v'_2 + \left(-\dfrac{R_C}{R_B}\right)v'_1 = v'_2 + \left(\dfrac{R_C}{R_B}\right)(v'_2 - v'_1)$$

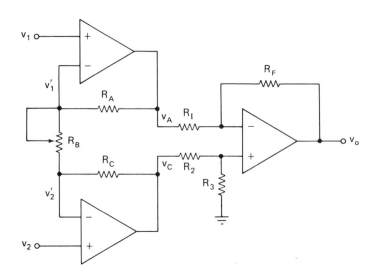

If $R_F/R_I = R_3/R_2$, then

$$v_o = \left(-\frac{R_F}{R_I}\right)(v_C - v_A) \qquad \text{by } \textbf{\textit{Rules 4 and 5}}$$

$$= \left(-\frac{R_F}{R_I}\right)\left[v_2' + \frac{R_C}{R_B}(v_2' - v_1') - v_1' + \left(\frac{R_A}{R_B}\right)(v_2' - v_1')\right]$$

$$= \left(-\frac{R_F}{R_I}\right)\left[\frac{R_B v_2' - R_B v_1' + R_C v_2' - R_C v_1' + R_A v_2' - R_A v_1'}{R_B}\right]$$

$$= \left(-\frac{R_F}{R_I}\right)\left[\left(\frac{R_B + R_C + R_A}{R_B}\right)v_2' - \left(\frac{R_B + R_C + R_A}{R_B}\right)v_1'\right]$$

$$= \left(-\frac{R_F}{R_I}\right)\left[\left(1 + \frac{R_A + R_C}{R_B}\right)(v_2' - v_1')\right]$$

and if $R_A = R_C$,

$$v_o = \left(1 + \frac{2R_A}{R_B}\right)\left(\frac{R_F}{R_I}\right)(v_1' - v_2')$$

But by **_Rule 1_**,

$$v_1 = v_1' \qquad \text{and} \qquad v_2 = v_2'$$

Thus

$$v_o = \left(1 + \frac{2R_A}{R_B}\right)\left(\frac{R_F}{R_I}\right)(v_1 - v_2) \qquad\qquad (6\text{-}2)$$

Appendix A10

Derivation of Transfer Admittance for Resistor "T" Network:

Derivation of Equation (6-4)

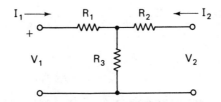

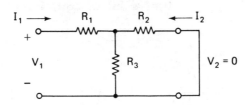

$$Y_{21} = \frac{i_2}{v_1}\bigg|_{v_2=0}$$

and

$$i_2 = i_{R_2} = \frac{v_{R2}}{R_2} = \frac{v_{in}\left(\dfrac{R_2\|R_3}{R_1 + R_2\|R_3}\right)}{R_2} = \frac{-v_{in}}{R_2}\left(\dfrac{\dfrac{R_2 R_3}{R_2 + R_3}}{R_1 + \dfrac{R_2 R_3}{R_2 + R_3}}\right)$$

$$i_2 = -\frac{v_{in}}{R_2}\left(\dfrac{\dfrac{R_2 R_3}{\cancel{R_2 + R_3}}}{\dfrac{R_1(R_2 + R_3) + R_2 R_3}{\cancel{R_2 + R_3}}}\right) = -\frac{v_{in}}{R_2}\left(\dfrac{R_2 R_3}{R_1 R_2 + R_1 R_3 + R_2 R_3}\right)$$

$$i_2 = -v_{in}\left(\dfrac{R_3}{R_1 R_2 + R_1 R_3 + R_2 R_3}\right)$$

So

$$\frac{i_2}{+v_{in}} = -\frac{R_3}{R_1 R_2 + R_1 R_3 + R_2 R_3} = Y_{21}$$

Therefore, if $Y_{21} = Y_{12}$ for passive networks,

$$Y_{12} = -\frac{R_3}{R_1 R_2 + R_1 R_3 + R_2 R_3} \qquad \text{transfer admittance}$$

and

$$\frac{1}{Y_{12}} = -\frac{R_1R_2 + R_1R_3 + R_2R_3}{R_3}$$

and

$$-\frac{1}{Y_{12}} = +\frac{R_1R_2 + R_1R_3 + R_2R_3}{R_3} \qquad \text{negative reciprocal of transfer admittance}$$

Appendix A11

Output Impedance of Constant Current Amplifier:

Derivation of Equation (6-8)

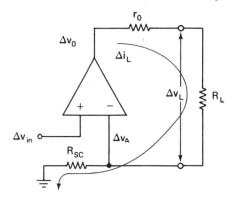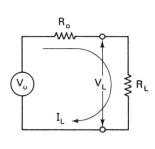

1. $v_o = i_L R_O + v_L$
2. $i_L R_O = v_o - v_L$
3. $R_o = \dfrac{v_o - v_L}{i_L}$
4. $\Delta v_o = i_L(r_o + R_L + R_{SC})$
5. $\Delta v_o = i_L r_o + \Delta i_L R_L + \Delta i_L R_{SC} = i_L r_o + v_L + v_A$
6. $v_o = (v_{in} - v_A)A$
7. $\Delta v_o = (v_{in} - \Delta v_A)A$
8. $\Delta v_{in} - A v_A = i_L r_o + \Delta v_L + \Delta v_A$
9. $\Delta v_{in} - (A + 1)\Delta v_A = A_L v_o + \Delta v_L$
10. $\Delta v_{in} - (A + 1)i_L R_{SC} = i_L r_o + v_L$
11. $\Delta v_{in} - v_L = i_L r_o + (A + 1)i_L R_{SC}$
    If i_L is constant and $\Delta V_A = 0$, then
12. $\dfrac{A v_{in} - v_L}{i_L} = r_o + (A + 1)R_{SC}$

    By comparison with (3),
    $R_o = r_o + (A + 1)R_{SC}$

Appendix A12

Grounded-Load Constant Current Amplifier:

Derivation of Equation (6-13)

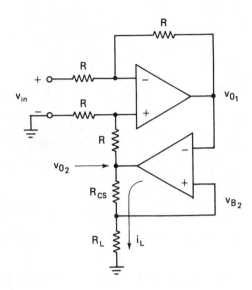

1. Ground reference the–circuit input to permit calculation.
2.

$$v_{o1} = \left(-\frac{R}{R}\right)v_{in} \;+\; \left(\frac{R}{2R}\right)\left(\frac{2R}{R}\right)v_{o2}$$

$$\uparrow$$

Rule 2   Rule 5   Rule 4

$v_{o1} = -v_{in} + v_{o2}$

$v_{in} = v_{o2} - v_{o1}$

3. $i_L = \dfrac{v_{o2} - v_{B2}}{R_{CS}}$

4. $v_{B2} = v_{o1}$ *Rule 1*

5. $i_L = \dfrac{v_{o2} - v_{o1}}{R_{CS}}$

   But $v_{o2} - v_{o1} = v_{in}$ from step (2); thus

6. $i_L = \dfrac{v_{in}}{R_{CS}}$

Appendix A13

AC Input to Integrator:

Derivation of Equation (6-23)

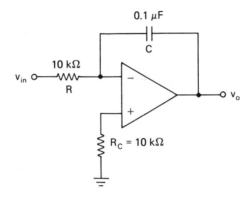

What is the output of an integrator when a sine wave is inserted at the input?

$$v_o = -\frac{1}{RC}\int_0^t v_{in}\, dt \qquad v_{in} = A\sin\omega t \qquad v_o = -\frac{A}{RC}\int_0^t \sin\omega t\, dt$$

To integrate this function, the integrand must be a perfect differential, or

$$\frac{d}{dt}(-\cos\omega t) = -\left[-\sin\omega t\,\frac{d(\omega t)}{dt}\right] = \omega\sin\omega t$$

Thus the integrand must be $\omega\sin\omega t$, and in order to balance out the ω in front of the sine function, a $1/\omega$ term must be placed in front of the integral sign:

$$v_o = -\frac{A}{RC}\left(\frac{1}{\omega}\right)\int\omega\sin\omega t\, dt = -\frac{A}{RC\omega}(-\cos\omega t) + v_{C(0)} \leftarrow \text{constant term}$$

$$= +\frac{A}{RC\omega}\cos\omega t + v_{C(0)=0}$$

But a trigonometric identity states that

$$\cos\omega t = +\sin\left(\frac{\pi}{2} - \omega t\right)$$

$$\sin\theta = -\sin(-\theta)$$

so

$$v_o = -\frac{A}{RC\omega}\sin\left(\omega t - \frac{\pi}{2}\right)$$

Example:

$$A = 4$$

$$f = 1000 \text{ Hz}$$

$$\omega = 2\pi f = 2\pi(1000) = 6280$$

$$v_{in} = 4\sin(2\pi 1000 t) = 4\sin(6280 t)$$

$$R_C = 10^4 \times 10^{-7} = 10^{-3}$$

$$v_o = -\frac{4}{(10^{-3})6280}\sin\left(6280 t - \frac{\pi}{2}\right)$$

$$v_o = -0.637\sin\left(6280 t - \frac{\pi}{2}\right)$$

When $t = 0$, $\theta = -\pi/2 = -90°$.

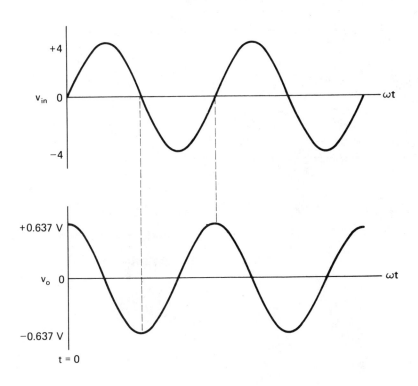

Derivation of Transfer Function For High-Pass *RC* Filter Section:

Derivation of Equation 6-24

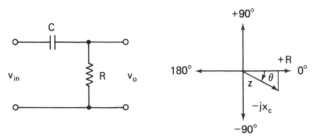

By the voltage-division rule,

$$v_o = \frac{R}{R + 1/j2\pi fC} v_{in} \qquad \frac{v_o}{v_{in}} = \frac{R}{R + 1/j2\pi fC}$$

Divide the top and bottom by R:

$$\frac{v_o}{v_{in}} = \frac{1}{1 + 1/j2\pi fRC}$$

Multiply the top and bottom by jf:

$$\frac{v_o}{v_{in}} = \frac{jf}{(1/2\pi RC) + jf} = \frac{f}{(1/2\pi RC) + jf} \underline{/+90°}$$

$$f \ll \frac{1}{2\pi RC} \qquad \frac{v_o}{v_{in}} = \frac{f}{1/2\pi RC} \underline{/+90°} = 2\pi fRC \underline{/90°} = (\omega RC) \underline{/+90°}$$

$$f = \frac{1}{2\pi RC} \qquad \frac{v_o}{v_{in}} = \frac{f}{f + jf} \underline{/+90°} = \frac{1}{\sqrt{2} \underline{/+45°}} \underline{/+90°} = 0.707 \underline{/+45°}$$

$$f \gg \frac{1}{2\pi RC} \qquad \frac{v_o}{v_{in}} = \frac{jf}{jf} = 1 \underline{/0°}$$

Appendix A15

Differentiator Transfer Function:

Derivation of Equation (6-27)

$$Z_I = R_I + \frac{1}{j\omega C_D} = R_I - \frac{j}{\omega C_D}$$

$$Z_F = \frac{R_D(-j/\omega C_I)}{R_D - j/\omega C_I} = \frac{R_D(-jX_C)}{R_D + (-jX_C)}$$

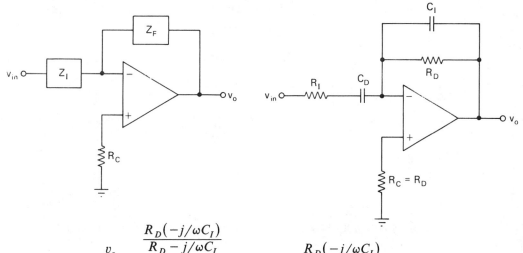

$$\frac{v_o}{v_{in}} = \frac{\dfrac{R_D(-j/\omega C_I)}{R_D - j/\omega C_I}}{R_I - j/\omega C_D} = -\frac{R_D(-j/\omega C_I)}{(R_I - j/\omega C_D)(R_D - j/\omega C_I)}$$

$$= -\frac{-j(R_D/\omega C_I)}{R_D R_I - jR_D/\omega C_D - jR_I/\omega C_I - 1/\omega^2 C_I C_D}$$

$$= -\frac{+jR_D(\omega C_D)}{1 - R_D R_I C_I C_D \omega^2 + j\omega R_D C_I + jR_I \omega C_D}$$

$$= -j\omega R_D C_D \left[\frac{1}{1 - (\omega R_I C_D)(\omega R_D C_I) + j(\omega R_D C_I + \omega R_I C_D)} \right]$$

If $R_D C_I = R_I C_D$, then

$$\frac{v_o}{v_{in}} = -j\omega R_D C_D \left[\frac{1}{1 - (\omega R_D C_I)(\omega R_I C_D) + j(\omega R_D C_I + \omega R_I C_D)} \right]$$

and

$$\frac{v_o}{v_{in}} = -j\omega R_D C_D \left[\frac{1}{1 - (\omega R_D C_I)^2 + 2j\omega R_D C_I} \right]$$

$$= -j\omega R_D C_D \left[\frac{1}{1 + 2j\omega R_D C_I - (\omega R_D C_I)^2} \right]$$

but the terms in the brackets form a perfect square, thus

$$\frac{v_o}{v_{in}} = -j\omega R_D C_D \left[\frac{1}{(1 + j\omega R_D C_I)^2} \right]$$

$$\underset{\substack{\text{basic} \\ \text{diff.}}}{} \qquad \underset{\text{modifier}}{}$$

Since the modifier term reduced to a perfect square only when $R_D C_I = R_I C_D$, v_o/v_{in} reduces to

$$\frac{v_o}{v_{in}} = -j\omega R_D C_D$$

only if $j\omega R_D C_I = j\omega R_I C_D$ and both are much less than unity ($j\omega R_D C_I \ll 1$).

Appendix A16

AC Input to Differentiator:

Derivation of Equation (6-30)

The output of a differentiator, with a sine-wave input, is

$$v_o = -RC\frac{d}{dt}(v_{in}) \qquad v_{in} = V_p\sin \omega t \qquad v_o = -RCV_p\frac{d}{dt}(\sin \omega t)$$

and after differentiation,

$$v_o = -RC\omega V_p\cos \omega t$$

But a trigonometric identity states that

$$\cos \theta = \sin(90° - \theta) \qquad \text{and} \qquad \sin(90° - \theta) = -\sin(\theta - 90°)$$

Thus

$$\cos \omega t = -\sin\left[\omega t - \frac{\pi}{2}\right]$$

Thus

$$v_O = \omega R_D C_D V_p\sin\left(\omega t - \frac{\pi}{2}\right) \qquad (6\text{-}30)$$

An example of the use of Eq. (6-37): Let $V_p = 0.6$ V, $f = 1$ kHz, $R = 10$ kΩ, and $C = 0.1$ μF. Then

$$v_{in} = 0.6\sin(2\pi \cdot 1000t)$$

$$v_o = (0.6)(2\pi)(1000)(10^4)(10^{-7})\sin\left(2\pi \cdot 1000t - \frac{\pi}{2}\right)$$

$$v_o = 3.77\sin\left(2\pi \cdot 1000t - \frac{\pi}{2}\right)$$

Because of the minus (−) sign in front of the argument (3.77), the sine wave begins going *downward*. Thus

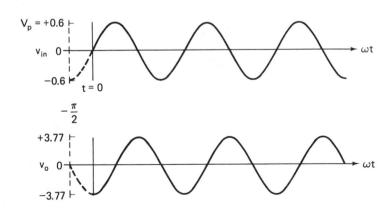

Appendix A17

Two-Op Amp Precision Rectifier

Explanation of Gain in Figure 7-2a

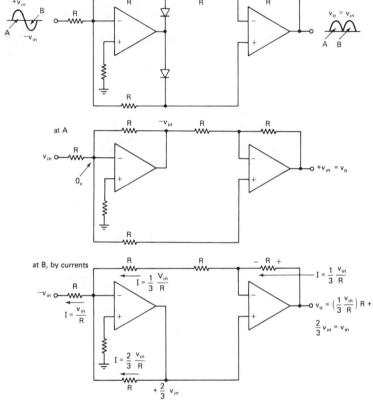

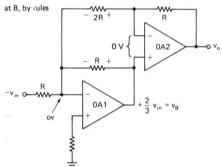

1. By Rule 1, voltage across ± of OA2 is 0 V.

2. Thus, 2R is in parallel with R.

3. R_F of OA1 is $\frac{2(1)}{3}R = \frac{2}{3}R$.

4. Gain of OA1 is $\left(-\frac{2}{3}\right)$ by Rule 2.

5. v_o of OA1 is $(-v_{in})\left(-\frac{2}{3}\right) = +\frac{2}{3}v_{in}$

6. Input to 2R is zero V by Rule 1.

7. By Rule 3, noninverting gain of OA2 is $\left(\frac{R + 2R}{2R}\right) = \frac{3}{2}$ and $v_o = \left(\frac{2}{3}v_{in}\right)\left(\frac{3}{2}\right) = v_{in}$

Four Types of Feedback

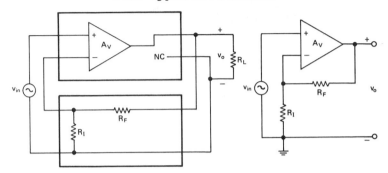

I. Voltage amplifier (voltage-series, series-shunt)

$$A_{OL} = A_V$$

$$\beta = \frac{R_I}{R_I + R_F}$$

$$\text{ideal gain} = \frac{R_F + R_I}{R_I}$$

$$\frac{v_o}{v_{in}} = \frac{A_V}{1 + A_V\left(\dfrac{R_I}{R_I + R_F}\right)}$$

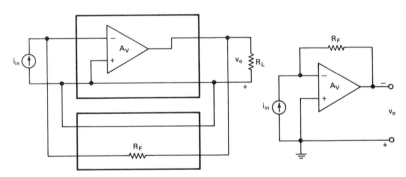

II. Transresistance amplifier (voltage-shunt, shunt-shunt)

$$A_{OL} = -A_V R_F$$

$$\beta = -\frac{1}{R_F}$$

$$\text{ideal gain} = -R_F$$

$$\frac{v_o}{v_{in}} = \frac{(-A_V R_F)}{1 + (-A_V R_F)\left(\dfrac{-1}{R_F}\right)}$$

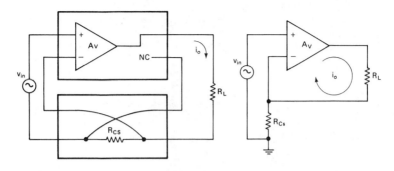

III. Transconductance amplifier (current-series, series-series)

$$A_{OL} = \frac{A_V}{R_{CS}}$$

$$\beta = R_{CS}$$

$$\text{ideal gain} = \frac{1}{R_{CS}}$$

$$\frac{i_o}{v_{in}} = \frac{A_V/R_{CS}}{1 + (A_V/R_{CS})R_{CS}}$$

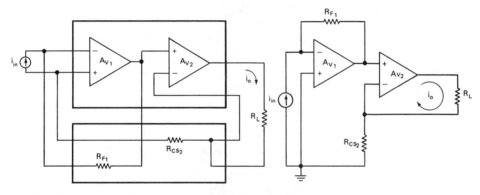

IV. Current amplifier (current-shunt, shunt-series)

$$A_{OL} = -A_{V_1}A_{V_2}\frac{R_{F_1}}{R_{CS_2}}$$

$$\beta = -\frac{R_{CS_2}}{R_{F_1}}$$

$$\text{ideal gain} = -\frac{R_{F_1}}{R_{CS_2}}$$

$$\frac{i_o}{i_{in}} = \frac{\left(-A_{V_1}A_{V_2}\dfrac{R_{F_1}}{R_{CS_2}}\right)}{1 + \left(-A_{V_1}A_{V_2}\dfrac{R_{F_1}}{R_{CS_2}}\right)\left(-\dfrac{R_{CS_2}}{R_{F_1}}\right)}$$

Appendix A19

Input Impedance to Transresistance Amplifier

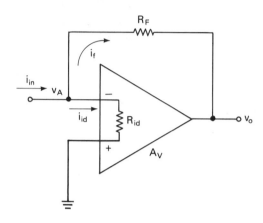

$$R_{\text{in}} = \frac{v_A}{i_{\text{in}}}$$

$$i_{\text{in}} = i_f + i_{id} = \frac{v_A - v_o}{R_F} + \frac{v_A}{R_{id}}$$

$$v_o = -A_V v_R \qquad i_{\text{in}} = v_A \left(\frac{1}{R_F} + \frac{A_V}{R_F} + \frac{1}{R_{id}} \right)$$

$$\frac{i_{\text{in}}}{v_A} = \frac{(1 + A_V)R_{id} + R_F}{R_F R_{id}}$$

$$R_{\text{in}} = \frac{v_A}{i_{\text{in}}} = \frac{R_F R_{id}}{(1 + A_V)R_{id} + R_F} = \frac{1}{\dfrac{1 + A_V}{R_F} + \dfrac{1}{R_{id}}}$$

$$R_{\text{in}} = \frac{1}{\dfrac{1}{R_F/(1 + A_V)} + \dfrac{1}{R_{id}}} = \frac{R_F}{(1 + A_V)} \left\| R_{id} \right.$$

When $R_{id} \to \infty$, $R_{\text{in}} \to \dfrac{R_F}{1 + A_V}$, which corresponds to Miller's Theorem.

Appendix A20

Transresistance Amplifier Gain and Feedback Ratio

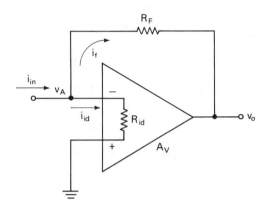

$$v_A = -\frac{v_o}{A_V}$$

$$i_{in} = i_f + i_{id} = \frac{v_A - v_o}{R_F} + \frac{v_A}{R_{id}} = \frac{-\dfrac{v_o}{A_V} - v_o}{R_F} + \frac{-\dfrac{v_o}{A_V}}{R_{id}}$$

$$i_{in} = -v_o \left[\frac{1 + A_V}{A_V R_F} + \frac{1}{A_V R_{id}} \right] = -v_o \left[\frac{(1 + A_V) R_{id} + R_F}{A_V R_F R_{id}} \right]$$

$$\frac{v_o}{i_{in}} = -\left[\frac{A_V R_F R_{id}}{(1 + A_V) R_{id} + R_F} \right] = -\left(\frac{A_V R_F}{1 + A_V + \dfrac{R_F}{R_{id}}} \right)$$

$$\lim_{R_{id} \to \infty} \frac{v_o}{i_{in}} \to \frac{-A_V R_F}{1 + A_V} \qquad \text{now arrange in } \frac{A_{OL}}{1 + A_{OL}\beta} \text{ form}$$

$$\frac{v_o}{i_{in}} = \frac{-A_V R_F}{1 + (-A_V R_F)\left(\dfrac{-1}{R_F} \right)} \qquad A_{OL} = -A_V R_F \qquad \beta = \frac{-1}{R_F}$$

$$\text{ideal gain} = \frac{1}{\beta} = -R_F$$

Appendix A21

Transconductance Amplifier Gain and Feedback Ratio

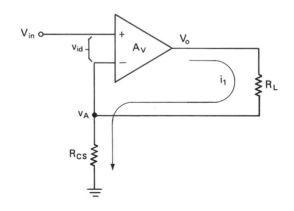

$$v_o = A_V v_{id} = A_V(v_{in} - v_A) = A_V v_{in} - A_V v_A$$

$$v_A = v_o\left(\frac{R_{CS}}{R_{CS} + R_L}\right) \qquad v_o = \frac{R_{CS} + R_L}{R_{CS}}$$

$$v_A = i_L R_{CS}$$

$$i_L R_{CS} = v_o\left(\frac{R_{CS}}{R_{CS} + R_L}\right) = (A_V v_{in} - A_V v_A)\left(\frac{R_{CS}}{R_{CS} + R_L}\right)$$

$$i_L = \frac{A_V v_{in}}{R_{CS} + R_L} - \frac{A_V i_L R_{CS}}{R_{CS} + R_L}$$

$$i_L\left(1 + \frac{A_V R_{CS}}{R_{CS} + R_L}\right) = v_{in}\left(\frac{A_V}{R_{CS} + R_L}\right)$$

$$\frac{i_L}{v_{in}} = \frac{A_V}{R_{CS} + R_L + A_V R_{CS}} = \frac{A_V/R_{CS}}{1 + A_V + \dfrac{R_L}{R_{CS}}}$$

$$\lim_{R_L \to 0} \frac{i_L}{v_{in}} \to \frac{A_V/R_{CS}}{1 + A_V} \qquad \text{now arrange in } \frac{A_{OL}}{1 + A_{OL}\beta} \text{ form}$$

$$\frac{i_L}{v_{in}} = \frac{A_V/R_{CS}}{1 + (A_V/R_{CS})(R_{CS})} \qquad A_{OL} \neq \frac{A_V}{R_{CS}} \qquad \beta = R_{CS}$$

$$\text{ideal gain} = \frac{1}{\beta} = \frac{1}{R_{CS}}$$

Appendix A22

Closed Loop Gain for Noninverting RGI Amplifier

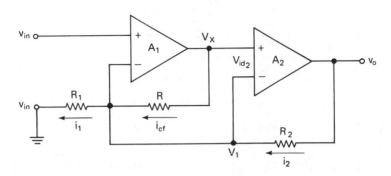

$$G = \frac{R_2 + R_1}{R_1} = \frac{R_2}{R_1} + 1$$

Because of large A_1, $V_1 \cong V_{ni}$, $V_{in} = 0$ (grounded)

$$i_{cf} = i_1 - i_2 \qquad i_1 = \frac{V_1 - 0}{R_1} \qquad i_2 = \frac{V_o - V_1}{R_2} \qquad i_{cf} = \frac{V_x - V_1}{R}$$

$$i_{cf} R = V_{id_2} = \frac{V_o}{A_2} \qquad i_{cf} = \frac{V_o}{A_2 R}$$

$$\frac{V_o}{A_2 R} = \frac{V_1}{R_1} - \frac{V_o - V_1}{R_2} \qquad V_o\left(\frac{1}{A_2 R} + \frac{1}{R_2}\right) = V_1\left(\frac{1}{R_1} + \frac{1}{R_2}\right)$$

$$V_o\left(\frac{R_2 + A_2 R}{A_2 R R_2}\right) = V_1\left(\frac{R_1 + R_2}{R R_2}\right) = V_{ni}\left(\frac{R_1 + R_2}{R R_2}\right)$$

$$\frac{V_o}{V_{ni}} = \frac{A_2 R (R_2 + R_1)}{R_1 R_2 + A_2 R R_1} \text{ factor } \left(\frac{R_2 + R_1}{R_1}\right) = G$$

$$\frac{V_o}{V_{ni}} = G\left(\frac{A_2 R}{R_2 + A_2 R}\right) = G\left(\frac{A_2}{\frac{R_2}{R} + A_2}\right)$$

$$\frac{V_o}{V_{ni}} = (G)\left(\frac{R}{R_2}\right)\left[\frac{A_2}{1 + A_2\left(\frac{R}{R_2}\right)}\right] = K\left(\frac{A}{1 + A\beta}\right)$$

$$K = G\frac{R}{R_2} \qquad A = A_2 = \frac{N_2}{D_2} \qquad \beta = \frac{R}{R_2} \qquad G = \frac{R_2}{R_1} + 1$$

$$\frac{V_o}{V_{ni}} = (G)\left(\frac{R}{R_2}\right)\left[\frac{\frac{N_2}{D_2}}{1 + \left(\frac{N_2}{D_2}\right)\left(\frac{R}{R_2}\right)}\right]$$

Appendix A23

Closed Loop Gain for Inverting RGI Amplifier

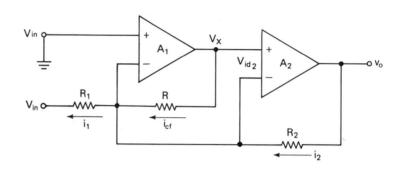

$$G = -\frac{R_2}{R_1}$$

Because of large A_1, $V_1 = V_{ni} = 0$

$$i_{cf} = i_1 - i_2 \qquad i_1 = \frac{V_1 - V_{in}}{R_1} = \frac{-V_{in}}{R_1} \qquad i_2 = \frac{V_o - V_1}{R_2} = \frac{V_o}{R_2}$$

$$i_{cf}R = v_{id_2} = \frac{V_o}{A_2} \qquad i_{cf} = \frac{V_o}{A_2 R}$$

$$\frac{V_o}{A_2 R} = \frac{-V_{in}}{R_1} - \frac{V_o}{R_2} \qquad V_o\left(\frac{1}{A_2 R} + \frac{1}{R_2}\right) = -V_{in}\left(\frac{1}{R_1}\right)$$

$$V_o\left(\frac{R_2 + A_2 R}{A_2 R R_2}\right) = -\frac{V_{in}}{R_1} \qquad \frac{V_o}{V_{in}} = -\frac{A_2 R R_2}{R_1 R_2 + A_2 R R_1}$$

Factor out $\dfrac{R_2}{R_1}$:

$$\frac{V_o}{V_{in}} = \left(-\frac{R_2}{R_1}\right)\left(\frac{A_2 R}{R_2 + A_2 R}\right) = \left(-\frac{R_2}{R_1}\right)\left(\frac{A_2}{\dfrac{R_2}{R} + A_2}\right)$$

$$\frac{V_o}{V_{in}} = \left(-\frac{R_2}{R_1}\right)\left(\frac{R}{R_2}\right)\left[\frac{A_2}{1 + A_2\left(\dfrac{R}{R_2}\right)}\right] = (G)\left(\frac{R}{R_2}\right)\left[\frac{\dfrac{N_2}{D_2}}{1 + \left(\dfrac{N_2}{D_2}\right)\left(\dfrac{R}{R_2}\right)}\right]$$

$$A = A_2 = \frac{N_2}{D_2} \qquad \beta = \frac{R}{R_2} \qquad G = -\frac{R_2}{R_1}$$

Appendix B1

Standard Resistor Values

Resistor Values (0.1 Ω to 22 MΩ)		Resistor Values, 1% (10.0 Ω to 1.00 MΩ)	
5%	10%		
10		100	316
		102	324
	11	105	332
		107	340
12		110	348
		113	357
	13	115	365
		118	374
15		121	383
		124	392
	16	127	402
		130	412
18		133	422
		137	432
	20	140	442
		143	453
22		147	464
		150	475
	24	154	487
		158	499
27		162	511
		165	523
	30	169	536
		174	549
33		178	562
		182	576
	36	187	590
		191	604
39		196	619
		200	634
	43	205	649
		210	665
47		215	681
		221	698
	51	226	715
		232	732
56		237	750
		243	768
	62	249	787
		255	806
68		261	825
		267	845
	75	274	866
		280	887
82		287	909
		294	931
		301	953
	91	309	976

Appendix B2

Nonpolarized Capacitors

Ceramic Disk (20% Tolerance)		Mylar Wrap (1% Tolerance) Film Capacitors	
Value (pF)	Voltage	Value (μF)	Voltage
5	1000	0.001	100
10	1000	0.0022	100
15	1000	0.0033	100
18	1000	0.0047	100
20	1000	0.0068	100
22	1000	0.01	100
25	1000	0.022	100
27	1000	0.033	100
30	1000	0.047	100
33	1000	0.050	100
39	1000	0.068	100
47	1000	0.1	100
50	1000	0.22	100
56	1000	0.33	100
68	1000	0.47	100
82	1000	0.50	100
100	1000	0.68	100
120	1000	1.00	100
150	1000	2.00	100
180	1000	4.00	100
200	1000	10.0	100
220	1000		
250	1000		
270	1000		
300	1000		
330	1000		
390	1000		
470	1000		
500	1000		
560	1000		
680	1000		
750	1000		
820	1000		
1,000	1000		
1,500	1000		
2,000	1000		
2,200	1000		
3,000	1000		
3,900	1000		
4,700	1000		
5,000	1000		
6,800	1000		
10,000	500		
15,000	500		
20,000	500		
30,000	500		
50,000	500		

Appendix B3

Polarized Capacitors

Miniature Computer-Grade Electrolytics		Computer-Grade Electrolytic Capacitors		
Value (µF)	Voltage	Value (µF)	Voltage	Max (ac) Ripple
1,000	10	12,000	10	4.1
1,600	10	76,000	10	12.1
5,000	10	280,000	10	23.6
10,000	10			
		5,500	15	3.0
600	15	8,900	15	4.0
1,200	15	25,000	15	6.9
2,500	15	38,000	15	10.6
4,000	15	210,000	15	23.6
8,000	15			
		4,700	25	3.4
200	25	5,600	25	5.3
500	25	8,900	25	4.7
1,000	25	13,000	25	5.8
2,200	25	20,000	25	9.7
4,000	25	29,000	25	11.8
150	30	2,200	40	2.6
250	30	2,700	40	2.9
500	30	5,100	40	4.0
1,100	30	11,000	40	8.8
3,000	30	40,000	40	15.3
		63,000	40	16.0
50	50			
250	50	2,200	50	2.8
500	50	4,100	50	5.5
1,100	50	6,100	50	6.7
2,300	50	20,000	50	12.2
25	150	1,100	75	2.9
50	150	2,100	75	4.1
100	150	4,800	75	6.8
250	150			
530	150	4,000	100	5.8
20	250	200	250	1.4
50	250			
100	250			
200	250			
20	450			

LVA Zener Diodes

Regulator and reference diodes for discrete and hybrid circuits with the sharpest breakdown below 10 volts

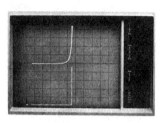

The LVA device exhibits considerably sharper breakdown characteristics than zeners, in the 4-10 volt range. Above 10 volts, the breakdown mechanism of zener regulators is avalanche, which produces a very sharp knee and provides good voltage regulation. Below 10 volts, the field emission phenomenon starts, and as the operating voltage decreases, field emission accounts for an increasingly higher percentage of the device breakdown mechanism.

The field emission breakdown is characterized by a semi-logarithmic relationship between the applied current and zener voltage which produces the soft knee in low voltage zener regulators.

In the LVA device the field emission breakdown mechanism is suppressed, producing a predominantly avalanche breakdown in the voltage range that has historically been characterized by soft, field emission knees.

The performance of the LVA is displayed in the photograph above; an unretouched scope comparison of 1N752 standard zener (upper), and 1N6085 LVA (Tektonix Type 576, 1V/div. horizontal 1mA/div. vertical.)

Maximum Ratings (Common to all Types)

Rating	Symbol	Value	Units
DC Power Dissipation Ta = 25°C (See Figure 5 for derating)	P_D	400	mW
Operating Temperature Range	T_j	-65 to +175	°C
Storage Temperature Range	T_{stg}	-65 to +200	°C

General Purpose LVA Zeners

TRW Type[1]	Nominal Zener Voltage @ Iz	Maximum Dynamic Impedance[2] Zz @ Iz		Maximum Noise Density[3] @ 250µA	Maximum Reverse Leakage I_R @ V_R	
	Vdc	Ohms	mA	µV/√Hz	µA	Vdc
LVA 43A	4.3	18	20	4	4.0	1.5
LVA 47A	4.7	15	10	4	4.0	2.0
LVA 51A	5.1	15	5	4	0.1	2.0
LVA 56A	5.6	40	1	4	0.05	3.0
LVA 62A	6.2	50	1	4	0.05	4.0
LVA 68A	6.8	50	1	4	0.05	5.0
LVA 75A	7.5	100	1	4	0.01	6.0
LVA 82A	8.2	100	1	4	0.01	6.5
LVA 91A	9.1	100	1	4	0.01	8.0
LVA 100A	10.0	100	1	4	0.01	9.0

V_F @ 200mA = 1.5V Max.
[1]A Suffix denotes ±5% V_Z tolerance.
 B Suffix denotes ±2% V_Z tolerance.
 C Suffix denotes ±1% V_Z tolerance.
[2]Measured @ DC test current with 10% AC superimposed (60Hz rms).
[3]1000Hz to 3000Hz, see Figure 1.

Reprinted courtesy of TRW Power Semiconductors, Lawndale, California.

Appendix B5

µA 741 Operational Amplifier

GENERAL DESCRIPTION — The µA741 is a high performance monolithic Operational Amplifier constructed using the Fairchild Planar* epitaxial process. It is intended for a wide range of analog applications. High common mode voltage range and absence of latch-up tendencies make the µA741 ideal for use as a voltage follower. The high gain and wide range of operating voltage provides superior performance in integrator, summing amplifier, and general feedback applications. Electrical characteristics of the µA741A and E are identical to MIL-M-38510/10101.

- NO FREQUENCY COMPENSATION REQUIRED
- SHORT CIRCUIT PROTECTION
- OFFSET VOLTAGE NULL CAPABILITY
- LARGE COMMON MODE AND DIFFERENTIAL VOLTAGE RANGES
- LOW POWER CONSUMPTION
- NO LATCH-UP

ABSOLUTE MAXIMUM RATINGS

Supply Voltage	
µA741A, µA741, µA741E	±22 V
µA741C	±18 V
Internal Power Dissipation (Note 1)	
Metal Can	500 mW
Molded and Hermetic DIP	670 mW
Mini DIP	310 mW
Flatpak	570 mW
Differential Input Voltage	±30 V
Input Voltage (Note 2)	±15 V
Storage Temperature Range	
Metal Can, Hermetic DIP, and Flatpak	−65°C to +150°C
Mini DIP, Molded DIP	−55°C to +125°C
Operating Temperature Range	
Military (µA741A, µA741)	−55°C to +125°C
Commercial (µA741E, µA741C)	0°C to +70°C
Lead Temperature (Soldering)	
Metal Can, Hermetic DIPs, and Flatpak (60 s)	300°C
Molded DIPs (10 s)	260°C
Output Short Circuit Duration (Note 3)	Indefinite

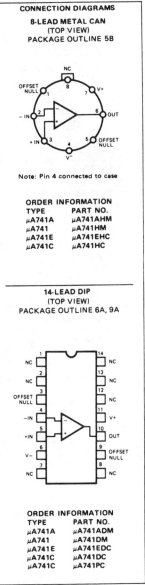

CONNECTION DIAGRAMS

8-LEAD METAL CAN
(TOP VIEW)
PACKAGE OUTLINE 5B

Note: Pin 4 connected to case

ORDER INFORMATION

TYPE	PART NO.
µA741A	µA741AHM
µA741	µA741HM
µA741E	µA741EHC
µA741C	µA741HC

14-LEAD DIP
(TOP VIEW)
PACKAGE OUTLINE 6A, 9A

ORDER INFORMATION

TYPE	PART NO.
µA741A	µA741ADM
µA741	µA741DM
µA741E	µA741EDC
µA741C	µA741DC
µA741C	µA741PC

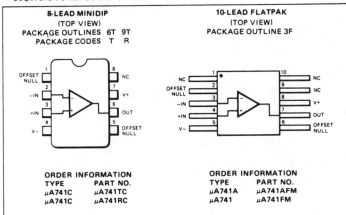

8-LEAD MINIDIP
(TOP VIEW)
PACKAGE OUTLINES 6T 9T
PACKAGE CODES T R

ORDER INFORMATION

TYPE	PART NO.
µA741C	µA741TC
µA741C	µA741RC

10-LEAD FLATPAK
(TOP VIEW)
PACKAGE OUTLINE 3F

ORDER INFORMATION

TYPE	PART NO.
µA741A	µA741AFM
µA741	µA741FM

Notes on following pages.

*Planar is a patented Fairchild process.

Reprinted courtesy of Fairchild Semiconductor Corp., Mt. View, California.

FAIRCHILD LINEAR INTEGRATED CIRCUITS • μA741

μA741A

ELECTRICAL CHARACTERISTICS (V_S = ±15V, T_A = 25°C unless otherwise specified)

PARAMETERS (see definitions)		CONDITIONS	MIN	TYP	MAX	UNITS
Input Offset Voltage		$R_S \leq 50\Omega$		0.8	3.0	mV
Average Input Offset Voltage Drift					15	μV/°C
Input Offset Current				3.0	30	nA
Average Input Offset Current Drift					0.5	nA/°C
Input Bias Current				30	80	nA
Power Supply Rejection Ratio		V_S = +10, −20; V_S = +20, −10V, R_S = 50Ω		15	50	μV/V
Output Short Circuit Current			10	25	35	mA
Power Dissipation		V_S = ±20V		80	150	mW
Input Impedance		V_S = ±20V	1.0	6.0		MΩ
Large Signal Voltage Gain		V_S = ±20V, R_L = 2kΩ, V_{OUT} = ±15V	50			V/mV
Transient Response	Rise Time			0.25	0.8	μs
(Unity Gain)	Overshoot			6.0	20	%
Bandwidth (Note 4)			.437	1.5		MHz
Slew Rate (Unity Gain)		V_{IN} = ±10V	0.3	0.7		V/μs
The following specifications apply for −55°C $\leq T_A \leq$ +125°C						
Input Offset Voltage					4.0	mV
Input Offset Current					70	nA
Input Bias Current					210	nA
Common Mode Rejection Ratio		V_S = ±20V, V_{IN} = ±15V, R_S = 50Ω	80	95		dB
Adjustment For Input Offset Voltage		V_S = ±20V	10			mV
Output Short Circuit Current			10		40	mA
Power Dissipation		V_S = ±20V, −55°C			165	mW
		+125°C			135	mW
Input Impedance		V_S = ±20V	0.5			MΩ
Output Voltage Swing		V_S = ±20V, R_L = 10kΩ	±16			V
		R_L = 2kΩ	±15			V
Large Signal Voltage Gain		V_S = ±20V, R_L = 2kΩ, V_{OUT} = ±15V	32			V/mV
		V_S = ±5V, R_L = 2kΩ, V_{OUT} = ±2 V	10			V/mV

NOTES
1. Rating applies to ambient temperatures up to 70°C. Above 70°C ambient derate linearly at 6.3mW/°C for the metal can, 8.3mW/°C for the DIP and 7.1mW/°C for the Flatpak.
2. For supply voltages less than ±15V, the absolute maximum input voltage is equal to the supply voltage.
3. Short circuit may be to ground or either supply. Rating applies to +125°C case temperature or 75°C ambient temperature.
4. Calculated value from: $\text{BW(MHz)} = \dfrac{0.35}{\text{Rise Time }(\mu s)}$

FAIRCHILD LINEAR INTEGRATED CIRCUITS • µA741

µA741

ELECTRICAL CHARACTERISTICS (V_S = ±15 V, T_A = 25°C unless otherwise specified)

PARAMETERS (see definitions)		CONDITIONS	MIN	TYP	MAX	UNITS
Input Offset Voltage		R_S ≤ 10 kΩ		1.0	5.0	mV
Input Offset Current				20	200	nA
Input Bias Current				80	500	nA
Input Resistance			0.3	2.0		MΩ
Input Capacitance				1.4		pF
Offset Voltage Adjustment Range				±15		mV
Large Signal Voltage Gain		R_L ≥ 2 kΩ, V_{OUT} = ±10 V	50,000	200,000		
Output Resistance				75		Ω
Output Short Circuit Current				25		mA
Supply Current				1.7	2.8	mA
Power Consumption				50	85	mW
Transient Response (Unity Gain)	Rise time	V_{IN} = 20 mV, R_L = 2 kΩ, C_L ≤ 100 pF		0.3		µs
	Overshoot			5.0		%
Slew Rate		R_L ≥ 2 kΩ		0.5		V/µs

The following specifications apply for −55°C ≤ T_A ≤ +125°C:

			MIN	TYP	MAX	UNITS
Input Offset Voltage		R_S ≤ 10 kΩ		1.0	6.0	mV
Input Offset Current		T_A = +125°C		7.0	200	nA
		T_A = −55°C		85	500	nA
Input Bias Current		T_A = +125°C		0.03	0.5	µA
		T_A = −55°C		0.3	1.5	µA
Input Voltage Range			±12	±13		V
Common Mode Rejection Ratio		R_S ≤ 10 kΩ	70	90		dB
Supply Voltage Rejection Ratio		R_S ≤ 10 kΩ		30	150	µV/V
Large Signal Voltage Gain		R_L ≥ 2 kΩ, V_{OUT} = ±10 V	25,000			
Output Voltage Swing		R_L ≥ 10 kΩ	±12	±14		V
		R_L ≥ 2 kΩ	±10	±13		V
Supply Current		T_A = +125°C		1.5	2.5	mA
		T_A = −55°C		2.0	3.3	mA
Power Consumption		T_A = +125°C		45	75	mW
		T_A = −55°C		60	100	mW

TYPICAL PERFORMANCE CURVES FOR µA741A AND µA741

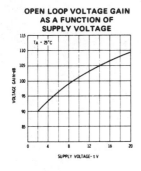

OPEN LOOP VOLTAGE GAIN AS A FUNCTION OF SUPPLY VOLTAGE

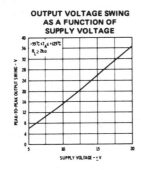

OUTPUT VOLTAGE SWING AS A FUNCTION OF SUPPLY VOLTAGE

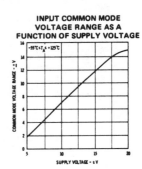

INPUT COMMON MODE VOLTAGE RANGE AS A FUNCTION OF SUPPLY VOLTAGE

Appendix B5 *(Continued)*

FAIRCHILD LINEAR INTEGRATED CIRCUITS • μA741

μA741E

ELECTRICAL CHARACTERISTICS ($V_S = \pm15V$, $T_A = 25°C$ unless otherwise specified)

PARAMETERS (see definitions)		CONDITIONS	MIN	TYP	MAX	UNITS
Input Offset Voltage		$R_S \leq 50\Omega$		0.8	3.0	mV
Average Input Offset Voltage Drift					15	μV/°C
Input Offset Current				3.0	30	nA
Average Input Offset Current Drift					0.5	nA/°C
Input Bias Current				30	80	nA
Power Supply Rejection Ratio		$V_S = +10, -20; V_S = +20, -10V, R_S = 50\Omega$		15	50	μV/V
Output Short Circuit Current			10	25	35	mA
Power Dissipation		$V_S = \pm20V$		80	150	mW
Input Impedance		$V_S = \pm20V$	1.0	6.0		MΩ
Large Signal Voltage Gain		$V_S = \pm20V$, $R_L = 2k\Omega$, $V_{OUT} = \pm15V$	50			V/mV
Transient Response	Rise Time			0.25	0.8	μs
(Unity Gain)	Overshoot			6.0	20	%
Bandwidth (Note 4)			.437	1.5		MHz
Slew Rate (Unity Gain)		$V_{IN} = \pm10V$	0.3	0.7		V/μs
The following specifications apply for $0°C \leq T_A \leq 70°C$						
Input Offset Voltage					4.0	mV
Input Offset Current					70	nA
Input Bias Current					210	nA
Common Mode Rejection Ratio		$V_S = \pm20V$, $V_{IN} = \pm15V$, $R_S = 50\Omega$	80	95		dB
Adjustment For Input Offset Voltage		$V_S = \pm20V$	10			mV
Output Short Circuit Current			10		40	mA
Power Dissipation		$V_S = \pm20V$			150	mW
Input Impedance		$V_S = \pm20V$	0.5			MΩ
Output Voltage Swing	$V_S = \pm20V$,	$R_L = 10k\Omega$	±16			V
		$R_L = 2k\Omega$	±15			V
Large Signal Voltage Gain		$V_S = \pm20V$, $R_L = 2k\Omega$, $V_{OUT} = \pm15V$	32			V/mV
		$V_S = \pm5V$, $R_L = 2k\Omega$, $V_{OUT} = \pm2 V$	10			V/mV

EQUIVALENT CIRCUIT

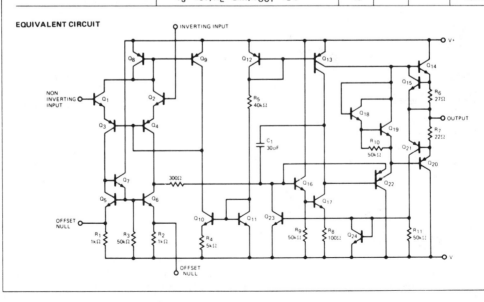

FAIRCHILD LINEAR INTEGRATED CIRCUITS • μA741

μA741C

ELECTRICAL CHARACTERISTICS ($V_S = \pm 15$ V, $T_A = 25°$C unless otherwise specified)

PARAMETERS (see definitions)	CONDITIONS		MIN	TYP	MAX	UNITS
Input Offset Voltage	$R_S \leqslant 10$ kΩ			2.0	6.0	mV
Input Offset Current				20	200	nA
Input Bias Current				80	500	nA
Input Resistance			0.3	2.0		MΩ
Input Capacitance				1.4		pF
Offset Voltage Adjustment Range				±15		mV
Input Voltage Range			±12	±13		V
Common Mode Rejection Ratio	$R_S \leqslant 10$ kΩ		70	90		dB
Supply Voltage Rejection Ratio	$R_S \leqslant 10$ kΩ			30	150	μV/V
Large Signal Voltage Gain	$R_L \geqslant 2$ kΩ, $V_{OUT} = \pm 10$ V		20,000	200,000		
Output Voltage Swing	$R_L \geqslant 10$ kΩ		±12	±14		V
	$R_L \geqslant 2$ kΩ		±10	±13		V
Output Resistance				75		Ω
Output Short Circuit Current				25		mA
Supply Current				1.7	2.8	mA
Power Consumption				50	85	mW
Transient Response (Unity Gain)	Rise time	$V_{IN} = 20$ mV, $R_L = 2$ kΩ, $C_L \leqslant 100$ pF		0.3		μs
	Overshoot			5.0		%
Slew Rate	$R_L \geqslant 2$ kΩ			0.5		V/μs

The following specifications apply for $0°$C $\leqslant T_A \leqslant +70°$C:

Input Offset Voltage					7.5	mV
Input Offset Current					300	nA
Input Bias Current					800	nA
Large Signal Voltage Gain	$R_L \geqslant 2$ kΩ, $V_{OUT} = \pm 10$ V		15,000			
Output Voltage Swing	$R_L \geqslant 2$ kΩ		±10	±13		V

TYPICAL PERFORMANCE CURVES FOR μA741E AND μA741C

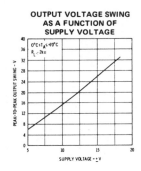

OPEN LOOP VOLTAGE GAIN AS A FUNCTION OF SUPPLY VOLTAGE

OUTPUT VOLTAGE SWING AS A FUNCTION OF SUPPLY VOLTAGE

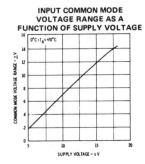

INPUT COMMON MODE VOLTAGE RANGE AS A FUNCTION OF SUPPLY VOLTAGE

FAIRCHILD LINEAR INTEGRATED CIRCUITS • μA741

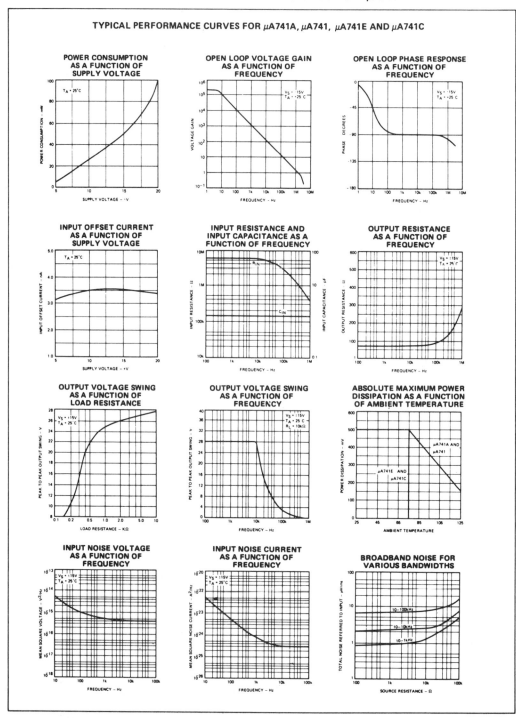

TYPICAL PERFORMANCE CURVES FOR μA741A, μA741, μA741E AND μA741C

Appendix B5 *(Continued)*

FAIRCHILD LINEAR INTEGRATED CIRCUITS • µA741

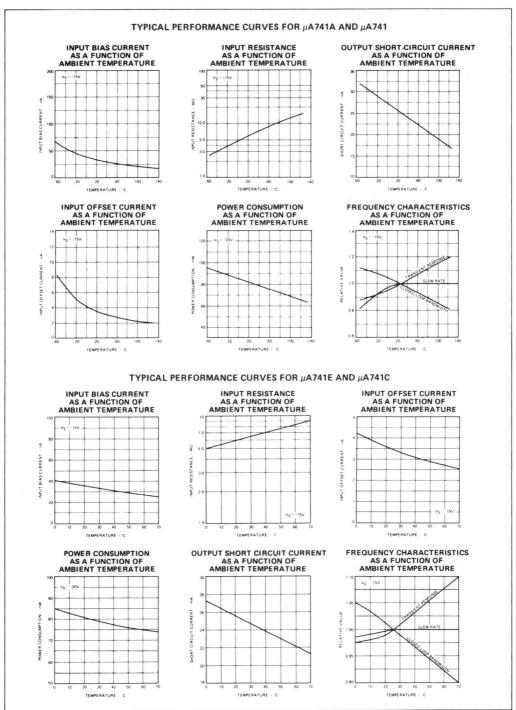

FAIRCHILD LINEAR INTEGRATED CIRCUITS • μA741

TRANSIENT RESPONSE

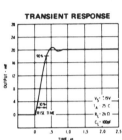

TRANSIENT RESPONSE TEST CIRCUIT

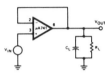

COMMON MODE REJECTION RATIO AS A FUNCTION OF FREQUENCY

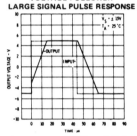

FREQUENCY CHARACTERISTICS AS A FUNCTION OF SUPPLY VOLTAGE

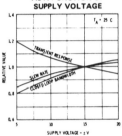

VOLTAGE OFFSET NULL CIRCUIT

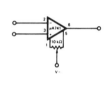

VOLTAGE FOLLOWER LARGE SIGNAL PULSE RESPONSE

TYPICAL APPLICATIONS

UNITY-GAIN VOLTAGE FOLLOWER

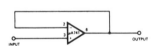

R_{IN} = 400 MΩ
C_{IN} = 1 pF
R_{OUT} << 1 Ω
B.W. = 1 MHz

NON-INVERTING AMPLIFIER

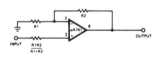

GAIN	R1	R2	B W	R_{IN}
10	1 kΩ	9 kΩ	100 kHz	400 MΩ
100	100 Ω	9.9 kΩ	10 kHz	280 MΩ
1000	100 Ω	99.9 kΩ	1 kHz	80 MΩ

INVERTING AMPLIFIER

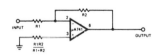

GAIN	R1	R2	B W	R_{IN}
1	10 kΩ	10 kΩ	1 MHz	10 kΩ
10	1 kΩ	10 kΩ	100 kHz	1 kΩ
100	1 kΩ	100 kΩ	10 kHz	1 kΩ
1000	100 Ω	100 kΩ	1 kHz	100 Ω

CLIPPING AMPLIFIER

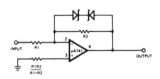

$$\frac{E_{OUT}}{E_{IN}} = \frac{R2}{R1} \text{ if } |E_{OUT}| < V_Z + 0.7 \text{ V}$$

where V_Z = Zener breakdown voltage

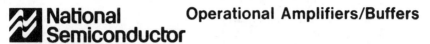

National Semiconductor

Operational Amplifiers/Buffers

LM108A/LM208A/LM308A, LM308A-1, LM308A-2 operational amplifiers

general description

The LM108/LM108A series are precision operational amplifiers having specifications about a factor of ten better than FET amplifiers over their operating temperature range. In addition to low input currents, these devices have extremely low offset voltage, making it possible to eliminate offset adjustments, in most cases, and obtain performance approaching chopper stabilized amplifiers.

The devices operate with supply voltages from ±2V to ±18V and have sufficient supply rejection to use unregulated supplies. Although the circuit is interchangeable with and uses the same compensation as the LM101A, an alternate compensation scheme can be used to make it particularly insensitive to power supply noise and to make supply bypass capacitors unnecessary. Outstanding characteristics include:

■ Offset voltage guaranteed less than 0.5 mV

■ Maximum input bias current of 3.0 nA over temperature

■ Offset current less than 400 pA over temperature

■ Supply current of only 300 µA, even in saturation

■ Guaranteed 5 µV/°C drift.

■ Guaranteed 1 µV/°C for LM308A-1

The low current error of the LM108A series makes possible many designs that are not practical with conventional amplifiers. In fact, it operates from 10 MΩ source resistances, introducing less error than devices like the 709 with 10 kΩ sources. Integrators with drifts less than 500 µV/sec and analog time delays in excess of one hour can be made using capacitors no larger than 1 µF.

The LM208A is identical to the LM108A, except that the LM208A has its performance guaranteed over a −25°C to 85°C temperature range, instead of −55°C to 125°C. The LM308A devices have slightly-relaxed specifications and performance guaranteed over a 0°C to 70°C temperature range.

compensation circuits

Standard Compensation Circuit

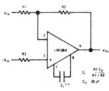

Alternate* Frequency Compensation

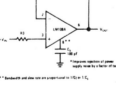

Feedforward Compensation

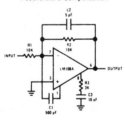

typical applications

Sample and Hold

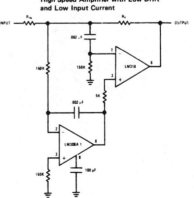

High Speed Amplifier with Low Drift and Low Input Current

LM108A/LM208A

absolute maximum ratings

Supply Voltage	$\pm$20V
Power Dissipation (Note 1)	500 mW
Differential Input Current (Note 2)	$\pm$10 mA
Input Voltage (Note 3)	$\pm$15V
Output Short-Circuit Duration	Indefinite
Operating Temperature Range LM108A	-55°C to 125°C
LM208A	-25°C to 85°C
Storage Temperature Range	-65°C to 150°C
Lead Temperature (Soldering, 10 sec)	300°C

electrical characteristics (Note 4)

PARAMETER	CONDITIONS	MIN	TYP	MAX	UNITS
Input Offset Voltage	$T_A = 25^{\circ}$C		0.3	0.5	mV
Input Offset Current	$T_A = 25^{\circ}$C		0.05	0.2	nA
Input Bias Current	$T_A = 25^{\circ}$C		0.8	2.0	nA
Input Resistance	$T_A = 25^{\circ}$C	30	70		MΩ
Supply Current	$T_A = 25^{\circ}$C		0.3	0.6	mA
Large Signal Voltage Gain	$T_A = 25^{\circ}$C, $V_S = \pm15$V $V_{OUT} = \pm10$V, $R_L > 10$ kΩ	80	300		V/mV
Input Offset Voltage				1.0	mV
Average Temperature Coefficient of Input Offset Voltage			1.0	5.0	μV/$^{\circ}$C
Input Offset Current				0.4	nA
Average Temperature Coefficient of Input Offset Current			0.5	2.5	pA/$^{\circ}$C
Input Bias Current				3.0	nA
Supply Current	$T_A = +125^{\circ}$C		0.15	0.4	mA
Large Signal Voltage Gain	$V_S = \pm15$V, $V_{OUT} = \pm10$V $R_L > 10$ kΩ	40			V/mV
Output Voltage Swing	$V_S = \pm15$V, $R_L = 10$ kΩ	$\pm$13	$\pm$14		V
Input Voltage Range	$V_S = \pm15$V	$\pm$13.5			V
Common Mode Rejection Ratio		96	110		dB
Supply Voltage Rejection Ratio		96	110		dB

Note 1: The maximum junction temperature of the LM108A is 150°C, while that of the LM208A is 100°C. For operating at elevated temperatures, devices in the TO-5 package must be derated based on a thermal resistance of 150°C/W, junction to ambient, or 45°C/W, junction to case. For the flat package, the derating is based on a thermal resistance of 185°C/W when mounted on a 1/16-inch-thick epoxy glass board with ten, 0.03-inch-wide, 2-ounce copper conductors. The thermal resistance of the dual-in-line package is 100°C/W, junction to ambient.

Note 2: The inputs are shunted with back-to-back diodes for overvoltage protection. Therefore, excessive current will flow if a differential input voltage in excess of 1V is applied between the inputs unless some limiting resistance is used.

Note 3: For supply voltages less than ±15V, the absolute maximum input voltage is equal to the supply voltage.

Note 4: These specifications apply for ±5V $< V_S < \pm20$V and -55°C $< T_A < 125^{\circ}$C, unless otherwise specified. With the LM208A, however, all temperature specifications are limited to -25°C $\leq T_A \leq 85^{\circ}$C.

application hints

A very low drift amplifier poses some uncommon application and testing problems. Many sources of error can cause the apparent circuit drift to be much higher than would be predicted.

Thermocouple effects caused by temperature gradient across dissimilar metals are perhaps the worst offenders. Only a few degrees gradient can cause hundreds of microvolts of error. The two places this shows up, generally, are the package-to printed circuit board interface and temperature gradients across resistors. Keeping package leads short and the two input leads close together help greatly.

Resistor choice as well as physical placement is important for minimizing thermocouple effects. Carbon, oxide film and some metal film resistors can cause large thermocouple errors. Wirewound resistors of evenohm or manganin are best since they only generate about 2 μV/°C referenced to copper. Of course, keeping the resistor ends at the same temperature is important. Generally, shielding a low drift stage electrically and thermally will yield good results.

Resistors can cause other errors besides gradient generated voltages. If the gain setting resistors do not track with temperature a gain error will result. For example a gain of 1000 amplifier with a con-

stant 10 mV input will have a 10V output. If the resistors mistrack by 0.5% over the operating temperature range, the error at the output is 50 mV. Referred to input, this is a 50 μV error. All of the gain fixing resistor should be the same material.

Offset balancing the LM308A-1 can be a problem since there is no easy offset adjustment incorporated into the circuit. These devices are selected for low drift with no offset adjustment to the internal circuitry, so any change of the internal currents will change the drift — probably for the worse. Offset adjustment must be done at the input. The three most commonly needed circuits are shown here.

Testing low drift amplifiers is also difficult. Standard drift testing technique such as heating the device in an oven and having the leads available through a connector, thermoprobe, or the soldering iron method — do not work. Thermal gradients cause much greater errors than the amplifier drift. Coupling microvolt signal through connectors is especially bad since the temperature difference across the connector can be 50°C or more. The device under test along with the gain setting resistor should be isothermal. The following circuit will yield good results if well constructed.

Offset Adjustment for Inverting Amplifiers

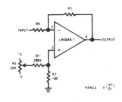

Offset Adjustment for Non-Inverting Amplifiers

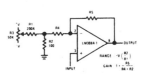

Offset Adjustment for Differential Amplifiers

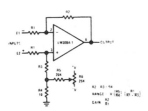

Drift Measurement Circuit

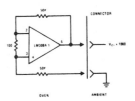

441

schematic diagram*

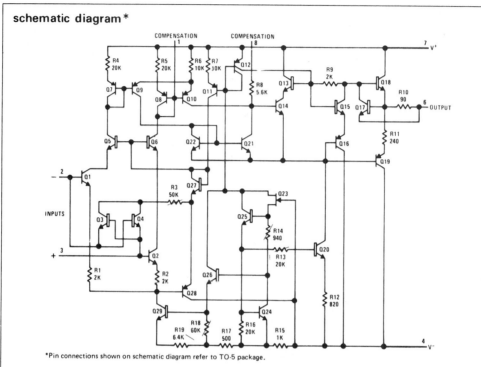

*Pin connections shown on schematic diagram refer to TO-5 package.

connection diagrams

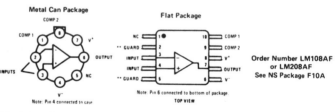

Metal Can Package

Order Number LM108AH, LM208AH,
LM308AH, LM308AH-1 or LM308AH-2
See NS Package H08C

Note: Pin 4 connected to case

Flat Package

Order Number LM108AF
or LM208AF
See NS Package F10A

Note: Pin 6 connected to bottom of package.
TOP VIEW

**Unused pin (no internal connection) to allow for input anti-leakage
guard ring on printed circuit board layout

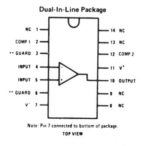

Dual-In-Line Package

Order Number LM108AD, LM208AD
or LM308AD
See NS Package D14E
Order Number LM108AJ, LM208AJ,
or LM308AJ
See NS Package J14A

Note: Pin 7 connected to bottom of package
TOP VIEW

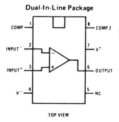

Dual-In-Line Package

Order Number LM108AJ-8,
LM208AJ-8 or LM308AJ-8
See NS Package J08A
Order Number LM208AN
or LM308AN
See NS Package N08B

TOP VIEW

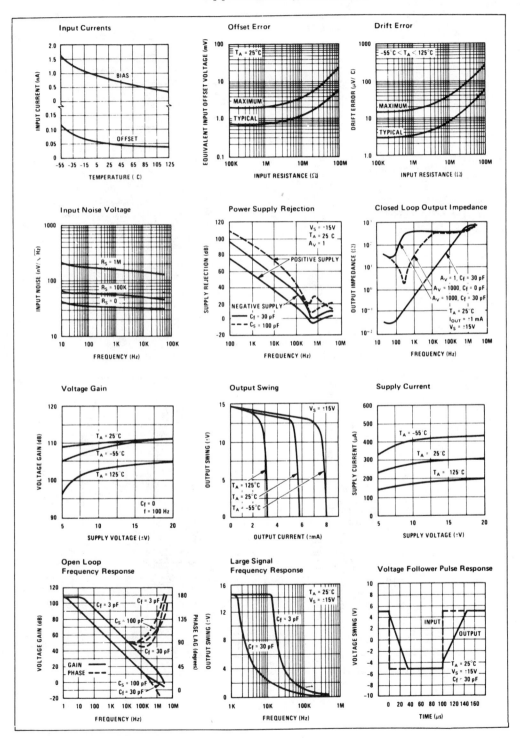

 **HARRIS**

HA-2539

Very High Slew Rate Wideband Operational Amplifiers

Features

- VERY HIGH SLEW RATE   600V/μs
- OPEN LOOP GAIN   15kV/V
- WIDE GAIN-BANDWIDTH   600MHz
- POWER BANDWIDTH   9.5MHz
- LOW OFFSET VOLTAGE   3mV
- INPUT VOLTAGE NOISE   6nV/$\sqrt{Hz}$
- OUTPUT VOLTAGE SWING   ±10V

Applications

- PULSE AND VIDEO AMPLIFIERS
- WIDEBAND AMPLIFIERS
- HIGH SPEED SAMPLE-HOLD CIRCUITS
- RF OSCILLATORS

Description

The Harris HA-2539 represents the ultimate in high slew rate wideband, monolithic, operational amplifiers. It has been designed and constructed with the Harris high frequency Bipolar dielectric isolation process and features dynamic parameters never before available from a truly differential device.

With a 600V/μs slew rate and a 600MHz gain-band-width-product, the HA-2539 is ideally suited for use in video and RF amplifier designs, in closed loop gains of 10 or greater. Full ±10V swing coupled with outstanding A.C. parameters and complemented by high open loop gain makes the devices useful in high speed data acquisition systems.

The HA-2539 is available in the 14 pin ceramic and epoxy packages, as well as a 20 pin LCC package. The HA-2539-2 denotes –55°C to +125°C operation while the HA-2539-5 operates over the 0°C to +75°C range.

Pinouts

TOP VIEWS

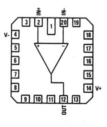

Schematic

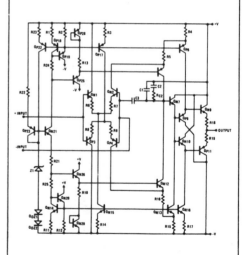

Courtesy of Harris Corporation.

HA-2539

Specifications

ABSOLUTE MAXIMUM RATINGS   (Note 1)

Voltage between V+ and V- Terminals	35V
Differential Input Voltage	6V
Output Current	50mA (Peak)
Internal Power Dissipation (Note 2)	870mW (Cerdip)
Operating Temperature Range: (HA-2539-2)	$-55^oC \leq T_A \leq +125^oC$
(HA-2539-5)	$0^oC \leq T_A \leq +75^oC$
Storage Temperature Range	$-65^oC \leq T_A \leq +150^oC$

ELECTRICAL CHARACTERISTICS $V_{SUPPLY} = \pm 15$ Volts; R_L = 1K ohms, unless otherwise specified.

PARAMETER	TEMP	HA-2539-2 -55°C to +125°C			HA-2539-5 0°C to +75°C			UNITS
		MIN	TYP	MAX	MIN	TYP	MAX	
INPUT CHARACTERISTICS								
Offset Voltage	+25°C		8	10		8	15	mV
	FULL		13	15			20	mV
Average Offset Voltage Drift	FULL		20			20		µV/°C
Bias Current	+25°C		5	20		5	20	µA
	FULL			25			25	µA
Offset Current	+25°C		1	6		1	6	µA
	FULL			8			8	µA
Input Resistance	+25°C		10			10		Kohms
Input Capacitance	+25°C		1.0			1.0		pF
Common Mode Range	FULL	± 10			± 10			V
Input Voltage Noise (f = 1kHz, R_g = 0Ω)	+25°C		6			6		nV/$\sqrt{Hz}$
TRANSFER CHARACTERISTICS								
Large Signal Voltage Gain (Note 3)	+25°C	10K	15K		10K	15 K		V/V
	FULL	5K			5K			V/V
Common-Mode Rejection Ratio (Note 4)	FULL	60			60			dB
Gain-Bandwidth-Product (Notes 5 & 6)	+25°C		600			600		MHz
OUTPUT CHARACTERISTICS								
Output Voltage Swing (Note 3)	FULL	± 10			± 10			V
Output Current (Note 3)	+25°C	10			10			mA
Output Resistance	+25°C		30			30		Ohms
Full Power Bandwidth (Note 3 & 7)	+25°C	8.7	9.5		8.7	9.5		MHz
TRANSIENT RESPONSE (Note 8)								
Rise Time	+25°C		7			7		ns
Overshoot	+25°C		15			15		%
Slew Rate	+25°C	550	600		550	600		V/µs
Settling Time: 10V Step to 0.1%	+25°C		200			200		ns
POWER REQUIREMENTS								
Supply Current	FULL		20	25		20	25	mA
Power Supply Rejection Ratio (Note 9)	FULL	60			60			dB

HA-2539

NOTES:

1. Absolute maximum ratings are limiting values, applied individually, beyond which the serviceability of the circuit may be impaired. Functional operability under any of these conditions is not necessarily implied.

2. Derate at 8.7mW/°C for operation at ambient temperatures above +75°C. Heat sinking required at temperatures above +75°C. $T_{JA} = 115°C/W$; $T_{JC} = 35°C/W$. Thermalloy model 6007 heat sink recommended.

3. $R_L = 1K\Omega$, VO = ± 10V

4. $V_{CM} = \pm 10V$

5. $V_0 = 90mV$.

6. $A_V = 10$.

7. Full power bandwidth guaranteed based on slew rate measurement using $FPBW = \dfrac{Slew\ Rate}{2\pi\ V_{peak}}$.

8. Refer to Test Circuits section of data sheet.

9. $V_{SUPPLY} = \pm5\ VDC\ to\ \pm15\ VDC$

Test Circuits

LARGE AND SMALL SIGNAL RESPONSE TEST CIRCUIT *

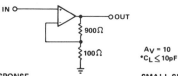

$A_V = 10$
*$C_L \leq 10pF$

LARGE SIGNAL RESPONSE
Vertical Scale: A=0.5V/Div., B=5.0V/Div.
Horizontal Scale: Time: 50ns/Div.

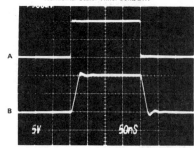

SMALL SIGNAL RESPONSE
Vertical Scale: Input=10mV/Div., Output=50mV/Div.
Horizontal Scale: Time: 20ns/Div.

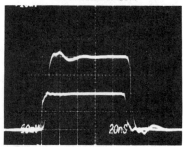

SETTLING TIME TEST CIRCUIT

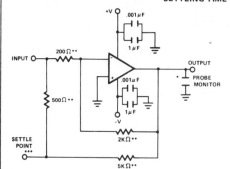

* Load Capacitance should be less than 10pF.

** It is recommended that resistors be carbon composition and that feedback and summing network ratios be matched.

*** SETTLE POINT (Summing Node) capacitance should be less than 10pF. For optimum settling time results, it is recommended that the test circuit be constructed directly onto the device pins. A Tektronix 568 Sampling Oscilloscope with S-3A sampling heads is recommended as a settle point monitor.

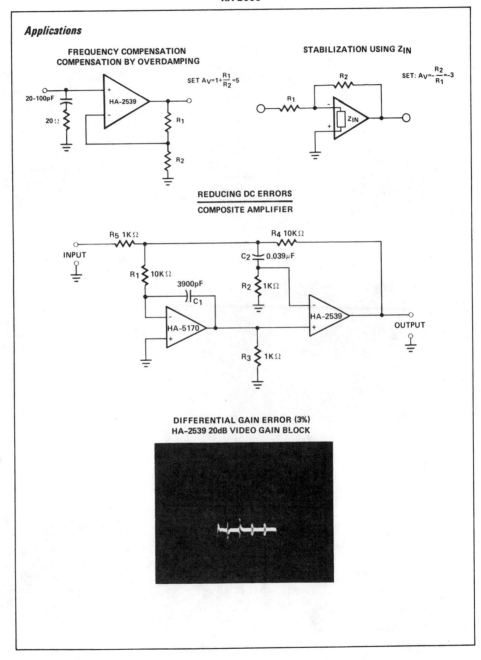

Applications

FREQUENCY COMPENSATION
COMPENSATION BY OVERDAMPING

$\text{SET } A_V = 1 + \dfrac{R_1}{R_2} = 5$

20-100pF
20 Ω
HA-2539
R_1
R_2

STABILIZATION USING Z_{IN}

$\text{SET: } A_V = -\dfrac{R_2}{R_1} = -3$

R_2
R_1
Z_{IN}

REDUCING DC ERRORS

COMPOSITE AMPLIFIER

INPUT
R_5 1KΩ
R_1 10KΩ
3900pF C_1
HA-5170
R_4 10KΩ
C_2 0.039μF
R_2 1KΩ
HA-2539
OUTPUT
R_3 1KΩ

DIFFERENTIAL GAIN ERROR (3%)
HA-2539 20dB VIDEO GAIN BLOCK

447

Performance Curves

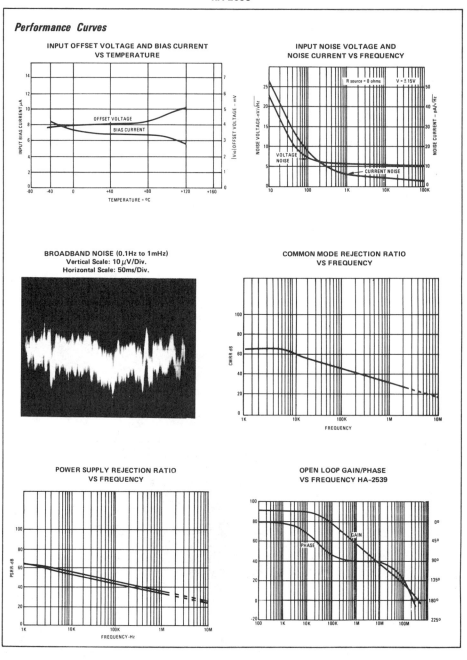

INPUT OFFSET VOLTAGE AND BIAS CURRENT
VS TEMPERATURE

INPUT NOISE VOLTAGE AND
NOISE CURRENT VS FREQUENCY

BROADBAND NOISE (0.1Hz to 1mHz)
Vertical Scale: 10 μV/Div.
Horizontal Scale: 50ms/Div.

COMMON MODE REJECTION RATIO
VS FREQUENCY

POWER SUPPLY REJECTION RATIO
VS FREQUENCY

OPEN LOOP GAIN/PHASE
VS FREQUENCY HA-2539

Appendix B7 *(Continued)*

HA-2539

Performance Curves *(Continued)*

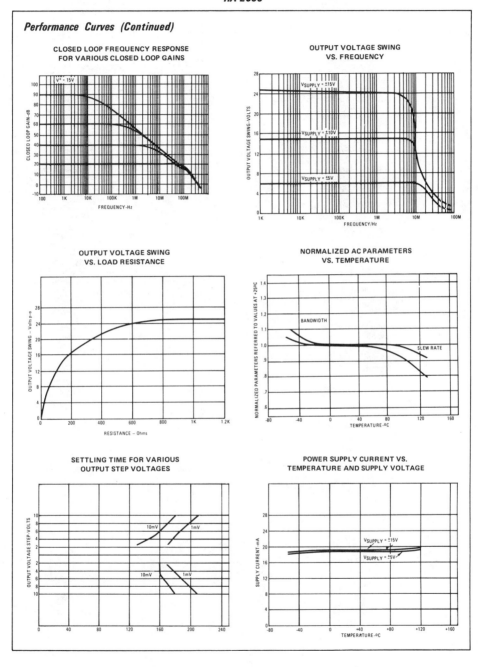

CLOSED LOOP FREQUENCY RESPONSE
FOR VARIOUS CLOSED LOOP GAINS

OUTPUT VOLTAGE SWING
VS. FREQUENCY

OUTPUT VOLTAGE SWING
VS. LOAD RESISTANCE

NORMALIZED AC PARAMETERS
VS. TEMPERATURE

SETTLING TIME FOR VARIOUS
OUTPUT STEP VOLTAGES

POWER SUPPLY CURRENT VS.
TEMPERATURE AND SUPPLY VOLTAGE

 **HARRIS**

PRELIMINARY

HA-2541

**Wideband, Fast Settling, Unity Gain Stable,
Operational Amplifier**

Features

- Unity Gain Bandwidth..40MHz
- High Slew Rate 280V/μs
- Fast Settling Time..90ns
- Power Bandwidth..4MHz
- Output Voltage Swing..±10V
- Unity Gain Stability
- Monolithic Bipolar Construction

Applications

- Pulse and Video Amplifiers
- Wideband Amplifiers
- High Speed Sample-Hold Circuits
- Fast, Precise D/A Converters
- High Speed A/D Input Buffer

Description

The HA-2541 is the first unity gain stable monolithic operational amplifier to achieve 40MHz unity gain bandwidth. A major addition to the Harris series of high speed, wideband op amps, the HA-2541 is designed for video and pulse applications requiring stable amplifier response at low closed loop gains.

The uniqueness of the HA-2541 is that its slew rate and bandwidth characteristics are specified at unity gain. Historically, high slew rate, wide bandwidth and unity gain stability have been incompatible features for a

monolithic operational amplifier. But features such as 280V/μs slew rate and 40MHz unity gain bandwidth clearly show that this is not the case for the HA-2541. These features, along with 90ns settling time, make this product an excellent choice for high speed data acquisition systems.

Packaged in a TO-8 metal can or 14 pin ceramic DIP, the HA-2541 is pin compatible with the HA-2540 and HA-5190 op amps. The HA-2541-2 is specified over the temperature range of -55°C to +125°C.

Pinouts

TOP VIEWS

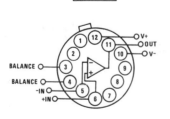

Schematic

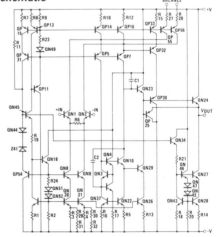

Courtesy of Harris Corporation.

Specifications HA-2541

Absolute Maximum Ratings (Note 1)

Voltage Between V+ and V-.. 35V
Differential Input Voltage... 6V
Output Current... 50mA (Peak)
Internal Power Dissipation (Note 2) TO-8............... 1.5W
          Dip................. 1.6W

Operating Temperature Range:

HA-2541-2......................................$-55^{\circ}C \leq T_A \leq +125^{\circ}C$
HA-2541-5...$0^{\circ}C \leq T_A \leq +75^{\circ}C$
Storage Temperature Range ....$-65^{\circ}C \leq T_A \leq +150^{\circ}C$
Maximum Junction Temperature ....$+175^{\circ}C$

Electrical Characteristics $V_{SUPPLY} = \pm 15$ Volts; $R_L = 2K$ Ohms, Unless Otherwise Specified

PARAMETER	TEMP	HA-2541-2 $-55^{\circ}C$ to $+125^{\circ}C$ MIN	TYP	MAX	HA-2541-5 $0^{\circ}C$ to $+75^{\circ}C$ MIN	TYP	MAX	UNITS
INPUT CHARACTERISTICS								
Offset Voltage (Note 11)*	+25°C			2			2*	mV
	Full			6			6*	mV
Average Offset Voltage Drift	Full		20			20		µV/°C
Bias Current	+25°C		6	35		6	35	µA
	Full			50			50	µA
Offset Current	+25°C		1	7		1	7	µA
	Full			9			9	µA
Input Resistance	+25°C		100			100		Kohms
Input Capacitance	+25°C		1			1		pF
Common Mode Range	Full	±10			±10			V
Input Noise Voltage (f = 1kHz, $R_g = 0\Omega$)	+25°C		10			10		nV/$\sqrt{Hz}$
TRANSFER CHARACTERISTICS								
Large Signal Voltage Gain (Note 3)	+25°C	10K			10K			V/V
	Full	5K			5K			V/V
Common-Mode Rejection Ratio (Note 5)	Full	70			70			dB
Unity Gain-Bandwidth (Note 6)	+25°C		40			40		MHz
OUTPUT CHARACTERISTICS								
Output Voltage Swing (Note 4)	Full	±10	±11		±10	±11		V
Output Current (Note 4)	+25°C	10			10			mA
Output Resistance	+25°C		2			2		Ohms
Full Power Bandwidth (Note 3 & 7)	+25°C	3	4		3	4		MHz
Differential Gain	+25°C		0.1			0.1		%
Differential Phase	+25°C		0.2			0.2		Degree
Harmonic Distortion (Note 10)	+25°C		<0.01			<0.01		%
TRANSIENT RESPONSE (Note 8)								
Rise Time	+25°C		4			4		ns
Overshoot	+25°C		40			40		%
Slew Rate	+25°C	200	280		200	280		V/µs
Settling Time:								
10V Step to 0.1%	+25°C		90			90		ns
POWER REQUIREMENTS								
Supply Current	+25°C		33			33		mA
	Full			40			45	mA
Power Supply Rejection Ratio (Note 9)	Full	70			70			dB

HA-2541

Notes:

1. Absolute maximum ratings are limiting values, applied individually, beyond which the serviceability of the circuit may be impaired. Functional operability under any of these conditions is not necessarily implied.

2. TO-8: θ_{jA} = 100°C/W, θ_{jC} = 15°C/W.
   Recommended heat sink: Thermalloy 2240A θ_{SA} = 27°C/W
   Cerdip: θ_{jA} = 91°C/W, θ_{jC} = 35°C/W
   Recommended heat sink: AAVID #5802 θ_{SA} = 15°C/W

3. V0 = ±10V

4. R_L = 1KΩ

5. V_{CM} = ±10V

6. V_0 = 90mV

7. Full Power Bandwidth guaranteed based on slew rate measurement using FPBW = $\dfrac{\text{Slew Rate}}{2\pi V_{PEAK}}$

8. Refer to Test Circuits section of data sheet.

9. V_{SUPPLY} = ±5VDC to ±15VDC

10. V_{IN} = 1VRMS; f = 10KHz; Av = 10

11. Relaxed Offset Voltage "C" version available in 1986.

Test Circuits

TEST CIRCUIT *

2KΩ

* $C_L \leq$ 10pF

LARGE SIGNAL RESPONSE

SMALL SIGNAL RESPONSE

SETTLING TIME TEST CIRCUIT

SETTLING POINT

5K   5K

2K

2K

V_{IN}

V_{OUT}

- A_V = ⁻1
- FEEDBACK AND SUMMING RESISTORS MUST BE MATCHED (0.1%)
- HP5082-2810 CLIPPING DIODES RECOMMENDED
- TEKTRONIX P6201 FET PROBE USED AT SETTLING POINT

HARRIS

PRELIMINARY

HA-2542
Wideband, High Slew Rate, High Output Current Operational Amplifiers

Features

- Stable at Gains of 2 or Greater
- Gain Bandwidth (A_{VCL} = 2) 120MHz
- High Slew Rate..300V/μs
- High Output Current..100mA
- Power Bandwidth 5.5MHz
- Output Voltage Swing ....±10V
- Monolithic Bipolar Construction

Applications

- Pulse and Video Amplifiers
- Wideband Amplifiers
- Coaxial Cable Drivers
- Fast Sample-Hold Circuits
- High Frequency Signal Conditioning Circuits

Description

The HA-2542 is a wideband, high slew rate, monolithic operational amplifier featuring an outstanding combination of speed, bandwidth, and output drive capability.

Utilizing the advantages of the Harris D. I. technology this amplifier offers 350V/μs slew rate, 120MHz gain bandwidth, and ±100mA output current. Application of this device is further enhanced through stable operation down to closed loop gains of 2.

For additional flexibility, offset null and frequency compensation controls are included in the HA-2542 pinout.

The capabilities of the HA-2542 are ideally suited for high speed coaxial cable driver circuits with gain. With 5.5MHz full power bandwidth, this amplifier is most suitable for high frequency signal conditioning circuits and pulse/video amplifiers. Other applications utilizing the HA-2542 advantages include wideband amplifiers and fast sample-hold circuits.

Packaged in a 12 pin (TO-8) can, the HA-2542 is pin compatible with the HA-2540, HA-2541, HA-5190, LH0032, and H0S-050C.

Pinout

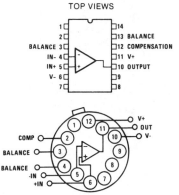

Schematic

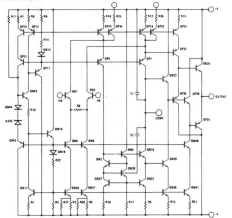

Courtesy of Harris Corporation, copyright © 1984 Harris Corporation.

Appendix B9 *(Continued)*

Specifications HA-2542

Absolute Maximum Ratings (Note 1)

Voltage between V+ and V− Terminals.................. 35V
Differential Input Voltage.. 6V
Output Current.. 125mA (Peak)
                        107mA$_{RMS}$ (Continuous)
Internal Power Dissipation (Note 2) TO-8........... 1.5W
                             Dip 1.6W

Operating Temperature Range:

HA-2542-2 .... −55°C $\leq$ T$_A$ $\leq$ +125°C
HA-2542-5 .... 0°C $\leq$ T$_A$ $\leq$ +75°C
Storage Temperature Range.....−65°C $\leq$ T$_A$ $\leq$ +150°C
Maximum Junction Temperature ....+175°C

Electrical Characteristics   V$_{SUPPLY}$ = ±15 Volts; R$_L$ = 1K ohms, unless otherwise specified.

PARAMETER	TEMP	HA-2542-2 −55°C to +125°C MIN	TYP	MAX	HA-2542-5 0°C to +75°C MIN	TYP	MAX	UNITS
INPUT CHARACTERISTICS								
Offset Voltage	+25°C			10			10	mV
	Full			20			20	mV
Average Offset Voltage Drift	Full		20			20		µV/°C
Bias Current	+25°C			35			35	µA
	Full			50			50	µA
Offset Current	+25°C		1	7		1	7	µA
	Full			9			9	µA
Input Resistance	+25°C		100			100		Kohms
Input Capacitance	+25°C		1			1		pF
Common Mode Range	Full	±10			±10			V
Input Noise Voltage (f = 1kHz, R$_g$ = 0Ω)	+25°C		10			10		nV/$\sqrt{Hz}$
TRANSFER CHARACTERISTICS								
Large Signal Voltage Gain (Note 3)	+25°C	10K			10K			V/V
	Full	5K			5K			V/V
Common-Mode Rejection Ratio (Note 4)	Full	70			70			dB
Gain-Bandwidth-Product (Note 5)	+25°C		120			120		MHz
OUTPUT CHARACTERISTICS								
Output Voltage Swing (Note 3)	Full	±10	±11		±10	±11		V
Output Current (Note 6)	+25°C	100			100			mA
Output Resistance	+25°C		5			5		Ohms
Full Power Bandwidth (Note 3 & 7)	+25°C	4.7			4.7			MHz
Differential Gain	+25°C		.1			.1		%
Differential Phase	+25°C		.2			.2		degree
Harmonic Distortion (Note 10)	+25°C		<0.04			<0.04		%
TRANSIENT RESPONSE (Note 8)								
Rise Time	+25°C		4			4		ns
Overshoot	+25°C		40			40		%
Slew Rate	+25°C	300	375		300	375		V/µs
Settling Time:								
10V Step to 0.1%	+25°C		100			100		ns
POWER REQUIREMENTS								
Supply Current	+25°C		31			31		mA
	Full			35			40	mA
Power Supply Rejection Ratio (Note 9)	Full	70			70			dB

454

Test Circuits *(continued)*

SETTLING TIME TEST CIRCUIT

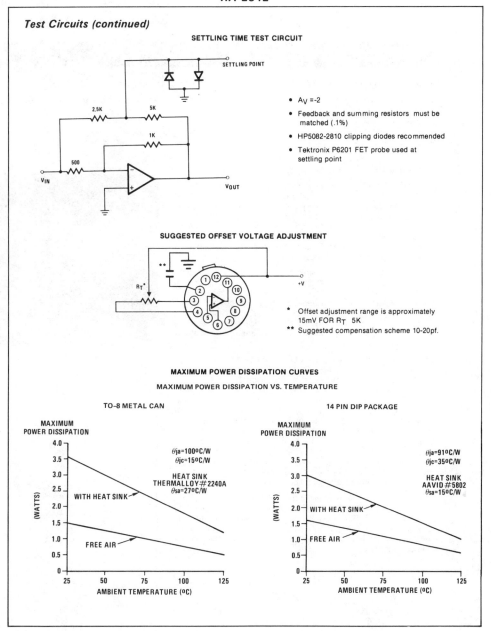

- $A_V = -2$
- Feedback and summing resistors must be matched (.1%)
- HP5082-2810 clipping diodes recommended
- Tektronix P6201 FET probe used at settling point

SUGGESTED OFFSET VOLTAGE ADJUSTMENT

* Offset adjustment range is approximately 15mV FOR R_T 5K
** Suggested compensation scheme 10-20pf.

MAXIMUM POWER DISSIPATION CURVES

MAXIMUM POWER DISSIPATION VS. TEMPERATURE

TO-8 METAL CAN

θja=100°C/W
θjc=15°C/W

HEAT SINK
THERMALLOY #2240A
θsa=27°C/W

WITH HEAT SINK

FREE AIR

14 PIN DIP PACKAGE

θja=91°C/W
θjc=35°C/W

HEAT SINK
AAVID #5802
θsa=15°C/W

WITH HEAT SINK

FREE AIR

Comlinear Corporation

Fast Settling, High Current Wideband Op Amps

CLC103AI, CLC103AM (Hi-Rel)

APPLICATIONS:

- coaxial line driving
- DAC current to voltage amplifier
- flash A to D driving
- baseband and video communications
- radar and IF processors

FEATURES:

- 80MHz full-power bandwidth (20V_{pp}, 100Ω)
- 200mA output current
- 0.4% settling in 10ns
- 6000V/μs slew rate
- 4ns rise and fall times (20V)

DESCRIPTION:

The CLC103 is a high-power, wideband op amp designed for the most demanding high-speed applications. The wide bandwidth, fast settling, linear phase, and very low harmonic distortion provide the designer with the signal fidelity needed in applications such as driving flash A to Ds. **The 80MHz full-power bandwidth and 200mA output current of the CLC103 eliminate the need for power buffers in most applications;** the CLC103 is an excellent choice for driving large high-speed signals into coaxial lines.

In the design of the CLC103 special care was taken in order to guarantee that the output settle quickly to within 0.4% of the final value for use with ultra fast flash A to D converters. This is one of the most demanding of all op amp requirements since settling time is affected by the op amp's bandwidth, passband gain flatness, and harmonic distortion. This high degree of performance ensures excellent performance in many other demanding applications as well.

The dynamic performance of the CLC103 is based on Comlinear's proprietary op amp topology. This new design provides performance far beyond that available from conventional op amp designs; unlike conventional op amps where optimum gain-bandwidth product occurs at a high gain, minimum settling time at a gain of -1, and maximum slew rate at a gain of $+1$, the Comlinear design provides consistent, predictable performance across its entire gain range. For example, the table below shows how **the$-$3dB bandwidth remains nearly constant over a wide range of gains.** And since the amplifier is inherently stable, no external compensation is required. The result is shorter design time and the ability to accommodate design changes (in gain, for example) without loss of performance or redesign of compensation circuits.

The CLC103 is constructed using thick film resistor/bipolar transistor technology. The CLC103AI is specified over a temperature range of $-25°C$ to $+85°C$, while the CLC103AM is specified over a range of $-55°C$ to $+125°C$ and is screened to Comlinear's M Standard for high reliability applications. Both devices are packaged in 24-pin ceramic DIPs.

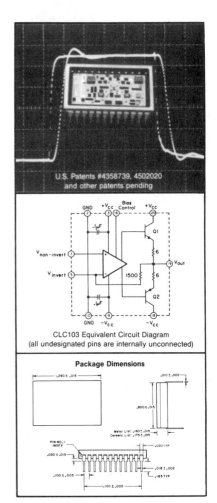

U.S. Patents #4358739, 4502020 and other patents pending

CLC103 Equivalent Circuit Diagram
(all undesignated pins are internally unconnected)

Package Dimensions

Typical Performance

parameter	gain setting						units
	+4	+20	+40	−4	−20	−40	
−3dB bandwidth	230	150	130	155	145	125	MHz
rise time (20V)	4	4	4	4	4	4	ns
slew rate	6	6	6	6	6	6	V/ns
settling time (0.4%)	10	10	12	10	10	12	ns

Comlinear Corporation • 4800 Wheaton Drive, Fort Collins, CO 80525 • (303) 226-0500 • TLX 45-0881 • FAX (303) 226-0564
DS103.01   June 1986

Courtesy Comlinear Corporation, Fort Collins, Colorado.

Appendix B10 *(Continued)*

Electrical Characteristics ($A_v = +20$, $V_{cc} = \pm15V$, $R_L = 100\Omega$)

Parameters	Conditions	typ	min and max ratings[1]			units	symbol
Ambient Temperature (AM)[1]:		+25°C	−55°C	+25°C	+125°C		
Ambient Temperature (AI)[1]:		+25°C	−25°C	+25°C	+85°C		
FREQUENCY DOMAIN RESPONSE							
* −3dB bandwidth	$V_{out} < 4V_{pp}$	150	>125	>135	>120	MHz	SSBW
gain flatness	$V_{out} < 4V_{pp}$						
* peaking	0.1 to 50MHz	0.1	<0.6	<0.3	<0.3	dB	GFPL
* peaking	>50MHz	0.2	<1.5	<0.6	<0.6	dB	GFPH
* rolloff	at 100MHz	—	<0.4	<0.6	<0.8	dB	GFR
group delay	to 75MHz	3.0±0.5	—	—	—	ns	GD
linear phase deviation	to 75MHz	1	<3	<2	<4	°	LPD
reverse isolation – non-inverting	to 150MHz	55	>45	>45	>45	dB	RINI
TIME DOMAIN RESPONSE							
rise and fall time	5V step	2.3	<2.8	<2.6	<2.9	ns	TRS
	20V step	4	<5	<5	<5	ns	TRL
settling time to 0.4%	10V step	10	<25	<20	<25	ns	TSP
overshoot	5V step	5	<15	<10	<10	%	OS
slew rate (overdriven input)		6	>5	>5	>5	V/ns	SR
overload recovery							
<50ns pulse, 200% overdrive		30	—	—	—	ns	OR
DISTORTION AND NOISE RESPONSE							
* 2nd harmonic distortion	$2V_{pp}$, 20MHz	−48	<−40	<−40	<−40	dBc	HD2
* 3rd harmonic distortion	$2V_{pp}$, 20MHz	−48	<−40	<−40	<−40	dBc	HD3
equivalent input noise							
noise floor	>100kHz	−158	<−152	<−152	<−152	dBm(1Hz)	SNF
integrated noise	1kHz to 100MHz	28	<56	<56	<56	μV	INV
* noise floor	>5MHz	−158	<−152	<−152	<−152	dBm(1Hz)	SNF
* integrated noise	5MHz to 100MHz	28	<56	<56	<56	μV	INV
STATIC, DC PERFORMANCE							
* input offset voltage		10	<30	<25	<30	mV	VIO
* average temperature coefficient		35	<80	<80	<80	μV/°C	DVIO
* input bias current	non-inverting	10	<40	<30	<40	μA	IBN
* average temperature coefficient		20	<80	<80	<80	nA/°C	DIBN
* input bias current	inverting	20	<110	<60	<110	μA	IBI
* average temperature coefficient		250	<500	<500	<500	nA/°C	DIBI
* power supply rejection ratio		54	>45	>45	>45	dB	PSRR
common mode rejection ratio		38	>30	>30	>30	dB	CMRR
* supply current	no load	30	<36	<34	<36	mA	ICC
MISCELLANEOUS PERFORMANCE							
non-inverting input	resistance	250	>100	>100	>100	kΩ	RIN
	capacitance	2.4	<3	<3	<3	pF	CIN
output impedance	at DC	—	<0.1	<0.1	<0.1	Ω	RO
	at 100MHz	2, 45	—	—	—	Ω, nH	ZO
output voltage range	no load	—	>±11	>±11	>±11	V	VO

MTBF is 2.21 million hours (AM version, GF @ 70°C case, per MIL-HDBK-217D).

Absolute Maximum Ratings

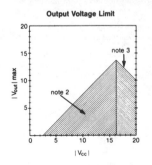

Output Voltage Limit

note 3
note 2

supply voltage (V_{cc})	±20V
output current	±200mA
thermal resistance (θ_{ca})	see thermal model
junction temperature	+175°C
operating temperature	AI: −25°C to +85°C
	AM: −55°C to +125°C
storage temperature	−65°C to +150°C
lead temperature (soldering 10s)	+300°C

* note 1: Parameters preceded by an * are the final electrical test parameters and are 100% tested. AM units are tested at −55°C (power stable), +25°C (low-duty cycle pulsed testing), and +125°C (low-duty cycle pulsed testing). AI units are tested only at +25°C although their performance is guaranteed at −25°C and +85°C as indicated above.

note 2: This rating protects against damage to the input stage caused by saturation of either the input or output stages. Under transient conditions not exceeding 1μs (duty cycle not exceeding 10%), maximum input voltage may be as large as twice the maximum. V_{cm} should never exceed ±5V. (V_{cm} is the voltage at the non-inverting input, pin 7.)

note 3: This rating protects against exceeding transistor collector-emitter breakdown ratings. Recommended V_{cc} is ±15V.

Appendix B10 *(Continued)*

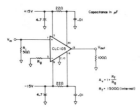

figure 1: recommended non-inverting gain circuit

figure 2: recommended inverting gain circuit

Test fixture layout artwork is available upon request.

CLC103 Operation

The CLC103 is based on Comlinear's proprietary op amp topology, a unique design which uses current feedback instead of the usual voltage feedback. This design provides dynamic performance far beyond that previously available, yet it is used basically the same as the familiar voltage-feedback op amp (see the gain equations above). A complete discussion of current feedback is given in application note AN300-1.

Layout Considerations

To obtain optimum performance from any circuit operating at high frequencies, good PC layout is essential. Fortunately, the stable, well-behaved response of the CLC103 makes operation at high frequencies less sensitive to layout than is the case with other wideband op amps, even though the CLC103 has a much wider bandwidth.

In general, a good layout is one which minimizes the unwanted coupling of a signal between nodes in a circuit. A continuous ground plane from the signal input to output on the circuit side of the board is helpful. Traces should be kept short to minimize inductance. If long traces are needed, use microstrip transmission lines which are terminated in their characteristic impedance. At some high-impedance nodes, or in sensitive areas such as near pin 5 of the CLC103, stray capacitance should be kept small by keeping nodes small and removing ground plane directly around the node.

The $\pm V_{cc}$ connections to the CLC103 are internally bypassed to ground with $0.1\mu F$ capacitors to provide good high-frequency decoupling. It is recommended that $1\mu F$ or larger tantalum capacitors be provided for low-frequency decoupling. The $0.01\mu F$ capacitors shown at pins 18 and 20 in figures 1 and 2 should be kept within 0.1″ of those pins. A wide strip of ground plane should be provided for a signal return path between the load-resistor ground and these capacitors.

Since the layout of the PC board forms such an important part of the circuit, much time can be saved if prototype amplifier boards are tested early in the design stage. Encased/connectorized amplifiers are available from Comlinear.

Settling Time, Offset, and Drift

After an output transition has occurred, the output settles very rapidly to the final value and no change occurs for several microseconds. Thereafter, thermal gradients inside the CLC103 will cause the output to begin to drift. When this cannot be tolerated, or when the initial offset voltage and drift is unacceptable, the use of a composite amplifier is advised.

A composite amplifier can also be referred to as a feed-forward amplifier. Most feed-forward techniques such as those used in the vast majority of wideband op amps, involve the use of a wideband AC-coupled channel in parallel with a low-bandwidth, high-gain DC-coupled amplifier. For the composite amplifier suggested for use with the CLC103, the CLC103 replaces the wideband AC-coupled amplifier and a low-cost monolithic op amp is used to supply high open-loop gain at low frequencies. Since the CLC103 is strictly DC coupled throughout, crossover distortion of less than 0.01dB and 1° results.

For composite operation in the non-inverting mode, the circuit in figure 1 should be modified by the addition of the circuit shown in figure 3. For inverting operation, modify the circuit in figure 2 by the addition of the circuit in figure 4. Keep all resistors which connect to the CLC103 within 0.2″ of the CLC103 pins. The other side of these resistors should likewise be as close to U1 as possible. For good overall results, U1 should be similar to the LF356; this gives $5\mu V/°C$ input offset drift and the crossover frequency occurs at about 2MHz. Since U1 has a feedback network composed of $R_a + R_b$ and a 15kΩ resistor, which is in parallel with R_g and the internal 1.5kΩ feedback resistor of the CLC103, R_b must be adjusted to match the feedback ratios of the two networks. This is done by driving the composite amplifier with a 70kHz square wave large enough to produce a transition from +5V to −5V at the CLC103 output and adjusting R_b until the output of U1 is at a minimum. R_a should be about $9.5R_g$ for best results; thus, R_b should be adjusted around the value of $0.5R_g$.

Bias Control

In normal operation, the bias control pin (pin 10) is left unconnected. However, if control over the bias of the amplifier is desired, the bias control pin may be driven with a TTL signal; a TTL high level will turn the amplifier off.

Distortion and Noise

The graphs of intercept point versus frequency on the preceding page make it easy to predict the distortion at any frequency, given the output voltage of the CLC103. First, convert the output voltage V_o to $V_{rms} = (V_{pp}/2\sqrt{2})$ and then to $P = (10\log_{10}(20V_{rms}^2))$ to get the output power in dBm. At the frequency of interest, its 2nd harmonic will be $S_2 = (I_2 - P)dB$ below the level of P. Its third harmonic will be $S_3 = 2(I_3 - P)dB$ below the level of P, as will the two-tone third order intermodulation products. These approximations are useful for $P < -1dB$ compression levels.

Approximate noise figure can be determined for the CLC103 using the Equivalent Input Noise graph on the preceding page. The following equation can be used to determine noise figure (F) in dB.

$$F = 10\log\left[1 + \frac{v_n^2 + \dfrac{i_n^2 \ R_F^2}{A_v^2}}{4kTR_s\Delta f}\right]$$

where v_n is the rms noise voltage and i_n is the rms noise current. Beyond the breakpoint of the curves (i.e., where they are flat), broadband noise figure equals spot noise, so Δf should equal one (1) and v_n and i_n should be read directly off the graph. Below the breakpoint, the noise must be integrated and Δf set to the appropriate bandwidth.

Application Notes and Assistance

Application notes that address topics such as data conversion, fiber optics, and general high-frequency circuit design are available from Comlinear or your Comlinear sales engineer.

Comlinear maintains a staff of highly-qualified applications engineers to provide technical and design assistance.

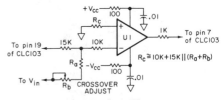

figure 3: non-inverting gain composite amp to be used with figure 1 circuit

figure 4: inverting gain composite amplifier to be used with figure 2 circuit

Typical Performance Characteristics ($A_v = +20$, $V_{cc} = \pm15V$, $R_L = 100\Omega$)

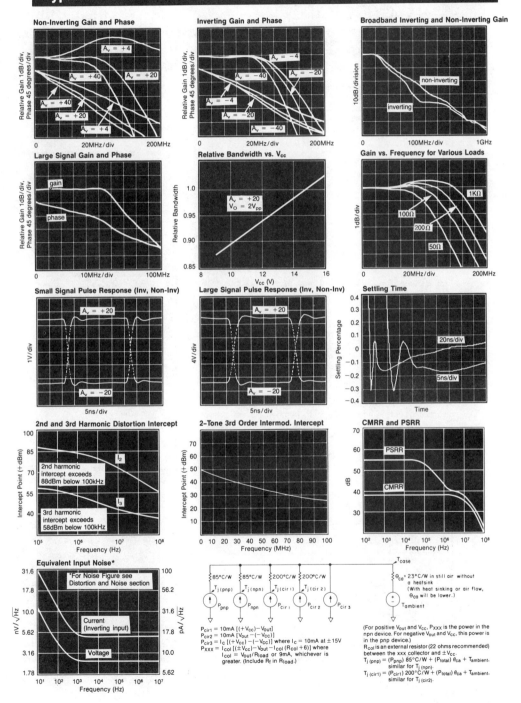

Non-Inverting Gain and Phase

Relative Gain 1dB/div, Phase 45 degrees/div

$A_v = +4$
$A_v = +40$
$A_v = +20$
$A_v = +40$
$A_v = +20$
$A_v = +4$

0   20MHz/div   200MHz

Inverting Gain and Phase

Relative Gain 1dB/div, Phase 45 degrees/div

$A_v = -4$
$A_v = -40$
$A_v = -20$
$A_v = -4$
$A_v = -20$
$A_v = -40$

0   20MHz/div   200MHz

Broadband Inverting and Non-Inverting Gain

10dB/division

non-inverting
inverting

0   100MHz/div   1GHz

Large Signal Gain and Phase

Relative Gain 1dB/div, Phase 45 degrees/div

gain
phase

0   10MHz/div   100MHz

Relative Bandwidth vs. V_{cc}

Relative Bandwidth

1.0
0.95
0.90
0.85

$A_v = +20$
$V_O = 2V_{pp}$

8   10   12   14   16
V_{cc} (V)

Gain vs. Frequency for Various Loads

1dB/div

$1K\Omega$
100Ω
200Ω
50Ω

0   20MHz/div   200MHz

Small Signal Pulse Response (Inv, Non-Inv)

1V/div

$A_v = +20$
$A_v = -20$

5ns/div

Large Signal Pulse Response (Inv, Non-Inv)

4V/div

$A_v = +20$
$A_v = -20$

5ns/div

Settling Time

Settling Percentage

0.4
0.3
0.2
0.1
0
-0.1
-0.2
-0.3
-0.4

20ns/div
5ns/div

Time

2nd and 3rd Harmonic Distortion Intercept

Intercept Point (+dBm)

100
85
70
55
40

I_2

2nd harmonic intercept exceeds 88dBm below 100kHz

I_3

3rd harmonic intercept exceeds 58dBm below 100kHz

10^5 10^6 10^7 10^8
Frequency (Hz)

2-Tone 3rd Order Intermod. Intercept

Intercept Point (+dBm)

70
60
50
40
30
20
10

0 10 20 30 40 50 60 70 80 90 100
Frequency (MHz)

CMRR and PSRR

dB

70
60
50
40
30

PSRR
CMRR

10^2 10^3 10^4 10^5 10^6 10^7 10^8
Frequency (Hz)

Equivalent Input Noise*

nV/$\sqrt{Hz}$

31.6
17.8
10.0
5.62
3.16
1.78

pA/$\sqrt{Hz}$

100
56.2
31.6
17.8
10.0
5.62

*For Noise Figure see Distortion and Noise section

Current (Inverting input)
Voltage

10^1 10^2 10^3 10^4 10^5 10^6 10^7
Frequency (Hz)

T_{case}

85°C/W   85°C/W   200°C/W   200°C/W

$T_{j\,(pnp)}$ $T_{j\,(npn)}$ $T_{j\,(cir\,1)}$ $T_{j\,(cir\,2)}$

P_{pnp} P_{npn} $P_{cir\,1}$ $P_{cir\,2}$ $P_{cir\,3}$

$\theta_{ca} = 23°C/W$ in still air without a heatsink (With heat sinking or air flow, θ_{ca} will be lower.)

$T_{ambient}$

$P_{cir1} = 10mA\,[(+V_{cc}) - V_{out}]$
$P_{cir2} = 10mA\,[V_{out} - (-V_{cc})]$
$P_{cir3} = I_c\,[(+V_{cc}) - (-V_{cc})]$ where $I_C = 10mA$ at $\pm15V$
$P_{xxx} = I_{col}\,[(\pm V_{cc}) - V_{out} - I_{col}\,(R_{col} + 6)]$ where $I_{col} = V_{out}/R_{load}$ or 9mA, whichever is greater. (Include R_f in R_{load}.)

(For positive V_{out} and V_{cc}, P_{xxx} is the power in the npn device. For negative V_{out} and V_{cc}, this power is in the pnp device.)
R_{col} is an external resistor (22 ohms recommended) between the xxx collector and $\pm V_{cc}$.
$T_{j\,(pnp)} = (P_{pnp})\,85°C/W + (P_{total})\,\theta_{ca} + T_{ambient}$, similar for $T_{j\,(npn)}$.
$T_{j\,(cir1)} = (P_{cir1})\,200°C/W + (P_{total})\,\theta_{ca} + T_{ambient}$, similar for $T_{j\,(cir2)}$.

459

 **Comlinear Corporation**

DC to 1.1GHz
Linear Amplifier

CLC104AI, CLC104AM (Hi-Rel)

APPLICATIONS:

- digital and wideband analog communications
- radar, IF and RF processors
- fiber optic drivers and receivers
- photomultiplier preamplifiers

FEATURES:

- −3dB bandwidth of 1.1GHz
- 325psec rise and fall times
- 14dB gain, 50Ω input and output
- low distortion, linear phase
- 1.4:1 VSWR (output, DC−1.1GHz)

DESCRIPTION

Comlinear Corporation's CLC104 Linear Amplifier represents a significant advance in linear amplifiers. Proprietary design techniques have yielded an amplifier with 14dB of gain and a −3dB bandwidth of DC to 1100MHz. Gain flatness to 750MHz of ±0.4dB coupled with excellent VSWR and phase linearity gives outstanding pulse fidelity and low signal distortion.

Designed for 50Ω systems, the CLC104 is very easy to use, requiring only properly bypassed power supplies for operation. This translates to time- and cost-savings in all stages of design and production.

Fast rise time, low overshoot and linear phase make the CLC104 ideal for high speed pulse amplification. These properties plus low distortion combine to produce an amplifier well suited to many communications applications. With a 1.1GHz bandwidth, the CLC104 can handle the fastest digital traffic, even when the demodulation scheme or the digital coding format requires that DC be maintained. It is also ideal for traditional video amplifier applications such as radar or wideband analog communications systems.

These same characteristics make the CLC104 an excellent choice for use in fiber optics systems, on either the transmitting or receiving end of the fiber. The low group delay distortion insures that pulse integrity will be maintained. As a photomultiplier tube preamp, its fast response and quick overload recovery provide for superior system performance.

The CLC104 is constructed using thin film resistor/bipolar transistor technology. The CLC104AI is specified over a temperature range of −25°C to +85°C, while the CLC104AM is specified over a range of −55°C to +125°C and is screened to Comlinear's M Standard for high reliability applications. Both devices are packaged in a 14-pin double-wide DIP package.

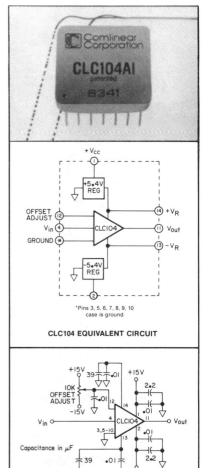

CLC104 EQUIVALENT CIRCUIT

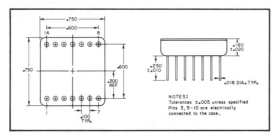

figure 1: Basic circuit

Comlinear Corporation • P.O. Box 20600, Fort Collins, CO 80522 • (303) 226-0500 • TLX 45-0881 • FAX (303) 226-0564
November 1984

Courtesy Comlinear Corporation, Fort Collins, Colorado.

Appendix B11 *(Continued)*

Electrical Characteristics (at 25°C, $R_S = R_L = 50\,\Omega$, $V_{cc} = \pm 15V$)

Parameter and Conditions	min	typ	max	units
FREQUENCY DOMAIN RESPONSE				
• −3dB bandwidth (0dBm out)	1000	1100		MHz
• −3dB bandwidth (10dBm out)		1050		MHz
• non-inverting gain (at 100MHz, note 1)	13.8	14.2	14.9	dB
• gain flatness (DC − 750MHz)		±.4		dB
• deviation from linear phase (DC − 600MHz)		1.5	3	degree
• group delay		600		ps
• reverse isolation DC − 750MHz		40		dB
750MHz − 1100MHz		35		dB
• input VSWR DC − 750MHz		1.2:1	1.35:1	
750MHz − 1100MHz		1.4:1	1.6:1	
• output VSWR DC − 750MHz		1.2:1	1.35:1	
750MHz − 1100MHz		1.4:1	1.6:1	
TIME DOMAIN RESPONSE				
• rise and fall time, 1V output step		325	375	ps
(10% to 90%) 2V output step		375	450	ps
• overshoot and aberrations (1V output step)		3		%
• settling time (to 0.8%, 1V step)		1.2		ns
• overload recovery ($V_{inpeak} = \pm 0.5V$)		1.2	1.6	ns
DISTORTION AND NOISE PERFORMANCE				
• max second/third harmonics at 100MHz 0dBm		47/53		−dBc
(see graph) 10dBm		30/35		−dBc
• third order intermodulation intercept 100MHz		26		+dBm
(two tone, 1MHz separation; see graph) 500MHz		17		+dBm
• equivalent input noise voltage, 10 Hz to 1200MHz		55		μV_{rms}
• usable dynamic range 100MHz		71		dB
500MHz		65		dB
GENERAL INFORMATION				
• input bias current (drift) note 2		80(0.6)	280(2.0)	$\mu A(\mu A/°C)$
• output offset voltage (drift) note 3		100(375)	250(625)	$mV(\mu V/°C)$
• supply current (no load)		54	60	±mA
• supply rejection (at 1KHz; see graph)		55		dB

Absolute Maximum Ratings

• supply voltage range		±9V to ±16V		
• input voltage		±0.5V		
• output current		40mA		
• output voltage (short circuit protection)		±3.5V		
• operating temperature (CLC104AI)		−25°C to 85°C		
(CLC104AM)		−55°C to 125°C		
• storage temperature		−65°C to 150°C		
• junction temperature		175°C		

note 1: nominal gain only — gain variation over temperature is ±0.1dB

note 2: (input bias current) $\times$ ($R_s \| 50\Omega$) = input offset voltage

note 3: output offset can be adjusted to zero with an external potentiometer — see "Reducing DC Offset"

Typical Performance Characteristics (at 25°C, R$_L$ = 50Ω, V$_{cc}$ = ±15V)

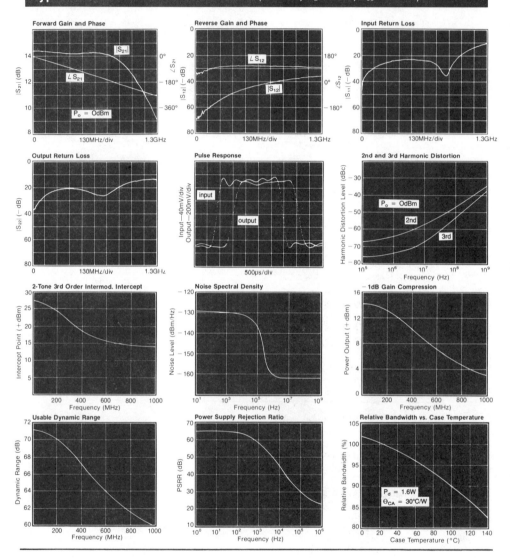

PC Board Layout Considerations

Proper layout of printed circuit boards is important to achieve optimum performance of a circuit operating in the 1GHz frequency range. Use of microstripline is recommended for all signal-carrying paths and low resistance, low inductance signal return and bypass paths should be used. To keep the impedance of these paths low, use as much ground plane as is possible. Ground plane also serves to increase the flow of heat out of the package.

The CLC104 has three types of connections: signal paths (input and output), DC inputs (supplies and offset adjust), and grounds. 50Ω microstrip is recommended for connection to the input (pin 4) and output (pin 11). Microstrip on a double-sided PC board consists of a ground plane on one side of the board and a constant-width signal-carrying trace on the other side of the board. For 1/16" G10 or FR-4 PC board material, a 0.1" wide trace will have a 50Ω characteristic impedance. The ground plane beneath the signal trace must extend at least one trace width on either side of the trace. Also, all traces (including ground) should be kept at least one trace width from the signal carrying traces.

To keep power supply noise and oscillations from appearing at the amplifier output, all supply pins should be capacitively bypassed to ground. The power supply pins (1 and 2) are the inputs to a pair of voltage regulators whose outputs are at pins 13 and 14. It is recommended that $0.01\mu F$ or larger ceramic capacitors be connected from pins 1, 2, 13 and 14 to ground, within 0.2" of the pins. A $1\mu F$ or larger solid tantalum capacitor to ground is required within 3" of pins 1 and 2, and for good low frequency performance, solid tantalum capacitors of at least $15\mu F$ should be connected from pins 13 and 14 to ground within 3" of the pins. Use 0.025" or wider traces for the supply lines. The offset adjust pin (12) also requires bypassing; a $0.01\mu F$ or larger ceramic capacitor to ground within 0.2" of the pin is recommended.

Grounding is the final layout consideration. Pins 3 and 5-10 should all be connected to a ground plane which should cover as much of one side of the board around the amplifier as possible.

Reducing DC Offset

DC offset of the CLC104 may be adjusted by applying a DC voltage to the amplifier's offset adjust pin (12). The simplest method is shown in Figure 1. Using this method of offset adjust it is possible to vary the output offset by approximately $\pm400mV$. This simple adjustment has no effect on the offset drift characteristics of the CLC104.

If lower offset and offset drift are required, a low frequency op amp may be used in conjunction with the CLC104 in a composite configuration. The suggested circuit appears in Figure 2. Its method of operation is to compare an attenuated version of the output signal to the input signal and apply a correcting voltage at the offset adjust pin. A compensation capacitor C_s reduces the bandwidth of the op amp correction circuit to limit the op amp's effect on the CLC104 to frequencies below f_{45}, the frequency at which the op amp has 45dB of open loop gain. Using an LM108, f_{45} is about 7Hz with $C_s = .1\mu F$. Thus the op amp can correct DC and low frequency errors below f_{45}, without affecting CLC104 performance above f_{45}. Also note that the noise performance of the op amp will dominate below f_{45}.

With an LM108 op amp in this composite configuration, input offset is typically 2mV and drift is $15\mu V/°C$. At frequencies well below f_{45}, the composite gain is equal to $(1 + 49.9K/(R_a + R_b))$ and the output impedance is very low. As the signal frequency increases beyond f_{45}, the op amp loses influence and the CLC104 gain and output impedance dominate. To ensure a smooth transition and matched gain at all frequencies, adjust R_b for a minimum op amp output swing with a $0.1V_{pp}$ sine wave input (to the CLC104) at the frequency f_{45}. Since the CLC104 has a 50Ω output impedance, its output voltage is a function of the load impedance $(A_v \simeq 10R_L/(R_L + 50))$, whereas the gain of the composite amplifier at low frequencies and DC is relatively independent of the load impedance, due to the high open-loop gain of the op amp. Thus, to avoid gain mismatching and phase nonlinearity, use the composite amplifier only if the load impedance is constant from DC to at least $10(f_{45})$.

Use of a composite amplifier reduces input offset voltage and its corresponding drift, but has no effect on input bias current. This current is converted to an input voltage by the resistance to ground seen at the amplifier input and the voltage appears, amplified, at the output. Typical input offset voltage due to the bias current is 2mV and input offset drift is approximately $15\mu V/°C$.

Thermal Considerations

The CLC104 case must be maintained at or below 140°C. Note that because of the amplifier design, power dissipation remains fairly constant, independent of the load or drive level. Therefore, standard derating is not possible. There are two ways to keep the case temperature low. The first is to keep the amount of power dissipated inside the package to a minimum and the second is to get the heat out of the package quickly by reducing the thermal resistance from case to ambient.

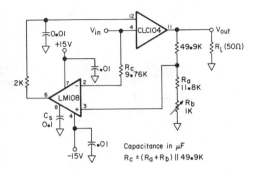

figure 2: Composite amplifier

A large portion of the heat dissipated inside the package is in the voltage regulators. At the minimum $\pm9V$ supply level the regulators dissipate 390mW and at the maximum $\pm16V$ supply level they dissipate 1.2W. The amplifier itself dissipates a fairly constant 600mW, so the total power dissipation can range from 1W to 1.8W. In cases where high supply voltages and high temperatures are likely to be encountered, the supply voltage to the CLC104 should be reduced. One possible circuit using a zener diode appears in Figure 3a. In selecting components for this circuit, remember that the maximum supply current is about 75mA and the minimum is about 45mA, and the minimum supply voltage $(V_{cc} - V_z)$ is 9V. To keep power dissipation in the zener diode within its limits, a resistor may be used in parallel with it to handle approximately 40mA of the current. The average power in the CLC104 will be reduced to $2 \times (V_{cc} - V_z) \times (55mA)$. Another circuit appears in Figure 3b and is simply a resistor in series with the supply. Again, since the maximum current is 75mA the maximum resistor is $R_s = (V_{cc} - 9V)/75mA$. The average power dissipation in the CLC104 is reduced to $2 \times [V_{cc} - (55mA \times R_s)] \times (55mA)$. In both of these circuits, bypassing should be placed on the amplifier side of the series element.

Several methods of decreasing the thermal resistance from case to ambient are possible. With no heat paths other than still air at 25°C, the thermal resistance from case to ambient for the CLC104 is about 40°C/W. When placed into a printed circuit board with all ground pins soldered into a ground plane $1" \times 1.5"$, the thermal resistance drops to about 30°C/W. In this configuration, the case rise will be 30°C for 9V supplies and 50°C for 16V supplies. This results in maximum allowable ambient temperatures of 110°C and 90°C, respectively. If higher operating temperatures are required, heat sinking of the package is recommended.

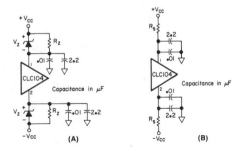

figure 3: Reducing Supply Voltages

Comlinear Corporation

Fast Settling, Wideband Operational Amplifiers

CLC200AI, CLC200AM (Hi-Rel)

APPLICATIONS:

- fast, precision A to D, D to A conversion
- baseband and video communications
- radar, sonar, IF processors
- laser drivers, photodiode preamps
- graphic CRT composite video drive amp

FEATURES:

- −3dB bandwidth of 95MHz
- .02% settling in 25ns
- 4000V/μs slew rate
- low distortion, linear phase
- 3.6ns rise and fall times

DESCRIPTION:

The CLC200 operational amplifier achieves performance far superior to that of other high performance op amps. A proprietary Comlinear design provides a **bandwidth of DC-95MHz and an unprecedented settling time of 25nsec to 0.02%.** And since thermal tail has been eliminated, the CLC200 can be depended upon to settle fast and solidly maintain its level. Drive capability is also impressive at 24V$_{pp}$ and 100mA.

Using the CLC200 is as easy as adding power supplies and a gain-setting resistor. The result is reliable, consistent performance because such characteristics as bandwidth and settling time are virtually independent of gain setting. Unlike conventional op amp designs where the optimum gain-bandwidth product occurs at a high gain, minimum settling time at a gain of −1, maximum slew rate at a gain of +1, et cetera, the CLC200 offers predictable response at gain settings from ±1 to ±50. This, coupled with consistent performance from unit to unit with **no external compensation,** makes the CLC200 a real time- and cost-saver in design and production situations alike.

Minimizing settling time was a design goal of the CLC200. Settling time is one of the most demanding of all op amp requirements since it is affected by the op amp's bandwidth, gain flatness, and harmonic distortion. The result of this effort is an amplifier fast enough for the most demanding high speed D to A converters and "flash" A to D converters.

The superior slew rate and rise and fall times of the CLC200 make it an ideal amplifier for a broad range of pulse, analog and digital applications. Flat gain and phase response from DC to beyond 50MHz ensure distortion levels well below those of other op amps. A **full power bandwidth of 20MHz** eliminates the need for power buffers in many applications.

The CLC200 is constructed using thin film resistor/bipolar transistor technology. The CLC200AI is specified over a temperature range of −25°C to +85°C, while the CLC200AM is specified over a range of −55°C to +125°C and is screened to Comlinear's M Standard for high reliability applications. Both devices are packaged in 12-pin metal TO-8 cans.

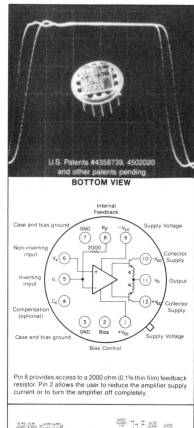

U.S. Patents #4358739, 4502020 and other patents pending

BOTTOM VIEW

Pin 8 provides access to a 2000 ohm (0.1% thin film) feedback resistor. Pin 2 allows the user to reduce the amplifier supply current or to turn the amplifier off completely.

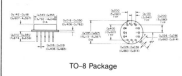

TO-8 Package

Typical Performance

parameter	+2	+20	+50	−2	−20	−50	units
−3dB bandwidth	150	95	75	100	95	90	MHz
rise time (20V)	—	4	5	4	4	4	ns
slew rate	4	4	4	4	4	4	V/ns
settling time (0.02%)	—	25	32	25	25	32	ns

(gain setting)

Comlinear Corporation • 4800 Wheaton Drive, Fort Collins, CO 80525 • (303) 226-0500 • TLX 45-0881 • FAX (303) 226-0564
DS200.02
December 1985

Courtesy Comlinear Corporation, Fort Collins, Colorado.

Appendix B12 *(Continued)*

Electrical Characteristics ($A_v = +20$, $V_{cc} = \pm15V$, $R_L = 200\,\Omega$, $R_f = 2000\,\Omega$)

Parameters	Conditions	typ +25°C	min and max ratings[1] −55°C	+25°C	+125°C	units	symbol
FREQUENCY DOMAIN RESPONSE							
* −3dB bandwidth	$V_{out} < 2V_{pp}$	95	>85	>85	>80	MHz	SSBW
gain flatness at	$V_{out} < 2V_{pp}$						
* peaking	<25MHz	0	<0.4	<0.3	<0.4	dB	GFPL
* peaking	>25MHz	0.2	<0.8	<0.6	<0.8	dB	GFPH
* rolloff	at 50MHz	—	<0.6	<0.4	<0.6	dB	GFR
group delay	to 50MHz	4.2±0.5	—	—	—	ns	GD
linear phase deviation	to 50MHz	1	<2	<2	<2	°	LPD
reverse isolation	to 50 MHz						
non-inverting		60	>50	>50	>50	dB	RINI
inverting		45	>35	>35	>35	dB	RIIN
TIME DOMAIN RESPONSE							
rise and fall time	2V step	3.6	<4.1	<4.1	<4.4	ns	TRS
	20V step	4	<5	<5	<6	ns	TRL
settling time to .02%	10V step	25	—	—	—	ns	TSP
to .1%	10V step	18	<25	<25	<25	ns	TS
overshoot	10V step	5	<12	<10	<10	%	OS
slew rate (overdriven input)		4	>3	>3	>3	V/ns	SR
overload recovery							
<50ns pulse, 200% overdrive		25	—	—	—	ns	OR
DISTORTION AND NOISE RESPONSE							
* 2nd harmonic distortion	$2V_{pp}$, 20MHz	−52	<−45	<−45	<−45	dBc	HD2
* 3rd harmonic distortion	$2V_{pp}$, 20MHz	−58	<−50	<−50	<−50	dBc	HD3
equivalent input noise							
noise floor	>100kHz	−156	<−150	<−150	<−150	dBm(1Hz)	SNF
integrated noise	1kHz to 100MHz	35	<70	<70	<70	μV	INV
* noise floor	>5MHz	−156	<−150	<−150	<−150	dBm(1Hz)	SNF
* integrated noise	5MHz to 100MHz	35	<70	<70	<70	μV	INV
STATIC, DC PERFORMANCE							
* input offset voltage		10	<25	<25	<25	mV	VIO
average temperature coefficient[1]		35	<80	<80	<80	μV/°C	DVIO
* input bias current	non-inverting	10	<40	<30	<40	μA	IBN
average temperature coefficient[1]		20	<80	<80	<80	nA/°C	DIBN
* input bias current	inverting	20	<70	<50	<70	μA	IBI
average temperature coefficient[1]		70	<200	<200	<200	nA/°C	DIBI
* power supply rejection ratio		55	>45	>45	>45	dB	PSRR
common mode rejection ratio		60	>40	>40	>40	dB	CMRR
* supply current	no load	29	<36	<34	<36	mA	ICC
MISCELLANEOUS PERFORMANCE							
non-inverting input	resistance	250	>100	>100	>100	kΩ	RIN
	capacitance	2.4	<3	<3	<3	pF	CIN
output impedance	at DC	—	<0.1	<0.1	<0.1	Ω	RO
	at 50MHz	1, 35	—	—	—	Ω, nH	ZO
output voltage range	no load	±12	>±11	>±11	>±11	V	VO

MTBF is 1.89 million hours (AM version, GF @ 70°C case, per MIL-HDBK-217D)

Absolute Maximum Ratings

Common Mode and Output Voltage Limits

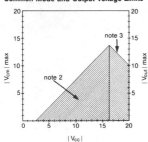

supply voltage (V_{cc})	±20V
output current	±100mA
thermal resistance (θ_{ca})	see thermal model
junction termperature	+175°C
operating temperature	AI: −25°C to +85°C
	AM: −55°C to +125°C
storage temperature	−65°C to +150°C
lead temperature (soldering 10s)	+300°C

* note 1: Parameters preceded by an * are the final electrical test parameters and are 100% tested. AM units are tested at −55°C, +25°C, and +125°C. AI units are tested only at +25°C although performance at −25°C and +85°C is guaranteed to be better than or equal to the performance specified for AM devices in the −55°C and +125°C ranges. Maximum temperature coefficient parameters apply only to AM devices.

note 2: This rating protects against damage to the input stage caused by saturation of either the input or output stages. Under transient conditions not exceeding 1μs (duty cycle not exceeding 10%), maximum input voltage may be as large as twice the maximum. V_{cm} should never exceed V_{cc}. (V_{cm} is the voltage at the non-inverting input, pin 6.)

note 3: This rating protects against exceeding transistor collector-emitter breakdown ratings. Recommended V_{cc} is ±15V.

Typical Performance Characteristics ($T_A = 25°C$, $A_v = 20$, $V_{cc} = \pm 15V$, $R_L = 200\Omega$)

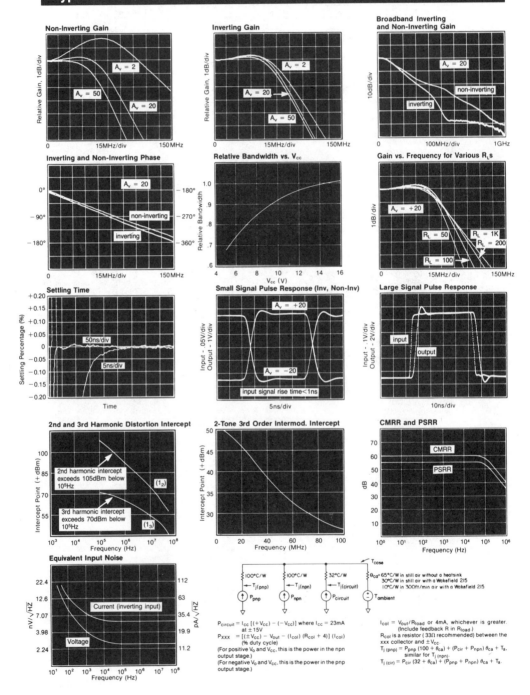

Appendix B12 *(Continued)*

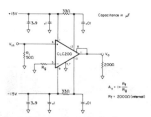

figure 1: suggested non-inverting gain circuit

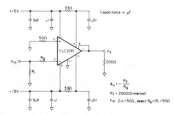

figure 2: suggested inverting gain circuit

Test fixture schematics are available upon request.

Controlling Bandwidth and Passband Response

As with any op amp, the ratio of the two feedback resistors R_f and R_g determines the gain of the CLC200. Unlike conventional op amps, however, the closed loop pole-zero response of the CLC200 is affected very little by the value of R_g. R_g scales the magnitude of the gain, but does not change the value of the feedback. This is possible due to a proprietary circuit topology. R_f does influence the feedback and so the CLC200 has been internally compensated for optimum performance with $R_f = 2000\Omega$, but any value of $R_f > 1K\Omega$ may be used with a single capacitor placed between pins 4 and 5 for compensation. See table 1. As R_f decreases C_c must increase to maintain flat gain. Slew rate will decrease slightly with increasing C_c but other parameters such as bandwidth, settling time, and phase linearity will improve. Large values of R_f and C_c can be used together or separately to reduce the bandwidth. This may be desirable for reducing the bandwidth in applications not requiring the full frequency response available although this may cause the output noise to increase at low gains.

table 1: Bandwidth versus R_f and C_c

R_f (kΩ)	C_c (pF)	$f \pm 0.3$dB (MHz)	$f - 3.0$dB (MHz)
10.0	0	5	15
5.0	0	10	30
3.0	0	20	60
2.0	0	50	100
1.5	0.25	70	130
1.0	0.50	120	170

Layout Considerations

To assure optimum performance the user should follow good layout practices which minimize the unwanted coupling of signals between nodes. During initial breadboarding of the circuit, use direct point to point wiring, keeping the lead lengths to less than .25". The use of solid, unbroken ground plane is helpful. Avoid wire-wrap type pc boards and methods. Sockets with small, short pin receptacles may be used with minimal performance degradation although their use is not recommended.

During pc board layout, keep all traces short and direct. The resistive body of R_g should be as close as possible to pin 5 to minimize capacitance at that point. For the same reason, remove ground plane from the vicinity of pins 5 and 6. In other areas, use as much ground plane as possible on one side of the board. It is especially important to provide a ground return path for current from the load resistor to the power supply bypass capacitors. Ceramic capacitors of .01 to .1µF (with short leads) should be less than .15 inches from pins 1 and 9. Larger tantalum capacitors should be placed within one inch of these pins. V_{cc} connections to pins 10 and 12 can be made directly from pins 9 and 1, but better supply rejection and settling time are obtained if they are separately bypassed as in figures 1 and 2. To prevent signal distortion caused by reflections from impedance mismatches, use terminated microstrip or coaxial cable when the signal must traverse more than a few inches.

Since the pc board forms such an important part of the circuit, much time can be saved if prototype boards of any high frequency sections are built and tested early in the design phase. Evaluation boards designed for either inverting or non-inverting gains are available from Comlinear at minimal cost.

Distortion and Amplification Fidelity

The graphs of intercept point versus frequency on the preceding page make it easy to predict the distortion at any frequency, given the output voltage of the CLC200. First, convert the output voltage (V_O) to $V_{RMS} = (V_{pp}/2\sqrt{2})$ and then to $P = (10log_{10}(20V_{RMS}^2))$ to get output power in dBm. At the frequency of interest, its 2nd harmonic will be $S_2 = (I_2 - P)$dB below the level of P. Its third harmonic will be $S_3 = 2(I_3 - P)$dB below P, as will the two-tone third order intermodulation products. These approximations are useful for $P < -1$dB compression levels.

Approximate noise figure can be determined for the CLC200 using the Equivalent Input Noise graph on the preceding page. The following equation can be used to determine noise figure (F) in dB.

$$F = 10log \left[1 + \frac{i_n{}^2 + \dfrac{R_F{}^2}{A_v{}^2}}{4kTR_s\Delta f} \right]$$

where v_n is the rms noise voltage and i_n is the rms noise current. Beyond the breakpoint of the curves (i.e., where they are flat), broadband noise figure equals spot noise figure, so Δf should equal one (1) and v_n and i_n should be read directly off the graph. Below the breakpoint, the noise must be integrated and Δf set to the appropriate bandwidth.

For linear operation of the CLC200 at large output voltage swings (DC component not included) and at high frequencies, observe the (AC output voltage) × (frequency) product specification of 400V · MHz. Exceeding this rating will cause the signal to be greatly distorted as the amplifier bias control circuit reduces the current available for slewing to prevent damage. At frequencies and voltages within this range the excess slew rate and bandwidth available will ensure the highest possible degree of amplified signal fidelity.

Operation with Reduced Bias Current

Placing a resistor between pins 1 and 2 will cause the CLC200 bias current to be reduced. A value of 20K will cause only a slight reduction, 3K will almost halve the current, while less than 1K will reduce bias to about 5mA and the amplifier will be off. In this condition, the input signal will be greatly attenuated. In the reduced bias, on condition, bandwidth will be roughly proportional to the reduction in bias current. A mechanical or semiconductor switch can be used to turn the amplifier off. Any connection which would cause current to flow out of pin 2 will result in increased bias current and may lead to device destruction from overheating and excessive current.

Thermal Considerations

At high ambient temperatures or large internal power dissipations, heat sinking is required to maintain acceptable junction temperatures. Use the thermal model on the previous page to determine junction temperatures. Many styles of heat sinks are available for TO-8 packages; the Wakefield 215 and the Thermalloy 2240 are good examples. Some heat sinks are the radial fin type which cover the pc board and may interfere with external components. An excellent solution to this problem is to use surface mounted resistors and capacitors. They have a very low profile and actually improve high frequency performance. For use of these heat sinks with conventional components, a .1" high spacer can be inserted under the TO-8 package to allow sufficient clearance.

Application Notes and Assistance

Application notes that address topics such as data conversion, fiber optics, and general high frequency circuit design are available from Comlinear or your Comlinear sales engineer.

Comlinear maintains a staff of highly qualified applications engineers to provide technical and design assistance.

Comlinear Corporation

Fast Settling, Wideband High Voltage Op Amps

CLC210AI, CLC210AM (Hi-Rel)

APPLICATIONS:

- precision, high speed D to A conversion
- graphic CRT composite video drive amp
- fast settling drivers for VCO
- high voltage pulse amplifier

FEATURES:

- ±30V output drive capability
- –3dB bandwidth of 50MHz
- settles to .02% in 40ns
- 7000V/μs slew rate
- low distortion, linear phase

DESCRIPTION:

The CLC210 employs proprietary Comlinear design techniques to achieve high voltage output at frequencies previously unavailable in a DC coupled, hybrid op amp. 60V$_{pp}$ from DC to over 5MHz has been attained without sacrificing other performance characteristics. 7000V/μsec slew rate and 40nsec to 0.02% settling time means that the CLC210 will get to its final value fast. And since thermal tail has been eliminated, the CLC210 will solidly maintain that level.

This combination of high voltage and high frequency is ideal for high-resolution display applications. The CLC210 allows designers to concentrate on improving display technology, rather than re-inventing the amplifier. The CLC210 is also ideal for intensity (Z–) modulation in more traditional CRT displays.

The CLC210 is ideally suited for driving varactors in VCO control loops and phase-correction loops. In a transimpedance mode, the CLC210 is fast enough to convert the current output from a high-speed DAC into a voltage large enough to drive the VCO frequency control circuitry. With 50mA current driving capability, the CLC210 can drive capacitive loads without severely degrading performance. In fact, the CLC210 will remain stable and oscillation-free with no external compensation.

Using the CLC210 is as easy as adding power supplies and gain-setting resistors. The result is reliable, consistent performance because such characteristics as bandwidth and settling time are virtually independent of gain setting. Unlike conventional op amp designs where the optimum gain-bandwidth product occurs at a high gain, minimum settling time at a gain of –1, maximum slew rate at a gain of +1, et cetera, the CLC210 offers predictable response at gain settings from 1 to 50. This, coupled with consistent performance from unit to unit with no external compensation, makes the CLC210 a real time- and cost-saver in design and production situations alike.

The CLC210 is constructed using thin film resistor/bipolar transistor technology. The CLC210AI is specified over a temperature range of –25°C to +85°C, while the CLC210AM is specified over a range of –55°C to +125°C and is screened to Comlinear's M Standard for high reliability applications. Both devices are packaged in 12-pin metal TO-8 cans.

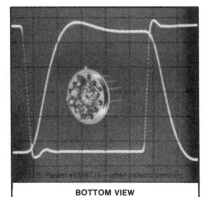

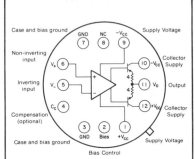

BOTTOM VIEW

Pin 4 can be used to reduce the bandwidth or to flatten frequency response when R$_F$ <3500 is used. Pin 2 allows the user to reduce the amplifier supply current, or to turn the amplifier off completely.

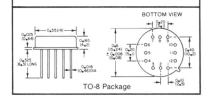

TO-8 Package

Comlinear Corporation • P.O. Box 20600, Fort Collins, CO 80522 • (303) 226-0500 • TLX 45-0881 • FAX (303) 226-0564
November 1984

Courtesy Comlinear Corporation, Fort Collins, Colorado.

Electrical Characteristics (25°C, $R_L = 1000\Omega$, $R_F = 3500\Omega$, $V_{cc} = \pm36V$ – see figures 1 and 2)

magnitude of gain [| V_{out}/V_{in} |]	2	20			50	
Parameter and Conditions	typ	min	typ	max	typ	units
FREQUENCY DOMAIN RESPONSE						
• −3dB bandwidth, $V_{out} \leq 5V_{pp}$	70	40	50		45	MHz
$V_{out} = 20V_{pp}$		15	20		20	MHz
• gain flatness, DC to 25MHz	±.5		±.3	±.5	±.3	dB
• group delay	6.5		7	8	7.5	ns
• deviation from linear phase, DC to 25MHz	1		1	1.5	1	degree
• reverse isolation	50		50		50	dB
• attenuation in shutdown mode	40		40		40	dB
• distortion – refer to graphs						
TIME DOMAIN RESPONSE						
• rise and fall time, 10V output step	7		7	9	8	ns
50V output step	—		9	12	10	ns
• settling time, 10V output step, to .1%	35		35	40	35	ns
to .02%	45		40	50	45	ns
• longterm output settling error (t < 10 seconds)	1		1	2	1	mV
• overshoot (input rise time ≤ 1ns), 5V output step	8		5	10	5	%
• slew rate		6.0	7.0			V/ns

GENERAL INFORMATION	min	typ	max	units
• input offset voltage (drift)		5(15)		mV(μV/°C)
• input bias current (drift), non-inverting input		±5(20)	±30(70)	μA(nA/°C)
inverting input		±20(50)	±50(100)	μA(nA/°C)
• equivalent input noise, 10Hz to 1MHz		4	5	μV
($R_s = 50\Omega$, gain = 20) 10Hz to 50MHz		40	55	μV
• input impedance, non-inverting input		100K/2.0		Ω/pF
• open loop transimpedance gain		1Meg		Ω
• power supply rejection ratio	40	45		dB
• common mode rejection ratio	70	80		dB
• output drive voltage/current		±33/±50		V/mA
• supply current		35		mA
• MTBF (AM suffix, per MIL-HDBK-217D, GF @ 70°C case)		1.93		10^6 hours

Absolute Maximum Ratings

- supply voltage (±V_{cc})   40V
- output current (I_O)   50mA
- input voltage (V_{imax})   (\| V_{cc} \| − 2.5)/gain
  (note 1)
- common mode input voltage   ±1/2 \| V_{cc} \|
- (AC output V_{pp}) × (frequency) product 300V·MHz

- junction temperature (T_J)   200°C
- operating temperature (T_A),   AI: −25 to +85°C
  (note 2)   AM: −55 to +125°C
- storage temperature   −65 to +150°C
- power dissipation, refer to graph
- still air thermal resistance (Θ_{CA}) 50°C/W

note 1: see figures 1 and 2. This rating protects against damage to the output stage caused by saturation, which occurs at \| V_o \| > \| V_{cc} \| − 2.5V. Under transient conditions not exceeding 1 μs (duty cycle not exceeding 10%), maximum input voltage may be as large as $2V_{imax}$.

note 2: heat sink is necessary at $T_A > 25$°C.

The collectors of the output transistors have been brought out separately to allow them to be protected from excess current and power dissipation, as well as to isolate current transients from the amplifier power supplies.

Typical Performance Characteristics

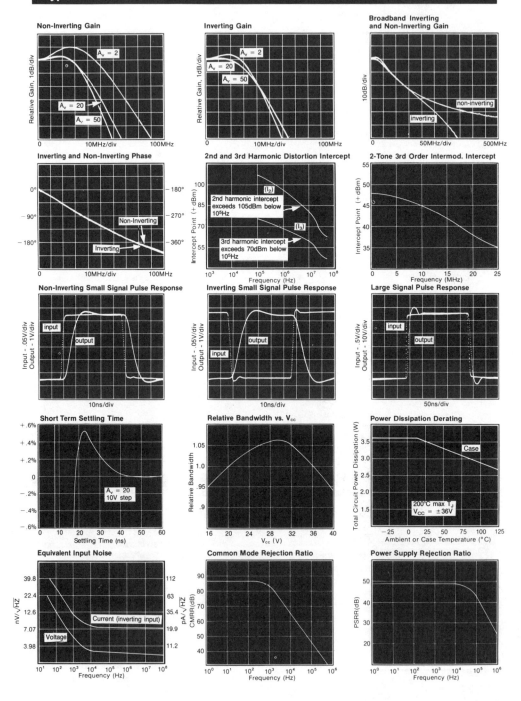

Appendix B13 *(Continued)*

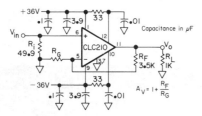

figure 1: non-inverting gain test fixture

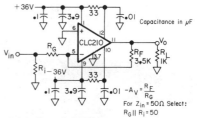

figure 2: inverting gain test fixture

Controlling Bandwidth and Passband Response

As with any op-amp, the ratio of the two feedback resistors R_f and R_g determines the gain of the CLC210. Unlike conventional op-amps, however, the closed loop pole-zero response of the CLC210 is affected very little by the value of R_g. R_g scales the magnitude of the gain, but does not change the value of the feedback. This is possible due to a proprietary circuit topology. R_f does influence the feedback and so the CLC210 has been internally compensated for optimum performance with $R_f = 3500\Omega$, but any value of $R_f > 1K\Omega$ may be used with a single capacitor placed between pins 4 and 5 for compensation. See table 1. As R_f decreases C_c must increase to maintain flat gain. Slew rate will decrease slightly with increasing C_c but other parameters such as bandwidth, settling time and phase linearity will improve. Large values of R_f and C_c can be used together or separately to reduce the bandwidth. This may be desirable for reducing the noise bandwidth in applications not requiring the full frequency response available.

table 1: Bandwidth versus R_f and C_C

R_f (kΩ)	C_C (pF)	$f \pm 0.3dB$ (MHz)	$f - 3.0dB$ (MHz)
10.0	0	2	10
5.0	0	15	35
3.5	0	25	50
2.0	1.2	35	65
1.5	4.7	50	75
1.0	20	60	95

Layout: Electrical and Thermal Considerations

To assure optimum performance the user should follow good layout practices which minimize the unwanted coupling of signals between nodes. During initial breadboarding of the circuit, use direct point to point wiring, keeping lead lengths to less than .25". The use of solid, unbroken ground plane is helpful. Avoid wire-wrap type pc boards and methods. Sockets with small, short pin receptacles may be used with minimal performance degradation although their use is not recommended.

During pc board layout keep all traces short and direct. The resistive bodies of R_f and R_g should be as close as possible to pin 5 to minimize capacitance at that point. For the same reason, remove ground plane from the vicinity of pins 4, 5 and 6. In other areas, use as much ground plane as possible on one side of the pc board. It is especially important to provide a ground return path for current from the load resistor to the power supply bypass capacitors. Ceramic capacitors of .01 to .1μF should be close to pins 1 and 9. Larger tantalum capacitors should also be placed within one inch of these pins. V_{cc} connections to pins 10 and 12 can be made directly from pins 9 and 1, but better supply rejection and settling time is obtained if they are separately bypassed as in figures 1 and 2. To prevent signal distortion caused by reflections from impedance mismatches, use terminated microstrip or coaxial cable when the signal must traverse more than a few inches. Since the pc board forms such an important part of the circuit, much time can be saved if prototype boards of any high frequency sections are built and tested early in the design phase.

At high ambient temperatures or large internal power dissipation, heat sinking is required to maintain acceptable junction temperatures. Many styles of heatsinks are available for TO-8 packages. Some of these, such as the radial fin type, cover the pc board and may interfere with external components. An excellent solution to this problem is to use surface mounted chip resistors and capacitors. They have a very low profile and actually improve high frequency performance. For use of these heatsinks with conventional components a .1" high spacer can be inserted under the TO-8 package to allow sufficient clearance.

Distortion and Amplification Fidelity

The graphs of intercept point versus frequency on the preceding page make it easy to predict the distortion at any frequency, given the output voltage of the CLC210. First, convert the output voltage (V_o) to $V_{RMS} = (V_{pp}/2\sqrt{2})$ and then to $P = (10\log_{10}(20V_{RMS}^2))$ to get output power in dBm. At the frequency of interest, its 2nd harmonic will be $S_2 = (I_2 - P)$dB below the level of P. Its third harmonic will be $S_3 = 2(I_3 - P)$dB below P, as will the two tone third order intermodulation products. These approximations are useful for $P < -1$dB compression levels.

For linear operation of the CLC210 at large output voltage swings (DC component not included) and at high frequencies, observe the (AC output voltage) $\times$ (frequency) product specification given under Absolute Maximum Ratings. Exceeding this rating will cause the signal to be greatly distorted as the amplifier bias control circuit reduces the current available for slewing to prevent damage. At frequencies and voltages within this range the excess slew rate and bandwidth available will insure the highest possible degree of amplified signal fidelity.

Operation with Reduced Bias Current

Placing a resistor between pins 1 and 2 will cause the CLC210 bias current to be reduced. A value of 20K will cause only a slight reduction, 3K will almost halve the current, while less than 1K will reduce bias to about 5mA and the amplifier will be off. In this condition, the input signal will be greatly attenuated and the amplifier will be protected from overloads. In the reduced bias, on condition, bandwidth will be roughly proportional to the reduction in bias current. A mechanical or semiconductor switch can be used to turn the amplifier off. Any connection which would cause current to flow out of pin 2 will result in increased bias current and may lead to device destruction from overheating and excessive current.

Application Notes

For information on the use of the CLC210 in a wide variety of applications, including operation with unequal supplies, refer to the application note AN200-1. Suggested pc board layouts are shown in the E Series data sheet.

 Comlinear Corporation

Fast Settling, Wideband Operational Amplifiers

CLC220AI, CLC220AM (Hi-Rel)

APPLICATIONS:

- very high speed D to A, A to D conversion
- high speed fiber optics systems
- baseband and video communications
- radar and IF processors
- very fast risetime pulse amplifiers

FEATURES:

- −3dB bandwidth of 190MHz
- .02% settling of 15ns
- 7000V/µs slew rate
- 1.9ns rise and fall times
- low distortion, linear phase

DESCRIPTION:

The CLC220 is a wide bandwidth DC-coupled operational amplifier that defines the state-of-the-art in high speed op amps. A −3dB bandwidth of DC to 190MHz is achieved using a proprietary Comlinear design. **Ultra-fast settling time (8nsec to 0.1%) and slew rate (7000V/µsec)** make the CLC220 a superior amplifier for pulsed and digital applications.

Since thermal tail has been eliminated, the CLC220 settles fast and remains solidly at the desired level. Flat gain and linear phase (1.2° deviation from linear) from DC to beyond 100MHz help the CLC220 to achieve distortion levels uncommonly low relative to conventional op amps.

Using the CLC220 is as easy as adding power supplies and a gain-setting resistor. The result is reliable, consistent performance because such characteristics as bandwidth and settling time are virtually independent of gain setting. Unlike conventional op amp designs where the optimum gain-bandwidth product occurs at a high gain, minimum settling time at a gain of −1, maximum slew rate at a gain of +1, et cetera, the CLC220 offers predictable response at gain settings from ±1 to ±50. This, coupled with consistent performance from unit to unit with **no external compensation,** makes the CLC220 a real time- and cost-saver in design and production situations alike.

This combination of features makes the CLC220 appropriate for a broad range of applications. The wide bandwidth, DC coupling, and fast settling lend themselves well to high speed D to A and "flash" A to D applications. Both receivers and transmitters in optical fiber systems have similar requirements. High **gain and phase linearity and corresponding low distortion** make the CLC220 ideal for many digital communication system applications, such as in the demodulator, where the need for both DC coupling and high frequency amplification creates requirements that are difficult to meet.

The CLC220 is constructed using thin film resistor/bipolar transistor technology. The CLC220AI is specified over a temperature range of −25°C to +85°C, while the CLC220AM is specified over a range of −55°C to +125°C and is screened to Comlinear's M Standard for high reliability applications. Both devices are packaged in 12-pin metal TO-8 cans.

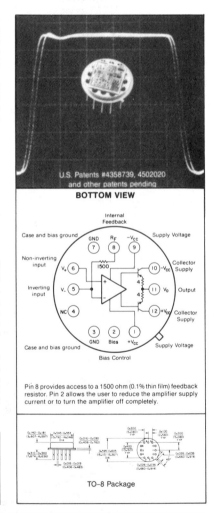

U.S. Patents #4358739, 4502020 and other patents pending

BOTTOM VIEW

Pin 8 provides access to a 1500 ohm (0.1% thin film) feedback resistor. Pin 2 allows the user to reduce the amplifier supply current or to turn the amplifier off completely.

TO-8 Package

Typical Performance

parameter	gain setting						units
	+4	+20	+50	−4	−20	−50	
−3dB bandwidth	250	190	120	200	190	150	MHz
rise time (2V)	1.6	1.9	2.3	1.6	1.9	2.3	ns
slew rate	7	7	7	7	7	7	V/ns
settling time (.02%)	15	15	18	15	15	18	ns

Comlinear Corporation • 4800 Wheaton Drive, Fort Collins, CO 80525 • (303) 226-0500 • TLX 45-0881 • FAX (303) 226-0564
DS220.02   December 1985

Courtesy Comlinear Corporation, Fort Collins, Colorado.

Electrical Characteristics (A$_v$ = +20, V$_{cc}$ = ±15V, R$_L$ = 200Ω, R$_f$ = 1500Ω)

Parameters	Conditions	typ +25°C	min and max ratings[1] −55°C	min and max ratings[1] +25°C	min and max ratings[1] +125°C	units	symbol
FREQUENCY DOMAIN RESPONSE							
* −3dB bandwidth	V$_{out}$ <2V$_{pp}$	190	>160	>170	>160	MHz	SSBW
gain flatness at	V$_{out}$ <2V$_{pp}$						
* peaking	<50MHz	0	<0.5	<0.3	<0.4	dB	GFPL
* peaking	>50MHz	0	<1.5	<0.6	<0.9	dB	GFPH
* rolloff	at 100MHz	0	<0.4	<0.6	<0.9	dB	GFR
group delay	to 100MHz	3.0±0.3	—	—	—	ns	GD
linear phase deviation	to 100MHz	1.2	<2	<2	<2	°	LPD
reverse isolation	to 100MHz						
non-inverting		60	>50	>50	>50	dB	RINI
inverting		45	>35	>35	>35	dB	RIIN
TIME DOMAIN RESPONSE							
rise and fall time	2V step	1.9	<2.2	<2.1	<2.2	ns	TRS
	5V step	2	<2.6	<2.5	<2.6	ns	TRL
settling time to .02%	5V step	15	—	—	—	ns	TSP
to .1%	5V step	8	<15	<12	<15	ns	TS
overshoot	5V step	7	<15	<12	<12	%	OS
slew rate (overdriven input)		7	>6	>6	>6	V/ns	SR
overload recovery							
<50ns pulse, 200% overdrive		25	—	—	—	ns	OR
DISTORTION AND NOISE RESPONSE							
* 2nd harmonic distortion	2V$_{pp}$, 20MHz	−58	<−50	<−50	<−50	dBc	HD2
* 3rd harmonic distortion	2V$_{pp}$, 20MHz	−62	<−50	<−50	<−50	dBc	HD3
equivalent input noise							
noise floor	>100kHz	−156	<−150	<−150	<−150	dBm(1Hz)	SNF
integrated noise	1kHz to 200MHz	50	<100	<100	<100	µV	INV
* noise floor	>5MHz	−156	<−150	<−150	<−150	dBm(1Hz)	SNF
* integrated noise	5MHz to 200MHz	50	<100	<100	<100	µV	INV
STATIC, DC PERFORMANCE							
* input offset voltage		10	<25	<25	<25	mV	VIO
average temperature coefficient[1]		35	<80	<80	<80	µV/°C	DVIO
* input bias current	non-inverting	10	<40	<30	<40	µA	IBN
average temperature coefficient[1]		20	<80	<80	<80	nA/°C	DIBN
* input bias current	inverting	20	<70	<50	<70	µA	IBI
average temperature coefficient[1]		70	<200	<200	<200	nA/°C	DIBI
* power supply rejection ratio		55	>45	>45	>45	dB	PSRR
common mode rejection ratio		60	>40	>40	>40	dB	CMRR
* supply current	no load	30	<36	<34	<36	mA	ICC
MISCELLANEOUS PERFORMANCE							
non-inverting input	resistance	250	>100	>100	>100	kΩ	RIN
	capacitance	2.4	<3	<3	<3	pF	CIN
output impedance	at DC	—	<0.1	<0.1	<0.1	Ω	RO
	at 100MHz	1, 35	—	—	—	Ω, nH	ZO
output voltage range	no load	—	>±10	>±10	>±10	V	VO

MTBF is 1.89 million hours (AM version, GF @ 70°C case, per MIL-HDBK-217D)

Absolute Maximum Ratings

Common Mode and Output Voltage Limits

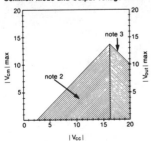

supply voltage (V$_{cc}$)	±20V
output current	±50mA
thermal resistance (θ$_{ca}$)	see thermal model
junction termperature	+175°C
operating temperature	AI: −25°C to +85°C
	AM: −55°C to +125°C
storage temperature	−65°C to +150°C
lead temperature (soldering 10s)	+300°C

* note 1: Parameters preceded by an * are the final electrical test parameters and are 100% tested. AM units are tested at −55°C, +25°C, and +125°C. AI units are tested only at +25°C although performance at −25°C and +85°C is guaranteed to be better than or equal to the performance specified for AM devices in the −55°C and +125°C ranges. Maximum temperature coefficient parameters apply only to AM devices.

note 2: This rating protects against damage to the input stage caused by saturation of either the input or output stages. Under transient conditions not exceeding 1µs (duty cycle not exceeding 10%), maximum input voltage may be as large as twice the maximum. V$_{cm}$ should never exceed V$_{cc}$. (V$_{cm}$ is the voltage at the non-inverting input, pin 6.)

note 3: This rating protects against exceeding transistor collector-emitter breakdown ratings. Recommended V$_{cc}$ is ±15V.

Typical Performance Characteristics ($T_A = 25°C$, $A_v = 20$, $V_{cc} = \pm15V$, $R_L = 200\Omega$)

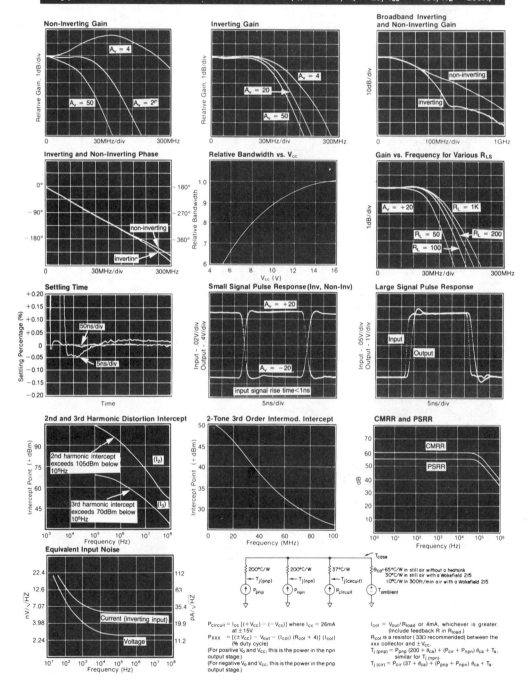

Appendix B14 *(Continued)*

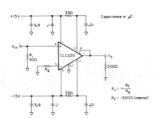

figure 1: suggested non-inverting gain circuit

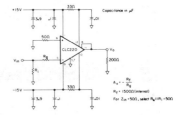

figure 2: suggested inverting gain circuit

Test fixture schematics are available upon request.

Controlling Bandwidth and Passband Response

As with any op amp, the ratio of the two feedback resistors R_f and R_g determines the gain of the CLC220. Unlike conventional op amps, however, the closed loop pole-zero response of the CLC220 is affected very little by the value of R_g. R_g scales the magnitude of the gain, but does not change the value of the feedback. This is possible due to a proprietary circuit topology. R_f does influence the feedback and so the CLC220 has been internally compensated for optimum performance with $R_f = 1500\Omega$. External R_f values greater than 1500Ω can be used with approximate results as listed in Table 1. Use of R_f values less than 1500Ω will result in extended bandwidth and peaking of the response at high frequencies. For example, $R_f = 1000\Omega$ will result in a $-3dB$ bandwidth of about 300MHz, with approximately 3.5dB of peaking above 200MHz. An RC network with a $-3dB$ bandwidth of about 250MHz could be used at the input to flatten the response, although it will reduce the bandwidth of the overall circuit.

table 1: Bandwidth versus R_f

R_f (kΩ)	$f \pm 0.3dB$ (MHz)	$f -3.0dB$ (MHz)
2	25	80
5	10	30
10	5	15

Layout Considerations

To assure optimum performance the user should follow good layout practices which minimize the unwanted coupling of signals between nodes. During initial breadboarding of the circuit, use direct point to point wiring, keeping the lead lengths to less than .25''. The use of solid, unbroken ground plane is helpful. Avoid wire-wrap type pc boards and methods. Sockets with small, short pin receptacles may be used with minimal performance degradation although their use is not recommended.

During pc board layout, keep all traces short and direct. The resistive body of R_g should be as close as possible to pin 5 to minimize capacitance at that point. For the same reason, remove ground plane from the vicinity of pins 5 and 6. In other areas, use as much ground plane as possible on one side of the board. It is especially important to provide a ground return path for current from the load resistor to the power supply bypass capacitors. Ceramic capacitors of .01 to .1μF (with short leads) should be less than .15 inches from pins 1 and 9. Larger tantalum capacitors should be placed within one inch of these pins. V_{cc} connections to pins 10 and 12 can be made directly from pins 9 and 1, but better supply rejection and settling time are obtained if they are separately bypassed as in figures 1 and 2. To prevent signal distortion caused by reflections from impedance mismatches, use terminated microstrip or coaxial cable when the signal must traverse more than a few inches.

Since the pc board forms such an important part of the circuit, much time can be saved if prototype boards of any high frequency sections are built and tested early in the design phase. Evaluation boards designed for either inverting or non-inverting gains are available from Comlinear at minimal cost.

Thermal Considerations

At high ambient temperatures or large internal power dissipations, heat sinking is required to maintain acceptable junction temperatures. Use the thermal model on the previous page to determine junction temperatures. Many styles of heat sinks are available for TO-8 packages; the Wakefield 215 and the Thermalloy 2240 are good examples. Some heat sinks are the radial fin type which cover the pc board and may interfere with external components. An excellent solution to this problem is to use surface mounted resistors and capacitors. They have a very low profile and actually improve high frequency performance. For use of these heat sinks with conventional components, a .1'' high spacer can be inserted under the TO-8 package to allow sufficient clearance.

Distortion and Amplification Fidelity

The graphs of intercept point versus frequency on the preceding page make it easy to predict the distortion at any frequency, given the output voltage of the CLC220. First, convert the output voltage (V_O) to $V_{RMS} = (V_{pp}/2\sqrt{2})$ and then to $P = (10\log_{10}(20V_{RMS}^2))$ to get output power in dBm. At the frequency of interest, its 2nd harmonic will be $S_2 = (I_2 - P)dB$ below the level of P. Its third harmonic will be $S_3 = 2(I_3 - P)dB$ below P, as will the two-tone third order intermodulation products. These approximations are useful for $P < -1dB$ compression levels.

Approximate noise figure can be determined for the CLC220 using the Equivalent Input Noise graph on the preceding page. The following equation can be used to determine noise figure (F) in dB.

$$F = 10\log\left[1 + \frac{v_n^2 + \dfrac{i_n^2\ R_F^2}{A_v^2}}{4kTR_s\Delta f} \right]$$

where v_n is the rms noise voltage and i_n is the rms noise current. Beyond the breakpoint of the curves (i.e., where they are flat), broadband noise figure equals spot noise figure, so Δf should equal one (1) and v_n and i_n should be read directly off the graph. Below the breakpoint, the noise must be integrated and Δf set to the appropriate bandwidth.

For linear operation of the CLC220 at large output voltage swings (DC component not included) and at high frequencies, observe the (AC output voltage) × (frequency) product specification of 600V · MHz. Exceeding this rating will cause the signal to be greatly distorted as the amplifier bias control circuit reduces the current available for slewing to prevent damage. At frequencies and voltages within this range the excess slew rate and bandwidth available will ensure the highest possible degree of amplified signal fidelity.

Operation with Reduced Bias Current

Placing a resistor between pins 1 and 2 will cause the CLC220 bias current to be reduced. A value of 20K will cause only a slight reduction, 3K will almost halve the current, while less than 1K will reduce bias to about 5mA and the amplifier will be off. In this condition, the input signal will be greatly attenuated. In the reduced bias, on condition, bandwidth will be roughly proportional to the reduction in bias current. A mechanical or semiconductor switch can be used to turn the amplifier off. Any connection which would cause current to flow out of pin 2 will result in increased bias current and may lead to device destruction from overheating and excessive current.

Application Notes and Assistance

Application notes that address topics such as data conversion, fiber optics, and general high frequency circuit design are available from Comlinear or your Comlinear sales engineer.

Comlinear maintains a staff of highly qualified applications engineers to provide technical and design assistance.

Comlinear Corporation

Wideband, High Speed Operational Amplifier

CLC300A

APPLICATIONS:

- digital communications
- baseband and video communications
- instrument input/output amplifiers
- fast A to D, D to A conversion
- graphic CRT video drive amp
- coaxial cable line driver

FEATURES:

- -3dB bandwidth of 85MHz
- new design topology eliminates gain-bandwidth trade-off
- 3000V/μsec slew rate
- 4ns rise and fall time
- 100mA output current
- low distortion, linear phase

DESCRIPTION:

The CLC300 operational amplifier represents a significant step forward in op amp performance. A unique, proprietary Comlinear design provides a **DC-85MHz -3dB bandwidth that is virtually independent of gain setting.** Rise and fall times of 4nsec and drive capability of 22V_{pp} and 100mA add to the CLC300's impressive specifications.

Ease-of-use is a design goal at Comlinear, and the CLC300 is a success in this area as well. Using the CLC300 is as easy as adding power supplies and a gain-setting resistor. And unlike conventional op amp designs in which optimum gain-bandwidth product occurs at a high gain, minimum settling time at a gain of -1, maximum slew rate at a gain of +1, et cetera, the CLC300 offers consistent performance at gain settings from 1 to 40 inverting or non-inverting. As a result, designing with the CLC300 is greatly simplified. And since **no external compensation is necessary,** "tweeks" on the production line have been eliminated, making the CLC300 an efficient component for use in production situations.

Flat gain and phase response from DC to 45MHz and superior rise and fall times make the CLC300 an ideal amplifier for a broad range of pulse, analog, and digital applications. A **45MHz full power bandwidth** (20V_{pp} into 100Ω) and 3000V/μsec slew rate eliminate the need for power buffers in many applications such as driving "flash" A to D converters or line-driving. For applications requiring lower power consumption, the CLC300 can operate on supplies as low as 5V. Fast overload recovery (20nsec) helps prevent loss of data in communications applications and flat phase response reduces distortion, even when data must be sent over extended lengths of line.

The CLC300 is specified over a temperature range of -25°C to +85°C and is packaged in a hermetic 24-pin ceramic DIP.

Layout Considerations

To assure optimum performance the user should follow good layout practices which minimize the unwanted coupling of signals between nodes. During initial breadboarding of the circuit, use direct point to point wiring, keeping lead lengths to less than .25". The use of solid, unbroken ground plane is helpful. Avoid wire-wrap type pc boards and methods. Sockets with small, short pin receptacles may be used with minimal performance degradation although their use is not recommended.

(continued on last page)

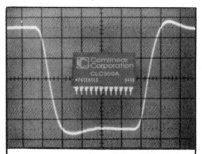

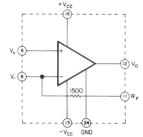

CLC300 Equivalent Circuit Diagram

Pin 11 provides access to a 1500Ω feedback resistor which can be connected to the output or left open if an external feedback resistor is desired. All undesignated pins are internally unconnected.

Package Dimensions

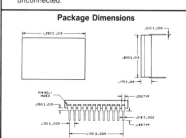

Comlinear Corporation • 4800 Wheaton Drive, Fort Collins, CO 80525 • (303) 226-0500 • TLX 45-0881 • FAX (303) 226-0564
DS300.01
March 1986

Courtesy Comlinear Corporation, Fort Collins, Colorado.

Electrical Characteristics (25°C, $R_L = 100\Omega$, $V_{cc} = \pm 15V$ – see figures 1 and 2)

magnitude of gain [$|V_{out}/V_{in}|$]	4*	20			40	
Parameter and Conditions	typ	min[2]	typ	max[2]	typ	units
FREQUENCY DOMAIN RESPONSE						
• −3dB bandwidth, $V_{out} < 4V_{pp}$	105	75	85		70	MHz
$V_{out} = 20V_{pp}$	45		45		45	MHz
• gain flatness, 100KHz to 20MHz	±0.25		±0.08	±0.3	±0.25	dB
20MHz to 45MHz	±0.5		±0.25	±0.6	±1	dB
• phase shift	1		1.6		2	deg/MHz
• deviation from linear phase, DC to 45MHz	2		3		5	degree
• reverse isolation	60		70		70	dB
• distortion—refer to graphs						
TIME DOMAIN RESPONSE						
• rise and fall time, 5V output step	3		4		5	ns
20V output step	7		7		7	ns
• settling time, 10V output step, to .8%	20		20		25	ns
• overshoot (input rise time ≤1ns), 5V output step	5		5		5	%
• slew rate	3		3		3	V/ns
• overload recovery (<50ns pulse width, 200% overdrive)	20		20		20	ns
GENERAL INFORMATION	min[2]		typ		max[2]	units
• input offset voltage (drift)			10(25)		32	mV(μV/°C)
• input bias current (drift), non-inverting input			10(20)		30	μA(nA/°C)
inverting input			30(50)		100	μA(nA/°C)
• equivalent input noise[1], integrated 0.1 to 100MHz			22		56	μV
($R_S = 50\Omega$, gain = 20)						
• second/third harmonic distortion (@ 20MHz, +10dBm)			48		38	−dBc
• input impedance, non-inverting input			100K/3			Ω/pF
• power supply rejection ratio (referred to input)	45		60			dB
• common mode rejection ratio (referred to input)			64			dB
• output drive voltage/current			10/100			V/mA
• supply current			24		33	mA

Absolute Maximum Ratings

- supply voltage ($\pm V_{cc}$)   16V ($\pm 5V$ min.)
- output current (I_O)   100mA
- input voltage (V_{imax})   ($|V_{cc}| - 2.5)/A_v$
- common mode input voltage $\pm \frac{1}{2}|V_{cc}|$
- power dissipation, refer to graph

- junction temperature (T_J)   150°C
- operating temperature (T_A)   −25 to +85°C
- storage temperature   −55 to +150°C
- still air thermal resistance (θ_{ca}) 25°C/W

[1] For Noise Figure, refer to Distortion and Noise Section in text.
[2] All min/max parameters are tested 100% at +25°C, $A_v = +20$, $R_L = 100\Omega$, and $V_{cc} = \pm 15V$.
*refer to Low Gain Operation section.

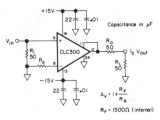

figure 1: recommended non-inverting gain circuit

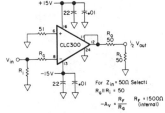

figure 2: recommended inverting gain circuit

Typical Performance Characteristics

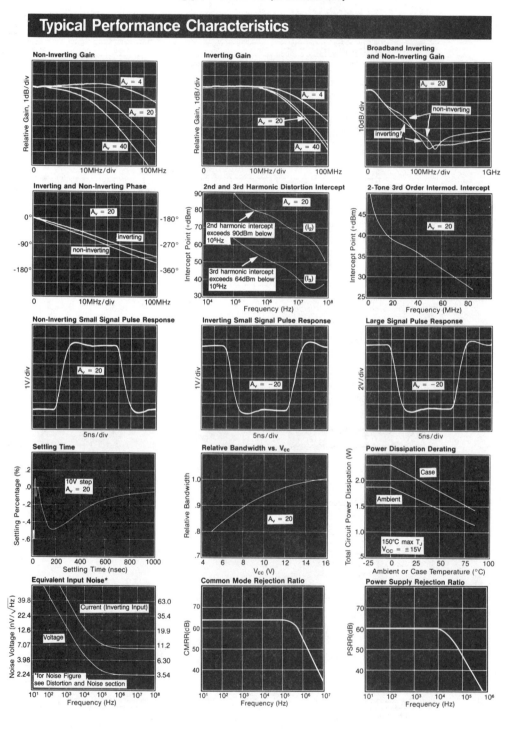

Layout Considerations (continued)

During pc board layout keep all traces short and direct. The resistive bodies of R_F and R_G should be as close as possible to pin 8 to minimize capacitance at that point. For the same reason, remove ground plane from the vicinity of pins 8 and 6. In other areas, use as much ground plane as possible on one side of the pc board. It is especially important to provide a ground return path for current from the load resistor to the power supply bypass capacitors. Ceramic capacitors of .01 to .1μF should be close to pins 13 and 16. Larger tantalum capacitors should also be placed within one inch of these pins. To prevent signal distortion caused by reflections from impedance mismatches, use terminated microstrip or coaxial cable when the signal must traverse more than a few inches. Since the pc board forms such an important part of the circuit, much time can be saved if prototype boards of any high frequency sections are built and tested early in the design phase.

Controlling Bandwidth and Passband Response

As with any op amp, the ratio of the two feedback resistors R_F and R_G determines the gain of the CLC300. Unlike conventional op amps, however, the closed loop pole-zero response of the CLC300 is affected very little by the value of R_G. R_G scales the magnitude of the gain, but does not change the value of the feedback. This is possible due to a proprietary circuit topology. R_F does influence the feedback and so the CLC300 has been internally compensated for optimum performance with $R_F = 1500\Omega$, but any value of $R_F > 500\Omega$ may be used with a single capacitor placed between pins 8 and 12 for compensation. See table 1. As R_F decreases, C_c must increase to maintain flat gain. Large values of R_F and C_c can be used together or separately to reduce the bandwidth. This may be desirable for reducing the noise bandwidth in applications not requiring the full frequency response available.

table 1: Bandwidth versus R_F and C_c ($A_V = +20$)

R_F (KΩ)	C_c (pF)	$f_{\pm 0.3dB}$ (MHz)	$f_{-3.0dB}$ (MHz)
10.0	0	2	5
5.0	0	3	12
2.0	0	8	40
1.5	0	45	85
1.0	0.3	90	115
.75	1.1	95	130
0.50	1.9	110	135

Low Gain Operation

The small amount of stray capacitance present at the inverting input can cause peaking which increases with decreasing gain. The gain setting resistor R_G is effectively in parallel with this capacitance and so a frequency domain pole results. With small R_G (Gain > 8), this pole is at a high frequency and it affects the closed loop gain of the CLC300 only slightly. At lower values of gain, this pole becomes significant. For example, at a gain of $+2$, the gain may peak as much as 3dB at 75MHz, and have a bandwidth exceeding 150MHz. The same behaviour does not exist for low inverting gains, however, since the inverting input is a virtual ground which maintains a constant voltage across the stray capacitance. Even at inverting gains $\ll 1$, the frequency response remains unchanged.

To avoid the peaking at low non-inverting gains, place a resistor R_p in series with the input signal path just ahead of pin 6, the non-inverting input. This forms a low pass filter with the capacitance at pin 6 which can be made to cancel the peaking due to the capacitance at pin 8, the inverting input. At a gain of $+2$, for example, choosing R_p such that the source impedance in parallel with R_i (see figure 1), plus R_p equals 175Ω will flatten the frequency response. For larger gains, R_p will decrease.

Settling Time, Offset, and Drift

After an output transition has occurred, the output settles very rapidly to final value and no change occurs for several microseconds. Thereafter, thermal gradients inside the CLC300 will cause the output to begin to drift. When this cannot be tolerated, or when the initial offset voltage and drift is unacceptable, the use of a composite amplifier is advised. This technique reduces the offset and drift to that of a monolithic, low frequency op amp, such as an LF356A. The composite amplifier technique is fully described in the CLC103 Data Sheet.

A simple offset adjustment can be implemented by connecting the wiper of a potentiometer, whose end terminals connect to ± 15V, through a 20K resistor to pin 8 of the CLC300. Variations of this technique are described in Application Note 200-1 Designer's Guide for 200 Series Op Amps.

Overload Protection

To avoid damage to the CLC300, care must be taken to insure that the input voltage does not exceed $(|V_{cc}| - 2.5)/A_V$. High speed, low capacitance diodes should be used to limit the maximum input voltage to safe levels if a potential for overload exists.

If in the non-inverting configuration the resistor R_i, which sets the input impedance, is large, the bias current at pin 6, which is typically a few μA but which may be as large as 18μA, can create a large enough input voltage to exceed the overload condition. It is therefore recommended that $R_i < [(|V_{cc}| - 2.5)/A_V]/(18\mu A)$.

Distortion and Noise

The graphs of intercept point versus frequency on the preceding page make it easy to predict the distortion at any frequency, given the output voltage of the CLC300. First, convert the output voltage (V_O) to $V_{RMS} = (V_{pp}/2\sqrt{2})$ and then to $P = (10 \log_{10}(20 V_{RMS}^2))$ to get output power in dBm. At the frequency of interest, its 2nd harmonic will be $S_2 = (I_2 - P)$dB below the level of P. Its third harmonic will be $S_3 = 2(I_3 - P)$dB below P, as will the two tone third order intermodulation products. These approximations are useful for $P < -1$dB compression levels.

Approximate noise figure can be determined for the CLC300 using the Equivalent Input Noise graph on the preceding page. The following equation can be used to determine noise figure (F) in dB:

$$F = 10 \log \left[1 + \frac{v_n^2 + \dfrac{i_n^2 R_F^2}{A_v^2}}{4kTR_s \Delta f} \right]$$

where v_n is the rms noise voltage and i_n is the rms noise current. Beyond the breakpoint of the curves (i.e., where they are flat), broadband noise figure equals spot noise figure, so Δf should equal one (1) and v_n and i_n should be read directly off of the graph. Below the breakpoint, the noise must be integrated and Δf set to the appropriate bandwidth.

Application Notes

For information on the use of the CLC300 in a wide variety of applications, refer to application note AN200-1.

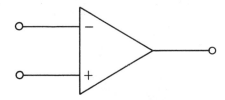

Answers to Selected Problems

CHAPTER 1

1. 1.5V
3. 2.5V
5. 5.7V
7. −1.5V
9. 0.6 Sin (100 wt)
11. −½
13. a. −6V
    b. 4.7V
15. a. +3.05
    b. +0.95
    c. +4V
    d. −4V
17. a. −14.2V
    b. +3.7V
19. a. −2.55V
    b. 0.9V
    c. 2.588V
    d. +1.95V
    e. +8.1V
    f. −8V
    g. −4V
21. a. −0.0975
    b. −120
    c. 0.936
23. 1×10^5

25. −119.05
27. $R_C = 9.167K\Omega, R_E = 4.58K\Omega$,
    $V_{RE} = 4.58V$, $V_{RC} = 9.167$ V,
    $V_{EE} = -5.28V$, $V_{CE} = 5.833V$
30. a. $R_1 = 7.07K\Omega$, $R_2 = 8.47K\Omega$
    b. $A V_{LS} = 0.545$
31. a. $V_{R1} = 6.1$, $R_1 = 4.067K\Omega$,
       $R_2 = 466.67\Omega$, $R_3 = 8.467K\Omega$
    b. Gain ≈ 1
32. CMRR = 47.133
34. 14,962.36
36. CMRR = 21.622

CHAPTER 2

1. $V_B = 978.05\mu V$
3. $V_O = -2.5089V$
5. a. 1.6mA
   b. −8V
7. a. −75
   b. 73.88
   c. 1.5%
9. a. −75
   b. 74.9
   c. For higher values of A_{VCL}, the error in
      the approximate A_{VCL} expression
      becomes less.

11. a. -26.4V

b. Rule 3

13. a. $R_I = 1.5K\Omega$, $R_F = 35.7K\Omega$

b. $A_{VEX} = 100$

c. $A_{VOL} = 2480$

15. -15V

17. $A_{VOL} = 12{,}520$, $H = \frac{1}{51}$, $V_O/V_{in} = 50.79$

19. $R1 = 10K\Omega$, $R2 = 14.3K\Omega$, $R3 = 20K\Omega$

21. $R_{IEFF} = 1.09K\Omega$, $A_{VOL} = 27.52$,

$A_{VEX} = 718.8$

23. $Z_{in} = 6.67M\Omega$

25. $R_O = 0.251\Omega$

CHAPTER 3

1. $+75$mV

3. $V_{OV} = 0.02$V, $V_{IO} = 4$mV

5. $V_{OI} = 120$mV

7. $V_{OV} = -150$mV, $V_{OI} = 112.5$mV,

$V_{OE} = 37.5$mV, $V_O = -11.2875$mV

9. $R_C = 2.5K\Omega$

11. $+2.625$mV

13. $R_O = 2.6M\Omega$

15. $R_O = 19.2M\Omega$

17. $R_O = 3.66M\Omega$

19. a. 10.00109 at ends, 10.000109 at center

b. 8.800176 at end, 8.800176 at center

CHAPTER 4

1. a. 16.975V

b. % ripple = 10.60%

c. 6,656 μF

3. a. 2.42

b. $V_{min} = 9.58$V

c. % Ripple = 20.2% No, ripple too high

d. Increase capacitor size

5. 31,955 μF, $V_{max} = 21.224$, $V_C = 1.224$,

% Ripple = 5.767%, $I_C = 5.25$A,

C = 31,955 μF

7. % Reg. = 4.96%

9. $V_{ops} = 141.6$ $V_{p\text{-}p}$

11. a. $R_F = 1.38R_I$

b. Let $R_I = 10.7K\Omega$, $R_F = 14.78K\Omega$, use 147KΩ

13. a. $V_C = 3.64$

b. $V_{min} = 3.36$V

c. No, ripple too high

d. Increase capacitance

15. a. 138.89μV drop

b. 0.002778%

CHAPTER 5

1. a. 169.3Hz

b. 0.707 $\underline{/-45°}$

c. 1 $\underline{/-0.573°}$

d. 0.01 $\underline{/-89.42°}$

e. 0.001 $\underline{/-89.94°}$

3. 1 $\underline{/+8.64°}$, 0.55 $\underline{/+56.45°}$, 0.006 $\underline{/+86.2°}$,

0.0066 $\underline{/+89.62°}$

5. a. 300 KHz

b. 98.74KHz

7. Oscillation

11. a. 60 dB

b. 10 $\underline{/90°}$

c. -9.94 $\underline{/6.27°}$

13. j54.0

15. 4.725 MHz

17. 2MHz

19. a. 7MHz

b. 70KHz

21. Initial SR = 0.0028V/μs, Average SR = 0.00102V/μs

23. -0.35V/μs or $+0.2$V/μs

$SR_{max} = 1.88$V/μs

25. $SR_1 = 0.030$V/μs, $SR_2 = 0.031$V/μs,

$SR_{AVG} = 0.011$V/μs

27. $V_B - V_A = 7.5V_{p\text{-}p}$

29. Op amp will not slew rate limit

33. 1.77 μV shot noise, 37.29 μV thermal noise, 37.31 μV total noise,

35. $V_{PSRR} = 66.6$ μV

CHAPTER 6

1. a. $V_{in} = 2$ $V_{p\text{-}p}$

b. CMRR = 20.1

5. 50 KΩ

7. $R_F = R_I = 28$ KΩ

9. a. $R_3 = 531.9$ Ω

b. $R_C = 24.5$ KΩ

11. a. $V_O = 19.44$ $(V_2 - V_1)$

b. Vary from 10 to 25

10:$R_3 = 2.83$ KΩ

25:$R_3 = 632.55$ Ω

13. $R_I = 3$ KΩ

15. a. $R_{CS} = 250$ Ω

b. $R_{Lmax} = 2.25$ KΩ

17. a. $R_{cs} = 2.61$ Ω

b. $F_h = 65.89$ KHz

19. $V_{O1} = 1.25$ V, $V_{O2} = 2.5$ V,

$V_{O3} = -3.75$ V, $V_{O4} = 2.5$ V,

$V_O = -3.75$ V, $V_{06} = 2.5$ V
21. Amp = 12.06 V, Phase = $-90°$
23. a. 0.267 $/74.56°$
    b. 0.94 $/19°$
    c. 10.93 dB/decade rise
    d. 0.999 $/2.08°$
    e. 1 $/0°$
25. $F_p = 7.24$ MHz
27. a. $F_I = 150$ Hz
    b. $R_I C_I = 1.06$ ms
    c. $F_{ID} = 2.82$ HZ
29. $V_O = -29.51 \sin(100 \, \eta \, \tau - \eta/2)$,
    Phase = $+90°$

CHAPTER 7

1. a. $R_F = 244.3$ KΩ
      $R_C = 14.13$ KΩ
3. a. R = 37.4 KΩ, 1%, RI = 18.7 KΩ,
      RC = 12.47 KΩ
5. a. T = 6.67 ms
      Time for rise = 12.5 μs
7. Output voltages:
   a. 5.9 V
   b. 5 V
9. A = 225 mV
   B = 275 mV
11. a.

V_{in}	V_o
0	0
1	0.2
2	0.8
3	1.8
4	3.2
5	5
6	7.2
7	9.8
8	12.8
9	16.2
10	20

   d. Lowest gain = 0.2, Divider ratio = 0.1
   e. R_A = 12 KΩ, R_B = 1.33 KΩ
13. $R_1 = 499$ Ω, $R_F = 39.9$ KΩ,
    $R_C = 493$ Ω, $R_O = 403.7$ KΩ,
    $R_P = 10$ KΩ
15. a. $V_P = 16.4$ V
    b. $R_F = 65.6$ KΩ
17. $R_F = 176$ KΩ
19. $V_{in} = -5.4$ V
    $V_P = 10$ V
    $V_S = \pm 12$ V

21. $V_{sw} = -2.22$ V, $V_{ch} = 132$ mV,
    -2.088 V, -2.352 V
23. $R_F/R_F = 2$, $V_R = 4.5$V
25. $R_F/R_I = 3.44$, $R_I = 1$ KΩ, $R_F = 4.5$ KΩ,
    $V_R = -4.85$ V
27. Eq (4-3): 0.0957 V/μs
    Eq (5-14): 156.67 μs
29. $F_{HZ} = 3.8$ KHz

CHAPTER 8

1. a. $R_{min} = 65$ KΩ
   b. Use 6.2K, 5%
3. a. $R_{min} = 910$ Ω
   b. use 910 Ω, 5%
5. a. $V_{OH} = 4.98$ V
   b. $V_{OH} = 4.99$ V
7. a. $I_O = 74.07$ μA, $I_2 = 148.89$ μA,
      $I_F = 222.95$ μA
   b. $V_O = -2.23$ V
11. a. $V_{LSB} = -117.19$ mV
    b. $V_{FC} = 10.117$ V
    c. $+15$ V, $+7.5$ V, $+1.875$ V, $+0.9375$,
       25.3125 V
13. a. -3.28125 V
    b. 0.0375 V off, therefore almost 5 steps
       off.
    c. 0010000000 − 0110000000
       0001111111 − 0010111111
    d. All steps not using bit 8
15. $R_{msb} = 18$ KΩ, $R_5 = 18$ KΩ,
    $R_4 = 36$ KΩ, $R_3 = 72$ KΩ
    $R_2 = 144$ KΩ, $R_1 = 288$ KΩ
17. $V_L = 0.244$ mV, $V_L = 0.208$ mV,
    $V_L = 0.833$ mV, $V_L = 3.33$ V
19. a. $V_{Lmsb} = 3.33$V, $V_N = 6.66$ V,
       V_L at full count (FC) = 6.61V
    b. $I_{Lmsb} = 66.667$ μA,
       $I_{Lvn} = 133.33$ μA,
       I_L at FC = 132.289μA
21. a. Gain = 7.2
    b. $V_O = 93.75$ mV
23. a. $R_F = 60$ KΩ
    b. $R_C = 20$ KΩ
    c. $V_{O(msb)} = -5$ V,
       $V_{O(LSB)} = -4.88$ mV
25. a. DAC $V_O = 7.266$ V
    b. 155 Clock Pulses
27. Between two counts
29. a. C = 2.05 μF

b. C = 2.00 mF, 1%, R = 10.0 KW
31. Digital magnitude = 100000001011011
33. a. Clock pulses, 12221
   Segments, 10001
   8 clock pulses, total
   b. Clock pulses, 21122
   Segments, 10011
   8 clock pulses, total

CHAPTER 9

1. Figure
3. Figure
5. a
6. a
7. e
8. e
9. d
11. a. Pole (x)
   b. Pole (x) at -5
13. a. 2 points
   b. poles
   c. $-2 \pm j7$
15. a. Four
   b. 2 poles, 2 zeros
   c. Quadratic, $S_1 = -4.25 + j6.16$ rad/s
   $S_2 = -4.25 - j6.16$ rad/s
17. a. $V_o/V_{in} = (S + 100)/(S^2 + 10^4 S + 10^8)$
   b. $\sigma = 100$, $\omega_z = O$, $\omega_o = 10^4$, $\zeta = 0.5$
   c. $S_1 = -5000 + j8660$ rad/s
   $S_2 = -5000 - j8660$ rad/s
19. a. $S_1 = -100 + j4999$ rad/s
   $S_2 = -100 - j4999$ rad/s
   b. $S^2 + 200 S + 25 \times 10^6$
21. a. $V_C = 165$ V
   b. $V_{C2} = 147$ V, $V_{C3} = 130$ V
   c. $V_{C2} = 137$ V, $V_{C3} = 113$ V
23. a. BW = 212 Hz, Q = 15
   BW = 21.2 Hz, Q = 150
   b. Q = 15; $S_1 = -667 + j20,000$
   $S_2 = -667 - J20,000$
   Q = 150; $S_1 = -66.7 + j20,011$ rad/s,
   $S_2 = -66.7 - j20,011$ rad/s
   c. Figure
   d. As poles move toward $j\omega$-axis,
   response becomes sharper
25. $R_C = 15$ KΩ
27. a. $V_O/V_{in} = (10.64 \times 10^6)/(S^2 + 25 S + 10.64 \times 10^6)$
   b. Constant term of denominator in
   numerator.

c. $\omega_o = 3261.9$, $\zeta = 0.003832$,
   $j\omega_p = j3261.88$, $jf_p = 519.4$
29. $S_1 = -740.75 + j1411.4$, $S_2 = -740.75 - j1411.4$
31. $L_{n1} = 1.3066$, $C_{n1} = 0.76536$
   $L_{n2} = 0.5412$, $C_{n2} = 1.84775$
33. 18° between poles
   9°
35. $C_3 = 0.01$ μF
35. $R_3 = 30.6$ KΩ, $R_C = 30$ KΩ, 5%
37. $C_2 = C_3 = 0.033$ μF, $R_1 = 19.7$ KΩ,
   $R_2 = 4.43$ KΩ, $R_3 = 24.1$ KΩ
39. $L_{n1} = 2.8514$, $C_{n1} = 0.3298$,
   $L_{n2} = 1.1811$, $C_{n2} = 1.5493$
41. Step 1. $S_{1,2} = -0.309 \pm j0.957$,
   $S_{1,2} = -0.809 \pm j0.5878$
   Step 2. $\delta = 0.2434$
   Sinh (0.2434) = 2.1207, $\beta_k = 0.4241$
   Step 3. Tanh (0.4241) = 0.4003
   $\delta = -0.4003$,
   $\delta_{1,2} = 0.1237$
   $\delta_{1,2} = 0.3238$
   $S = -0.4003$, $S_{1,2} = -0.1237 + j0.951$,
   $S_{1,2} = -0.3238 \pm j0.5878$
   Step 4. Cosh (0.4241) = 1.0913,
   $S = -0.4368$,
   $S_{1,2} = -0.1350 \pm j1.0378$
   $S_{1,2} = -0.3534 \pm j0.6415$
   Step 5 $(S + 0.4368) (S^2 + 0.2700S + 1.0952)$
   $(S^2 + 0.7068S + 0.5364)$
43. Chebyshev
45. Optimal
47. a. Faster rise, less delay, less overshoot
   b. Slower rise, more delay, more overshoot
49. $C_2 = C_3 = 0.033$ μF, f = 200 Hz,
   $C_{n1} = 1.613$, $C_{n2} = 0.7429$
   $R_1 = 13.2$ KΩ, $R_2 = 4.76$ KΩ,
   $R_3 = 38.9$ KΩ
51. 0.1789, 1.05925, 0.1895
53. $\omega_o = 1.6751$, $L_n C_n = 0.3564$
55. Skirt bcomes more shallow
57. $R_3 = 39.6$ KΩ, $R_C = 39.6$ Ω
59. $V_o/V_{in} = (S^2)/(S^2 + 25S + 25)$
61. a. $R_{11} = 5.95$ KΩ, $R_{21} = 39.9$ KΩ,
   $R_{12} = 8.44$ KΩ, $R_{22} = 9.89$ KΩ
   b. $R_{01} = R_{11} = 5.95$ KΩ,
   $R_{02} = R_{12} = 8.44$ KΩ,
   $L_1 = 3.9172$, $L_2 = 1.3773$,
   $C_{11} = 0.066$ μF, $C_2 = 0.066$ μF

c. $\omega_{o1} = 1967$ rad/s, $\sigma_{p1} = 759$ rad/s,
$\zeta_1 = 0.386$, $\omega_{o2} = 3317$ rad/s,
$\sigma_{p2} = 3064$ rad/s, $\zeta = 1.0825$
d. 4 zeros

63. 4 Pole
1. $L_n = 0.7236$, $C_n = 0.4231$
2. $L_n = 0.5107$, $C_n = 1.7421$
Four zeros at the origin.

65. Denominator is same as low pass. Numerator is same as single power of S (center term) term in denominator; one zero at origin. Gain related to Q.

67. a. $R_F/R_I = 2.5$
b. BW = 70 Hz
c. $f_{CH} = 175$ Hz, $f_{CL} = 105$ Hz

69. $R_{PL} = 22.7$ KΩ, $R_F/R_I = 2.31$, G = 3.31
$R_O = 223$ KΩ, $R_F/R_I = 1.76$, G = 2.76
$R_{PH} = 21.9$ KΩ, $R_F/R_I = 2.31$, G = 3.31
For $\omega_2 < 1$, the high and low pole positions exchange places, thus $f_{PL} = 450$ Hz and $f_{PH} = 466.7$ Hz.
The gain at resonance is 51.1
$R_B/(R_A + R_B) = 0.0196$ for voltage divider.

70. Y = 2.4

73. R = 5.46 KΩ, R/2 = 2.73 KΩ,
$R_F/R_I = 4$

74. a. $R_B/R_A + R_B) = 0.4288$
b. $R_B = 10$ KΩ, $R_A = 13.3$ KΩ,
use a 12.7 KΩ and a 1 KΩ pot

75. It has a biquadratic transfer function.

77. Q enhancement is the effect produced by the changing of circuit Q as the amplifier excess loop gain changes with frequency.

79. Q enhancement.

81. a. $R_1 = 5.42$ KΩ, R = 22.0 KΩ,
$R_2 = R_Q = QR = 22.3$ KΩ,
$R_3 = R_4 = R = 22.0$ KΩ,
b. $R = R_3 = R_4 = 22.1$ KΩ 1%,
$R_1 = 5.49$ KΩ 1%,
$R_2 = 22.1$ KΩ 1%,
c. $f_o = 738.8$ Hz, $f_z = 1.57$ KHz,
$f_p = 629.6$ Hz, $f_p = 785$ Hz at $\Omega_S = 2$,
$f_{j1} = 573.7$ Hz

83. a. $R_1 = 9.31$ KΩ 1%,
$R_3 = R_4 = R = 11.3$ KΩ,
$R_2 = R_Q = QR = 11.5$ KΩ
b. Use: $R = R_3 = R_4 = 11.3$ KΩ 1%,
$R_1 = 9.31$ KΩ 1%, R2 = 11.5 KΩ 1%,
c. $f_o = 174$ Hz, $f_z = 80.36$, $f_p = 151.3$ Hz,

$f_p = 160.72$ at $\Omega_S = 2$,
$f_{j1} = 219.5$ Hz

85. a. Circuit b. R = 10.05 KΩ

87. a. at dc, Vo/Vin = $1\underline{/0°}$
at 50 Hz, Vo/Vin = $1\underline{/-13.79°}$
at 100 Hz, Vo/Vin = $1\underline{/-172.3°}$
at 200 Hz, Vo/Vin = $1\underline{/-343.9°}$
at 500 Hz, Vo/Vin = $1\underline{/-355°}$
at 1000 Hz, Vo/Vin = $1\underline{/-357.6°}$
b. The maximum phase change occurs below 100 Hz
c. The minimum phase change occurs above 500 Hz

89. a. GRAPH; Effective closed loop gain at 500 Hz is 110.
b. At 500 Hz, A_{VEX} is $\approx$ 90 or 39 dB
c. % Error = 0.00154%
d. Error = 0.845%, $f_{co} = 3000$ Hz

CHAPTER 10

1. Yes
3. Waveform is clipped at the top and bottom.
5. Amplitude limiting, soft clipping, or reduction of A_{VEX} at large amplitudes.
7. $F_o = 2.23$ KHz, $R_F/R_I = 1.32$
9. R = 1.63 KΩ, $R_F = 19.44$ KΩ,
$V_z = 2.5$ V
11. $C_o = 0.38$ μF, $R_o = 4.04$ KΩ,
$R_F = 48.24$ KΩ,
$R_F/2 = 24.12$ KΩ, use 24.3 KΩ
and a 50 KΩ pot set to 24.3 KΩ,
$R_C = 51$ KΩ 5%, $C_p = 0.13$ μF,
$R_p = 20$ KΩ
13. R = 53.6 KΩ
14. a. 150 KHz is the highest frequency
b. $V_{in} = \pm 15$ V, $V_o = \pm 7$ V;
Actual period is 20 μS min.
$t_{min} = 10$ μS, $RC_{min} = 21.4$ μS,
$R_{min} = 25$ KΩ, C_{min} 857.14 pF
16. a. C = 0.569 μF, R = 63.9 Ω,
I = 125 mA
b. $R_I = 12$ KΩ, $R_F = 24$ KΩ,
$V_R = 7.5$ V, $R_1 = 2$ KΩ
c. $f_L = 0.003515$, DR = 135.1 dB,
$V_{in} = 1.28$ μV

CHAPTER 11

1. $I_O = 1.974$ μA
2. $R_{ABC} = 146.45$ KΩ
5. $R_O = 37.5$ Ω

7. $A_{VCL} = 11,500$
9. $A_{VCL} = 10$, $R_I = 10 \ K\Omega$, $R_B = 200 \ K\Omega$
11. 2000 Hrs, 11.9 weeks
12. a. Circuit diagram
    b. $R_I = (200 \ K\Omega)/2.5 = 80 \ K\Omega$, $R'_F/R'_I = 4$
15.

PARAMETER	LM 208	TL 089
V10 (m V)	0.5	0.15
Drift (V/μs)	5	0.25

17. a. $A_{CE} = -1$, ACB $= (+)$ so no Miller effect
    b. $A_{CASCODE} = -267$
    c. $A_{CE} = -267$, $f_o = 65 \ KHz$
    d. $f_o = 4 \ MHz$, 61 times as great
23. Let $R_1 = 10 \ K\Omega$, $C_1 = 0.48 \ \mu F$,
    $C_2 = 9.86 \ \mu F$, $C_3 = 20 \ \mu F$

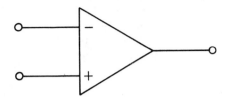

Index

Acronyms (cont.):
MOS-FET (metal oxide semi-conductor field effect transistor), 28, 207
MSB (most significant bit), 208
OTA (operational transconductance amplifier), 357
PLL (phase-locked loop), 387–400
PSRR (power-supply rejection ratio), 91, 96–99, 102, 103, 133, 369, 379, 385, 425–26
RGI (Response-gain independent), 385–87
SI (International System of Units), 2
SR (slew rate), 127–32
THD (total harmonic distortion), 184
TRW (Thompson, Ramo, Woolridge), 209
TTL (transistor-transistor logic), 201, 397
TTY (teletype), 400
VCO (voltage-controlled oscillator), 350–53, 388, 400
Active filter (*see* Filter)
Active load, 25
Adder (*see* Inverting summer)
Adjacent channel frequencies, 310
Akerberg-Mossberg (A-M), 300–309, 315
All-pass equalizer, 293, 311–17, 341–43
A-M (Akerberg-Mossberg) (*see* Akerberg-Mossberg)
Amplifiers:
antilog, 184
cascode, 373–76
chopper stabilized, 370–73
common-mode rejection quotient (CMRQ), 14, 18, 28
common-mode rejection ratio (CMRR), 28–30, 96–99, 374
commutating auto-zero (CAZ), 373–74
constant current: (*see also* Transconductance)

floating load, 55, 146–47
grounded load, 150–51
current, 56–57
current differencing (Norton), 362–69
current-series, 55, 421
current-shunt, 56, 421
difference, 48, 141
differential, 2–23, 91, 111
common-mode, 3–5, 14–16
common-mode range, 22–23
deviation, 6–13
differential, 6–13
input voltage, 3–10
output voltage, 3–10
differential-mode, 49, 140–42, 146–47
input impedance, 140
instrumentation, 138–140
line receiver, 137
gain:
common-mode, 5–6, 14–16
current, 56–57, 421
deviation, 7
differential, 8, 14, 17–18
ideal (*see also* β), 39, 42, 54–56, 420–24
transconductance, 55, 148–150
transresistance, 55, 420
voltage, 41, 54, 420
ideal, 114
input impedance:
common-mode, 15, 61–63
constant current, 412
difference, 407
noninverting, 402
transconductance (*see* Constant current)
transresistance, 422
voltage (*see* Noninverting)
instrumentation, 141–42
inverting op amp, 38–40, 145, 175, 184, 212–13, 215, 217, 300, 317, 339
bias current, 74–75, 357
bias current compensation, 76–78

Flash encoder, (*see also* ADC) 228
Flicker noise (*see* Noise)
Flip-flop, 205
FM (frequency modulation), 391
Folding, (*see also* s-plane) 284
Forward biased diode, 181
Four quadrant devices, 175
Frequency (Hz), 2
  audio, 334
  beat, 388, 391
  center, 391
  cutoff, 252
    high, 289
    low, 289
  dominate pole, 118, 121, 128, 317
  free running, 391
  lower limit, 353
  mark, 400
  measureable, 238
  − 3 dB, 265
  peak amplitude frequency, 303
  pole frequency, 114, 115, 303
  radian, 246
  reflected, 238
  resonant frequency, 242, 250, 305
  ring frequency, 238
  rolloff, 28, 117, 118, 317, 387
  space, 400
  stable, 344
  touch-tone, 399–400
  upper limit, 353
  zero frequency, 303
Frequency, synthesized, 397, 398
Frequency axis, (*see also* Axis) 236,
    239, 249
Frequency compensation, (*see also*
    Compensation) 28, 186
Frequency modulation (FM), 391
Frequency response, 112–18, 123–25,
    237–325, (*see also* Bode plot)
Frequency shift keying (FSK), 391, 400
Frequency spectrum, 235–36, 334
FSK (frequency shift keying), 391, 400
Fulcrum:
  differential amplifier, 9

inverting amplifier, 40
noninverting amplifier, 44
Full count, 209, 221, 228
Function generator (*see also*
    Waveform):
  FG 501, 349
  sine, 183–84
  square, 346
  triangular, 346

G

G (*see* Bandpass filter; Control systems;
    State variable biquad)
Gain, 38
  current, 23
  voltage (*see also* Voltage follower;
      Subtractor; and Amplifiers
      inverting op amp):
    closed loop, 39
      apparent, 59
      divider, 45
      effective, 317
      inverting, (*see also* Inverter) 38–
        41, 51
        actual at ac, 121
        actual at dc, 38
        ideal, 39, 54
        unity, 118
      noninverting, 41–45, 97,
        290–91, 341–42 (*see also*
        Voltage follower) 288–289
        actual at ac, 121
        actual at dc, 42
        ideal, 42, 57, 72
        unity, 49
      excess loop, 49–52, 64, 65, 118,
        207, 317–21
        apparent, 59
    open loop:
      ac, 114, 117, 123
      compensated, 117, 118, 123,
        124
      dc, 36, 37, 41, 59, 103
    unity, 57

Monolithic circuit element, 23–24, 56
Monolithic op amp, 69, 357
Monolithic R/2R ladder network, 212
Monotonic:
    active filter, 263, 267, 270
    DAC staircase, 206, 207
Monotonic L, 265 (*see also* Optimal
    filter)
MOS-FET (metal oxide semi-conductor
    field effect transistor), 28, 207
Most significant bit (MSB), 210, 216
Motorola, 184, 205
MSB (most significant bit), 207
Multiplier:
    frequency, 398
    numerator; Chebyshev, elliptic, 275–
    78
    voltage, 184, 272, 278
Multivibrator:
    astable (relaxation oscillator), 343–
    46, 351
    bistable (flip-flop), 351
    monostable (one-shot), 345–46

N

National Semiconductor Corporation,
    27, 80, 142–43, 365–68
Natural logarithm, 181, 269, 343
    base of (ϵ), 269
Negative feedback (*see* Feedback)
Negative voltage (*see* Voltage regulator)
Nelson, David A., xvi
Newcomb (*see* Kerwin, Huelsman and
    Newcomb)
Nodal network, 60
Noise, 98, 133–34, 334
    burst, 133
    combination of, 134
    flicker, 134
    in differential mode amplifiers, 139
    Johnson, 133
    1/f, 134
    popcorn, 134
    Schottky, 134
    shot, 134

thermal, 134
Noninverting amplifier (*see* Amplifiers)
Noninverting averager, 59–61
Noninverting gain, (*see* Gain)
Noninverting input (*see* Input)
Nonminimum phase function:
    all pass equalizer, 293, 311
    definition, 252
Normalized, high pass transfer function,
    276
Normalized capacitance, 258
Normalized characteristic equation,
    257–59, 267, 268, 270, 272,
    275
Normalized frequency, 258, 304
Normalized inductance, 257–58
Normalized low pass polynomials
    (factored form), 259, 268, 270,
    272, 275
    Bessel, 272
    biquadratic, 276
    Butterworth, 265–68
    Chebyshev, 268
    elliptic (Cauer), 275, 304, 308
    high pass, 277, 279, 283
    low pass, 257–59, 289
    optimal, 272
    parabolic, 272
Normalized resistance, 41
Norton amplifier (*see* Current
    differencing amplifier)
Norton equivalent circuit:
    common-base amplifier, 18–19
    constant current amplifier, 146
    constant current tail, 19
Notation, 1, 2
Notch depth amplifier, 296–98
Notch filter, 293–95
Null, terminals, 28, 81–85
Numerator, 57, 276–77, 293, 295, 303

O

Octal, 227
Octave, 113
Offset, 69–85, 135, 184

true image, 239–41, 245, 246, 251, 285, 289, 303
Pole-zero cancellation, 124, 295
Polynominal equations, 267, 280
Position-type feedback system (Type O), 52
Position voltage, 193
Positive feedback, 117, 191, 334, 348
Potential:
    common (*see* Common; Bus)
    different (*see* Voltage)
    ground (*see* Ground; Bus)
    zero volt (*see* Zero)
Power supply (*see also* Voltage regulator):
    dual wave rectifier, 92
    dual tracking (slaved), 104–7
    feedback, 101–7
    negative voltage, 91, 350
    pass transistor, 101–2
    positive voltage, 91, 349
    rectifier, dual full wave, 92
    regulation, 99–108
        line, 108
        load, 108
    regulator, 92, 99–108
    sources, 2
    three terminal, 103
    variation, 300
    zener, 99–101
Power supply filter, 91–96
Power supply rail voltages, 28, 364–66, 369
Power supply rejection (PSR), 96
Power supply rejection ratio (PSRR), 8–26, 91, 96–99, 101–3, 108, 135, 186, 373
Power supply transformer, 92
Preamplifier, 132
Precision:
    capacitor, 428 (*see also* Capacitor)
    resistor, 427 (*see also* Resistor)
    voltage (*see* Voltage)
Precision matched resistors, 140
Precision rectifier:
    full wave, 172–77

half wave, 171–72
Preregulator, 103
Priority encoder, 227
Programmable calculator (*see* Calculator)
PSRR (power supply rejection ratio), 91, 96–99, 102, 103, 133, 369, 379, 385, 425–26
Pull-up resistor:
    comparator, 186
    phase locked loop, 394, 398
    TTL gate, 203 (*see also* Open collector)
Pulse response (*see* Step response; Voltage follower, pulse response)
Pump, 92

Q

q (charge on an electron), 134
Q (*see* Quality factor)
Q-enhancement, 299–300
    definition, 299
Quad op amp, 300
Quadrant (*see* Four quadrant devices; Two quadrant circuit)
Quadratic equation, 245, 246, 261, 282, 293
Quadrature phase, 391, 393
Quality factor (Q), 27
    biquadratic filter, 299–309
    bandpass filter, 285–90
    definition of, 248
    oscillator, for, 335–36
    minimum value, 289
    relation to bandwidth, 251
    relation to damping factor (σ), 289
    relation to damping ratio (ζ), 248

R

R (resistor) (*see* Resistor)
r_e (emitter resistance), 20
Radian frequency, 246 (*see also* Frequency)